S0-CAT-702

Elevated Work Platforms and Scaffolding

Job Site Safety Manual

Elevated Work Platforms and Scaffolding

Job Site Safety Manual

Matthew J. Burkart, PE

Michael McCann, PhD, CIH

Daniel M. Paine, CSE

McGraw-Hill

New York Chicago San Francisco Lisbon London Madrid
Mexico City Milan New Delhi San Juan Seoul
Singapore Sydney Toronto

The McGraw·Hill Companies

2 3 4 5 6 7 8 9 0 IBT / IBT 0 1 0 9 8

ISBN 0-07-141493-2

The sponsoring editor for this book was Larry S. Hager and the production supervisor was Cheryl Souffrance. It was set in Weiss by Lone Wolf Enterprises, Ltd.

Printed and bound by IBT Global.

This book is printed on recycled, acid-free paper containing a minimum of 50% recycled, de-inked fiber.

Preface

This publication, which we consider to be the first of a series of construction safety books, is of particular importance because its subject matter is found on all construction building sites. Falls account for many construction site accidents and is one of four major categories that OSHA examines for compliance when inspecting a site.

We feel this comprehensive book, with illustrations, references to standards and regulations, inspection requirements, checklists, hazards, appropriate fall protection information, and listing of typical malfunctions and accidents, will be of interest to:

- Architects/engineers in their planning processes
- Owners, as responsible parties in the safety process
- Controlling contractors, as enforcers of safety processes, second only to the owner
- Subcontractors, as those who will have to make safety planning part of their work process
- Labor unions, as part of their training process for their members
- Other contractor/labor associations, as part of their training process for their members
- Equipment manufacturers, as part of their manufacturing processes and in keeping their equipment requirements current with standards and regulations
- Attorneys, as a reference in their litigation library

This book is intended to be comprehensive in scope, written for the novice in easily understood terms. It will present the latest innovations and practices in the industry, including, not only scaffolds, but aerial lift devices, ladders, etc, as they increase in popularity.

It is also intended to provide current information to the experienced user, erector, and designer of scaffold systems and aerial work platforms of the regulatory requirements, industry standards, and innovations in the industry, as well as to provide examples of the most common failures which result in injury and death to construction workers every day. The authors have attempted, through statistical research and their sum experience, to point to these factors in an effort to help eliminate future occurrences.

We would like to give our thanks to those who have provided help and assistance in providing materials or input to this endeavor including:

- Jim Lapping, for providing encouragement and inspiration as we worked on this book
- The Scaffold, Shoring, and Forming Institute for permitting us to use their publications in entirety in this book
- The OSHA and NIOSH sites, for the wealth of information they make available
- Jim Palmer, the Scaffold Training Institute, for allowing us to utilize some of his material in the text
- Ron Cox, Miller Equipment, for providing information and input, as well as allowing us to use information and pictures from Miller Equipment publications
- The USDA Forest Service for providing us with electronic formatting for the publication, "Calculating Apparent Reliability of Wood Scaffold Planks" by David S. Gromala, as well as allowing us to use this publication in this text
- The Canadian Centre for Occupational Health and Safety for allowing us to use drawings and materials in the interest of ladder safety.
- And to the entire office staff at AEGIS CORP, Innovative Safety, Michael McCann for his outstanding input and resultant chapters,

and Denise Almonte who worked with Mike McCann in co-authoring the chapter on aerial lifts, we extend our deepest appreciation for their nimble minds and their nimble fingers.

Michael McMann wishes to acknowledge Bill Wiehegen of NIOSH Pittsburgh Research Laboratories for reviewing the ladder chapter and making some recommendations; Alan Kline, President of Lynn Ladder and Scaffolding Co., Inc. for reviewing the ladder chapter and providing several figures; Camille Villanova of OSHA Washington office for reviewing the ladder and stairways chapters. Also Denise Almonte as co-author of the aerial lift chapter.

Table of Contents

Preface v

Introduction xxv

Chapter 1

OSHA Regulations and Industry Standards 1

History of Construction Legislation and How Subpart L Developed 1

Development of Subpart L 3

The Structural Work Act 5

Chapter 2

Accidents, Injuries, and Fatality Statistics 7

Hazards of Elevated Work Platforms 7

Statistics 9

Why Do Scaffolds Fall? 14

Engineering Design Analysis 15

Case Studies 16

FACE 9012 Case Study # 1 16

Introduction 17

Investigation 17

Cause of Death 19

Recommendations/Discussion 19

Failure to Understand and/or Comply with Regulatory Requirements 20

FACE 890 Case Study # 2 20

Contacts and Activities. 21
Overview of Employer's Safety Program 21
Synopsis of Events 21
Cause of Death. 23
Recommendations/Discussion. 23
Proper Assembly of Scaffolds 26
FACE 9013 Case Study #3 26
Summary 26
Case Study # 4 27
Introduction. 27
Investigation 28
Cause of Death. 29
Recommendations/Discussion. 29
FACE 8827 Case Study #5. 30
Dry Wall Finisher Dies in Fall from Ladder on Scaffold 30
Contacts/Activities 30
Overview of Employers' Safety Program 30
Synopsis of Events 31
Cause of Death. 31
Recommendations/Discussion. 31
Listing of Citations 33

Chapter 3 (contributed by Daniel Paine, CSE)
Preplanning — Selection & Use 35

Preplanning is Essential! 35
Beginning 35
The Pre-Bid Process 36
Job Hazard Analysis and Other Planning Instruments 36
Equipment and Training 37

Special Equipment 39
OSHA Standards 40
Capacity 54
Scaffold Platform Construction 54
Criteria for Supported Scaffolds 55
Criteria for Suspended Scaffolds 57
Hoists 60
Access Requirements 61
Use Requirements 62
Fall Protection Requirements 62
Falling Object Protection 63
Aerial Lift Requirements 1926.453 63

Chapter 4
Training 65

Specific Requirements 65
Training Materials 65
Competent Person 72
Competent, Qualified, and Authorized Designations 72
Scaffold Tags 73
Competent and Qualified Persons on the Jobsite 75

Chapter 5
Supported Scaffolds — Materials and Methods 85

Soil Bearing Capacities 85
Foundation/Support 86
Scaffolds on Elevated Structures 94
Planks 94

Lumber Grading. . . . 96
Evaluating the Condition of Scaffold Planking 98
Evaluating the Span of a Scaffold Plank 99
Platforms. . . . 99
Falling Object Protection 102
Causes of Scaffold Accidents. . . . 104
Wire Suspension Ropes. . . . 104
Fiber Ropes 108

Chapter 6 (contributed by Daniel Paine, CSE)
Fall Protection During Erection and Use of Scaffolds 115

Introduction 115
Identifying Hazards – The Job Hazard Analysis Form 115
Fall Protection Solutions. . . . 118
Passive Solutions 118
Active Solutions and Equipment 119
Configurations for Active Fall Protection. . . . 123
Rescue Plans 126
Summary. . . . 128

Chapter 7
Electrocution on Scaffolds. . . . 129

Introduction 129
Safety Factors 130
Determining Location of Power Lines 131
Preventing Contact. . . . 132
Use of Electrical Power Tools on Scaffolds. . . . 133
GFCI. . . . 134

Construction Applications of GFCI 134
AEGCP 135
Suspended Scaffolds and Welding 136
Case Studies 137
Death Due to Lack of Ground-Fault Protection 137
Frayed Insulation Causes Arc, Damages Suspension Rope 138
Employee Dies After Contacting Power Line 138
Summary — Working in the Vicinity of Power Lines 138

Chapter 8
Emergency Response and Rescue 141

Introduction 141
The Contract 142
Possible Scenarios 143
Workplace Emergencies 145
Alarms 146
Evacuation 147
Resolution 148

PART II

Supported Scaffolds 151

General Definition and Requirements 151
Maintaining Stability 152
Guy Wires 152
Tie Wires 153
Rigid Braces 153
Preventing Contact with Electrical Lines 154

Access 155
Square and Level 158
Criteria for Fall Protection on Supported Scaffolds 158
Guardrail Systems 160

Chapter 9
Tubular Welded-Frame Scaffolds - Mason's Frames 163

Introductory Description 163
The Half-Ladder 165
The Walk-Through Frame 167
End Frames 169
The Pad 170
Jacks 170
Planning 170
Pre-Erection Inspection 171
OSHA Regulations 172
Climbing of Tubular Welded-Frame Scaffolds 172
Outrigger Brackets 173

Chapter 10
Tube and Coupler Scaffolding 175

Definition 175
Important Features of Tube and Coupler Scaffolding 175
Standard Tables for Minimum Constructions 179

Chapter 11
Wood Pole Scaffolds 183

Use of Wood Pole Scaffolds 183
Wood Pole Scaffold Requirements 183

Chapter 12
Pump Jack Scaffolds and Ladder Jack Scaffolds 187

Definition 187
The Importance of the Structure and Poles 187
Structure of Pump Jack Scaffolds 189
Mending Plates 189
Poles 189
Guardrails 190
Ladder Jack Scaffolds 190

Chapter 13
Job-Manufactured Scaffolds and Form Scaffolds 193

The Need for Forming Systems 193
Roof Brackets 196

PART III

Suspended Scaffolds — General Information 199

Introduction 199
Support 199
Structure Attachment Points 200
Loads 200
Weight and Stability 201
Hoisting Devices 201
Stall Force 202
Fall Protection on Suspended Scaffolds 202
Guardrail Systems 204
Erectors and Dismantlers 205
Competent Person 205

Chapter 14
Single-Point and Two-Point Suspension Scaffolds 207

Introduction 207

Methods of Support 207

Preventing Scaffold Sway 210

Platforms. 211

Single-Point Suspension Scaffolds (Boatswains' Chair). 213

Definition and Requirements 213

Powered Platform 216

Chapter 15
Multiple Point Suspension Scaffolds. 217

Introduction 217

The Interior Hung Scaffold 217

Float Scaffolds. 219

Catenary Scaffolds. 220

Chapter 16
Adjustable Multi-Point Suspension Scaffolds
(Stone Setters — Mason's). 223

Mason's Multi-Point Scaffolds. 223

Chapter 17
Outrigger Type Scaffolding 227

Introduction 227

Needle Beam Scaffolds. 227

Outrigger Beam Scaffolds 229

Window-Jack Scaffolds 230

Chapter 18
Manually Propelled Rolling Towers 233

Various Uses 233
Guidelines for Use. 233
Construction and Safety 233
Casters. 234
Moving a Scaffold 235
Outriggers 236
Access 237

Chapter 19 (contributed by Michael McCann, PhD and Denise Almonte)
Aerial Work Platforms 239

Introduction 239
Types of Aerial Work Platforms 239
OSHA Regulations 242
Industry Standards 246
Specifications for Aerial Work Platforms 246
Power Source 247
Tire Type 247
Maximum Platform Height 247
Wheel Base Width 248
Load Capacity 248
Safety Features 249
Uses 249
Rental/Lease and Purchase 250
Training Requirements. 250
Operator Training 251
Training of Mechanics 252
Inspection Requirements 253

Annual Inspections 253
Frequent Inspections 253
Pre-Operation Inspection 253
Common Hazards and Precautions 254
Electrocution Hazards 254
Electrocution Precautions 255
Fall Hazards 255
Fall Precautions for Bucket Trucks and Boom Lifts 256
Fall Precautions for Scissor Lifts 256
Collapses/Tipovers 257
Hazards of Boom Lifts and Bucket Trucks 257
Precautions with Bucket Trucks and Boom Lifts 258
Hazards with Scissor Lifts 259
Precautions with Scissor Lifts 260
Other Hazards and Precautions 260
Fall Protection 261
Bucket Trucks 262
Boom Lifts 262
Scissor Lifts 262
Typical Malfunctions and Injuries 262
Bucket Truck Case Study 1: Electrocution (8) 263
Bucket Truck Case Study 2: Electrocution (9) 263
Bucket Truck Case Study 3: Electrocution (10) 263
Bucket Truck Case Study 4: Fall (11) 263
Bucket Truck Case Study 5: Tipover (12) 264
Boom Lift Case Study 1: Electrocution/Fall (13) 264
Boom Lift Case Study 2: Fall (14) 264
Boom Lift Case Study 3: Boom Collapse (15) 265
Boom Lift Case Study 4: Caught Between (16) 265

Scissor Lift Case Study 1: Electrocution (17) 265
Scissor Lift Case Study 2: Fall (18) 266
Scissor Lift Case Study 3: Tipover (19) 266
Scissor Lift Case Study 4: Tipover (20) 266
Scissor Lift Case Study 5: Caught Between (21) 267
References 267
Chapter 19 Useful Forms and Checklists 270
Checklist 1 Sample Pre-Operation Checklist for Bucket Trucks 271
Checklist 2 Sample Pre-Operation Checklist for Boom Lifts 273
Checklist 3 Sample Pre-Operation Checklist for Scissor Lifts 275
Checklist 4 Sample Pre-Delivery and Frequent Inspection Form for Boom Lifts 277
Checklist 5 Sample Frequent Inspection Checklist for Scissor Lifts 279
Checklist 6 Sample Annual Inspection Checklistfor Bucket Trucks 282
Checklist 7 Sample Annual Inspection Form for Boom Lifts 285
Checklist 8 Sample Annual Inspection Form for Scissor Lifts 293

Chapter 20 (contributed by Michael McCann, PhD)
Stairways 301

Types of Stairways 301
OSHA Regulations 301
Industry Standards 303
Installation 304
Training Requirements 305
Common Hazards 305
Common Precautions 308
Fall Protection 309
Typical Malfunctions and Injuries 309
Case History No. 1: Fall through Stairway Opening (11) 309

Case Study No. 2: Fall while Installing Staircase (12). 310
Case Study No. 3: Fall from Stairway Landing (13) 310
References. 310

Chapter 21 (contributed by Michael McCann, PhD)
Ladders 313

Types of Ladders. 313
Portable Ladders 314
Non-Self-Supporting Ladders 314
Job-Made ladders. 314
Self-Supporting Portable Ladders 315
Fixed Ladders. 317
OSHA Regulations 317
Ladder Construction. 318
Ladder Use 319
Industry Standards 320
Installation and Removal of Portable Ladders. 320
Ladder Selection 321
Moving Ladders 322
Setting Up All Portable Ladders 322
Setting Up Straight and Extension Ladders 322
Leveling and Stabilizing the Ladder 323
Setting Up Self-Supporting Ladders. 325
Removing Ladders 328
Ladder Care 328
Installation of Fixed Ladders 329
Training Requirements 329
Inspection Checklists. 330
Common Hazards 330
Common Precautions 332

Working from a Ladder. 332
Stepping On/Off a Ladder at Height 334
Descending Ladders 334
Fall Protection. 335
Fixed Ladders 335
Fall Protection on Portable Ladders 336
Typical Malfunctions and Injuries 337
Case History No. 1: Fall from Extension Ladder (16) 337
Case History No. 2: Collapse of Extension Ladder Section (17) 337
Case History No. 3: Collapse of Extension Ladder (18). 337
Case History No. 4: Movement of a Step Ladder (19) 338
Case History No. 5: Fall from a Fixed Ladder (20) 338
Case History No. 6: Electrocution While Carrying Metal Ladder (21). 339
Case Study No. 7: Roofer Falls from Ladder After Contacting Overhead Power Line (22). 339
References. 340
Portable Ladder Inspection Checklist. 343
General 343
Inspection Item 343
References 344
Guidelines for Extension Ladder Safety Set-up and Repositioning 345
Select the Ladder 345
Scanning the Work Area. 345
On-Site Ladder Inspection 345
Ladder Extension/ Set-up 346
Ladder Testing/ Securing 346
General Ladder Handling. 347
Guidelines for Extension Ladder Safety – Use 348
Recheck Setup at Start of Shift and After Breaks. 348
Climbing. 348
Raising or Lowering Materials and Tools. 348

Working From a Ladder 349
Stepping On/Off a Ladder at Height. 349

Appendix A: General Requirements for Scaffolding. 351
Appendix B: Safety Standards for Aerial Lifts. 377
Appendix C: Safety Standards for Specific Scaffolds. 383
Appendix D: Scaffold Construction. 401
Appendix E: Scaffold Hazards. 411
Appendix F: Scaffold Specifications 415
Appendix G: Scaffold Shoring and Forming Institute. 429
Appendix H: Most Frequent Citations. 477
Appendix I: Calculating Scaffold Planks. 485
Appendix J: OSHA-Approved Safety and Health Plans 497
Appendix K: MultiEmployer Directive 507
Appendix L: OSHA Regional Offices 525
Appendix M: Glossary 529
Index. 539

Introduction

This text is designed for both the experienced user of scaffolding, ladders, and elevated work platforms, as well as the novice. The ultimate objectives for those who use this text are to become familiar with:

- All types of scaffolding and elevated work platforms commonly used in the construction industry
- The regulations which are promulgated under federal law to control the use of such scaffoldings, elevated work platforms, ladders, and stairways
- The consensus standards and other recommended industry practices which further improve the safety and utilization of scaffolding and other elevated work platforms, as well as stairways and ladders
- A rational basis to develop an effective safety program
- Aid in developing a site-specific safety program for scaffold, ladder, and electrical work platforms

A scaffold is defined as a temporary elevated work platform and its support structure. For purposes of this text, specialized equipment such as ladders, aerial lifts, and other elevated platforms are included in this definition and will be presented in light of the same standards and regulations.

Scaffolds result in numerous fatal accidents and injuries. The construction industry gives rise to very intensive use of scaffolds. These scaffolds are moved and/or dismantled more frequently and are used under more adverse conditions. This situation causes the majority of accidents.

According to the Bureau of Labor Statistics, some 65% or 2.3 million construction workers are regularly involved in the use of scaffolds and other elevated work platforms. The Bureau estimates that this results in some 50 deaths and 4500 injuries per year, which costs the construction industry around $90 million dollars. Estimates project that the total consensus approximation cost of these accidents and injuries is $360 million to $1.5 billion dollars. Documented injury accidents are only a small portion of the total number of accidents and costs. The goal is to assist in preventing even a minor part of this injury, death, and property damage.

72% of all fatal scaffold accidents can be related to one of four primary causes:

- A failure of the planking
- A failure of the support or support structure of the scaffold
- Slipping or falling scaffold
- Being struck by falling objects while working on a scaffold

The 1992 – 1999 average number of falls from scaffolds resulting in deaths, is documented by the Center to Protect Workers' Rights, (The Construction Chart Book, 2002, p. 37). Figure 1 indicates that 17.8% of all fatal falls in construction occur in falls from scaffolds and 14% result from falls from ladders. Therefore, 31.8%, or just under one third, of total fatal falls are a result of falls from scaffolds and ladders and other work platforms. Unclassified locations account for an additional 25%.

Figure 2, from the same publication and page, indicates that fatal falls associated with the collapse of the structure are a large percentage of the fatalities associated with aerial lifts and suspended scaffolds.

These basic statistics give a good indication of the magnitude of the problem and help identify problem areas.

According to the Center to Protect Workers' Rights, (The Construction Chart Book, 2002, p. 38), in 1999, non-fatal falls are indicated to account for 40,061 injuries. The construction industry has the largest number of falls (Figure 3) when compared to other industries, resulting in days away from work. That translates into 69 falls per 10,000 workers.

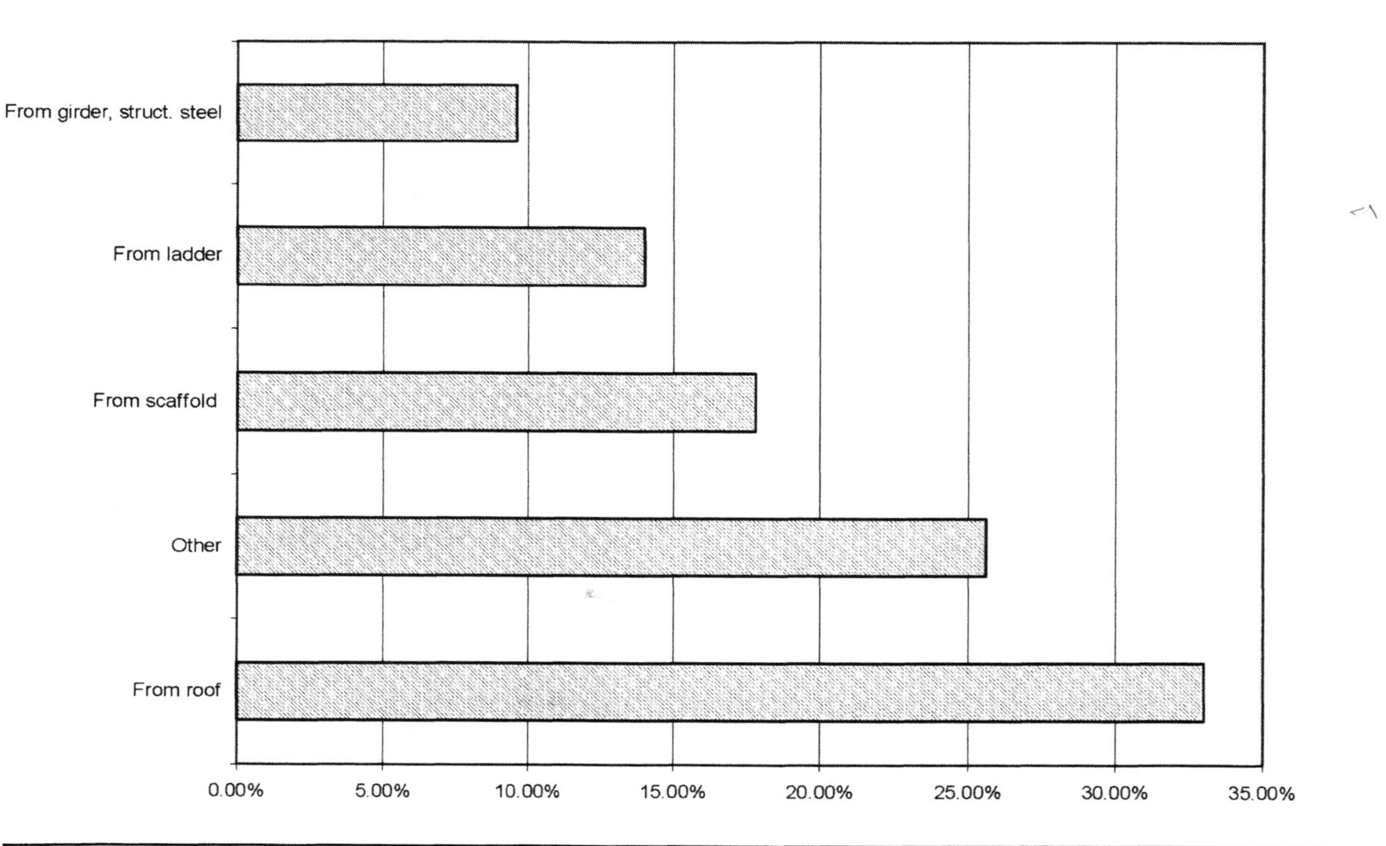

FIGURE 1 Distribution of causes of deaths from falls in construction, 1992-99 Average.

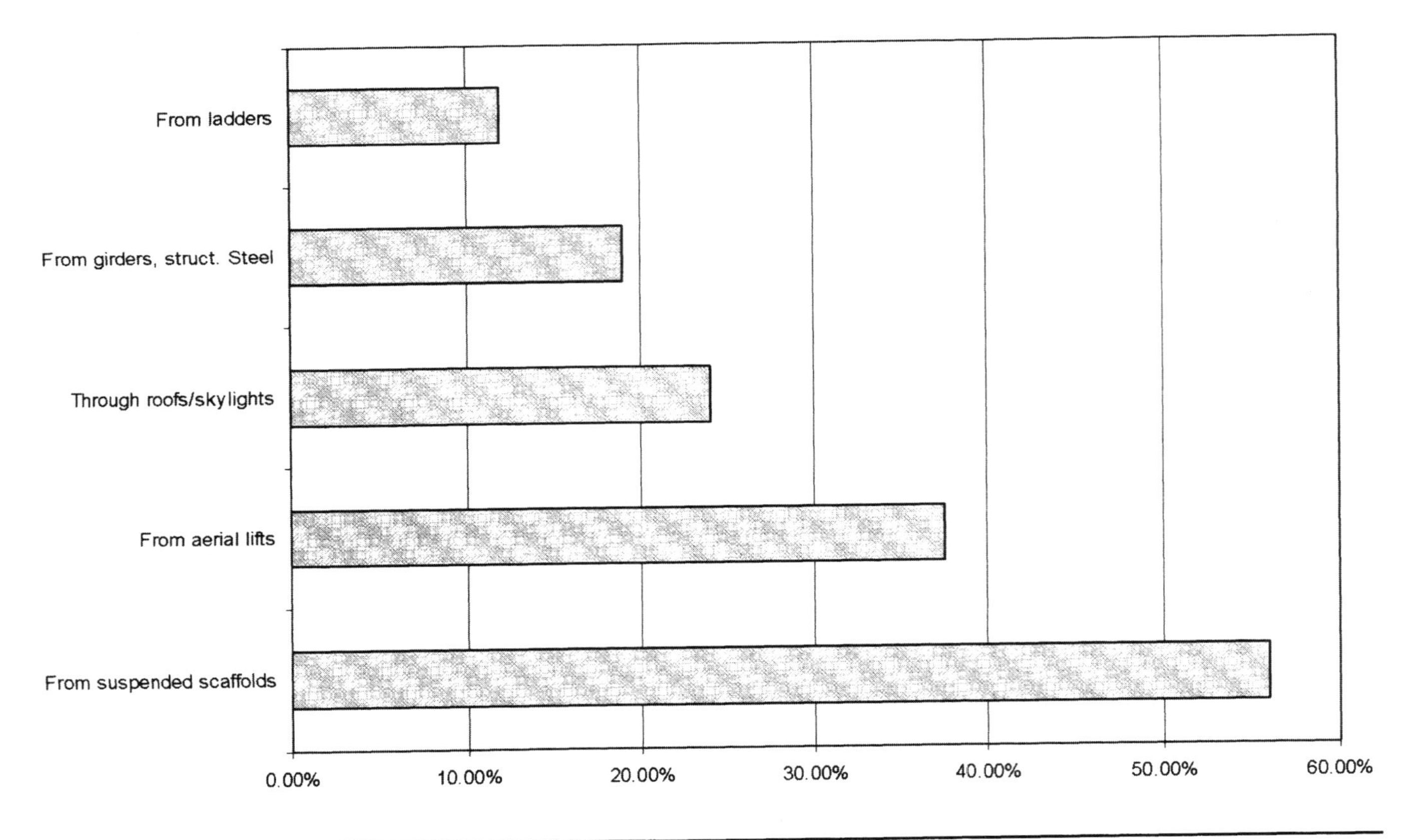

FIGURE 2 Percentage of collapses in selected fall categories in construction, (1992-99 average).

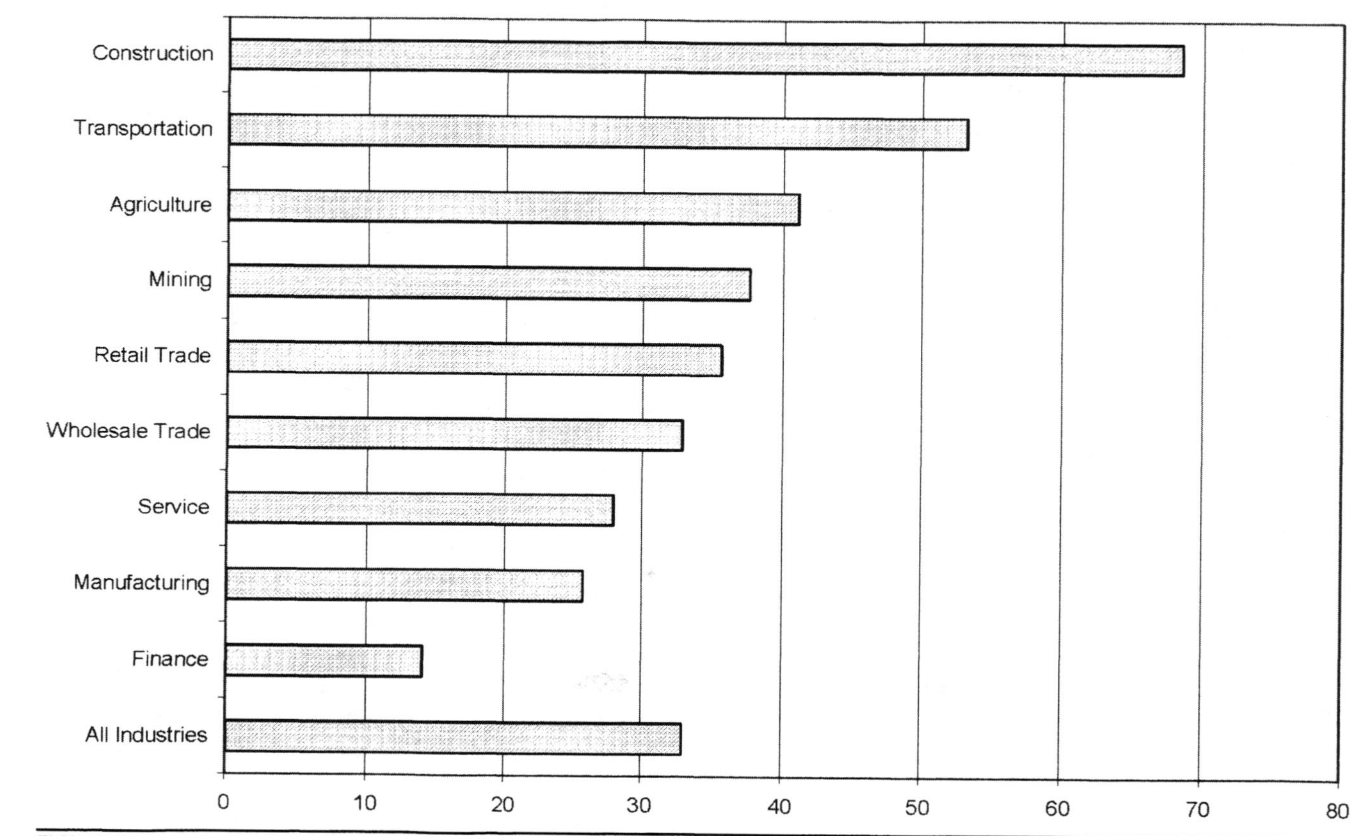

FIGURE 3 Rate of work-related nonfatal falls, by industry, 1999.

Using the same time period, falls from scaffolding account for 8.9% of lost work days, while falls from ladders account for 20.7% of lost work days. We are looking at a total of 29.6%, or again, almost 1/3 of lost works days as a result of injuries from falls. Far more fatal falls occur from scaffolds than do injuries. Because of the methods of collecting data and the reporting system, fatality statistics are considered more reliable.

The Occupational Safety and Health Regulations for Scaffolds and Aerial Lifts were most recently revised in 1996 and re-published by the Occupational Safety and Health Administration. These regulations for construction are published in CFR 29, SubPart L, Section 1926.450. These subpart regulations, in addition to providing general requirements for all scaffolding, in Section 1926.452, list requirements for specific types of scaffolds. Ladders and Stairways are covered in CFR 29, SubPart X, Section 1926.1050. Although the regulations do not change frequently, current regulations may be viewed and printed from the OSHA website, www.osha.gov, clicking on "regulations" and selection "construction".

Although some states maintain their own state OSHA plans, each of these plans is required to be at least as stringent as Federal OSHA. Readers are urged to check with these statutes for more stringent regulations.

The premise of this text is that the Occupational Safety and Health requirements for construction, as promulgated by the Occupational Safety and Health Act, are the minimum requirements for the erection, use, and dismantling of scaffolds. We will endeavor to provide industry standard practices, as well as state of the art developments, to promote safety and efficiency in the use of scaffolds.

Prior to the August 30, 1996 revision, the Occupational Safety and Health Regulations divided scaffolds into sections based upon the type and use of the scaffold. The OSHA regulations divide scaffolds into a small general section and 25 separate categories of scaffolds:

- Pole scaffolds
- Tube and coupler scaffolds
- Fabricated frame scaffolds

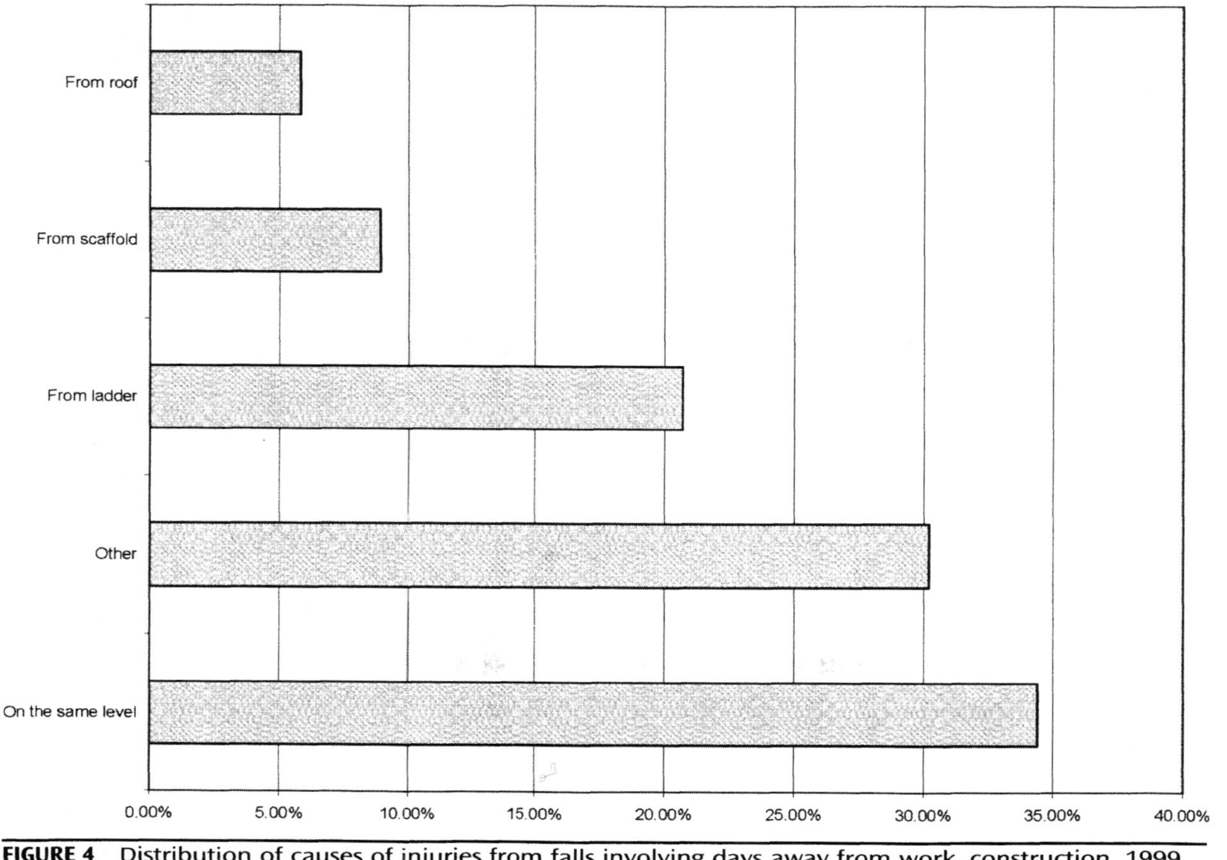

FIGURE 4 Distribution of causes of injuries from falls involving days away from work, construction, 1999

- Plasterers, decorators and large area scaffolds
- Brick layer square scaffolds
- Horse scaffolds
- Form scaffolds and carpenters brackets scaffolds
- Roof bracket scaffolds
- Outrigger scaffolds
- Pump jack scaffolds
- Ladder jack scaffolds
- Window jack scaffolds
- Crawling boards
- Step platform and trestle ladder scaffolds
- Single-point adjustable scaffolds
- Two-point adjustable suspension scaffolds
- Multi-point suspension scaffolds, stone-setters multi-point adjustable scaffolds, and masons multi-point adjustable suspension scaffolds
- Catanary scaffolds
- Float (ship) scaffolds
- Interior hung scaffolds
- Needle-beam scaffolds
- Multi-level suspension scaffolds
- Mobile scaffolds
- Repair-bracket scaffolds
- Stilts

The 1996 revision of SubPart L is divided into general requirements, scaffold platform criteria for supported scaffolds, criteria for suspended platform access, use, fall protection, falling object protection, and then in 1926.452, specific requirements for the individual types of support. Scaffold users are well advised to become familiar with the regulations, that relate to the specific type of scaffold or elevated platform they utilize.

In this text we will discuss the general requirements and group the 25 scaffold types into five major categories:

- Supported scaffolds
- Suspended scaffolds
- Mobile scaffolds
- Aerial lifts
- Stairways and ladders

In addition to these specific sections on scaffold regulations, this book contains sections on:

- Specific individual scaffolding components that can present hazards. These are, specifically, soil support, fiber rope, wire rope, wood and wood planks; in addition, separate sections are presented on fall protection during erection use and dismantling.
- Preplanning
- Accident statistics
- Selection & use of scaffolds
- Emergency response & rescue
- Electrical hazards

Specific emphasis is devoted to the regulatory requirement for a competent person. Although this request has been in force since the standard was removed, this requirement is recently being strictly enforced.

OSHA has implemented a multi-employer policy in which they may cite multiple contractors for OSHA violations that occur on the site. The impact of the multi-employer policy on contractors is significant.

Of major importance to the user is the training required, the specific requirements for training in the various processes, and also the duties and requirements of the competent person.

In addition to the Occupational Safety and Health Regulations, we also present industry recommended practices, as promulgated in the industry consensus standards and also other industry publications from contractor associations and trade associations.

Each section contains discussion of major hazards and accidents or incidents associated with those hazards for a specific type of scaffolding, as well as sketches and diagrams to identify the scaffold component, types, and specific hazards. Helpful hints for best practices are included to assist in complying with the regulation and to provide the optimum level of safety for workers and others.

Chapter 1

OSHA Regulations and Industry Standards

History of Construction Legislation and How Subpart L Developed

Job site and public safety did not become a concern of society until the time of King Charles II of England, who passed the first recorded safety regulations. These regulations addressed the fencing of quarries. They are some of the earliest documented regulations by a government for the safety of its citizens.

The Commonwealth of Massachusetts was the first in this country to pass safety legislation. Although this was primarily industrial safety, there was concern for workers. In 1877, Massachusetts passed factory acts, patterned after the acts passed in England. These were aimed at the inspection of factories and public places. Items affected were dangerous machines, such as those involving gears, weights, and pulleys. It required that machines were to be turned off when cleaning, except for steam engines, and mandated fire escapes for all buildings over three stories in height. This was a major step in job safety, but did not address many other conditions that affected the worker.

Workers' Compensation was applied to federal workers in 1908, when Theodore Roosevelt and Congress passed a compensation bill. Today, the Federal Employees Compensation Act (FECA) covers federal workers. Other workers were not covered until 1911, when Wisconsin passed the first effective workers' compensation law.

Previous to the Wisconsin law, New York attempted to put workers' compensation laws into effect, but the state legislature failed to pass such legislation. Workers' compensation must have been a radical idea at this time, because it apparently was challenged and, in 1916, the Supreme Court declared Workers' Compensation to be constitutional. Workers' Compensation laws and benefits for workers other than federal workers are determined by each state.

In 1936, the Walsh-Healey Public Contracts Act was passed. This act required that contractors performing work for the federal government in excess of $10,000 follow certain safety requirements. In 1969 the Construction Safety Act was enacted to provide safety regulations for workers' protection on federally funded construction projects. This act was published as the code of Federal Regulations, Part 29 CFR 1926. The regulations did not provide protection to the workers on the majority of construction sites; only federally funded sites were required to comply with these regulations.

Many individual states were ahead of the federal government in developing labor laws that were intended to protect workers from unsafe conditions on worksites. These state regulations varied in content and were not uniform from state to state. This variation in the regulations caused much confusion. This was particularly true of large contractors who worked in many states, but on an infrequent basis. Construction is a unique field of endeavor. Employers and employees go from job to job in a relatively short period of time and these jobs can be in various locations within a state or from state to state. Awareness of safety requirements became burdensome and the requirements were difficult to follow.

On December 29, 1970, The Williams-Steiger Occupational Safety and Health Act (OSH), Public Law 91-596, 91st Congress, S. 2193, was enacted. This law was enacted for many reasons, but primarily because of the concern for the large numbers of fatalities, injuries, and illnesses that were occurring to workers in this country and the effect that these illnesses and injuries were having on the economy. This act was signed by President Nixon and went into effect in April of 1971. The OSH Act established, for the first time, a nationwide, federal program to protect almost the entire work force from work-related deaths, injuries, and illnesses, as well as providing a central organization to enforce these safety regulations.

The Occupational Safety and Health Administration (OSHA) promulgated regulations under the act. Parts of the Construction Act incorporated sections of 29 CFR 1926, which applied to the construction industry. In addition, the Act also promulgated additional regulations for General Industry - 29 CFR 1910, Maritime - 29 CFR 1915 and Longshoring - 29CFR 1918.

The Occupational Safety and Health Act of 1970 gave the Secretary of Labor authority to adopt established federal standards, including the Construction Safety Act, as OSHA standards. These regulations, as they applied to construction, are designated as 29 CFR 1926. With 29 CFR 1926, protection from death, injury, and illness became mandatory and uniform for the construction industry. Standards applying to scaffolds were adopted in SubPart L as a part of this process.

Development of Subpart L

During the first two years of the OSH Act, amendments involving scaffold provisions, planking grades, wood pole scaffold construction, overhead protection, scaffold loading, and plank spans were revised. Then, in 1977, OSHA began a complete review of SubPart L. In November 1986 a notice of proposed rulemaking on scaffolds in construction was issued.

In January 1988 OSHA announced it would convene an informal public hearing in March to hear comments on specific issues related to scaffolds, fall protection, and stairways and ladders. An Administrative Law Judge oversaw the hearing and in August 1988 the rulemaking record was certified, including all comments and submissions, which closed the record for this proceeding.

In 1988 The American National Standards Institute (ANSI), a voluntary consensus standards group, through its accredited standards committee "Safety in Construction and Demolition Operations", A10, approved a revision of their standard, A10. 8-1977,"Scaffolding Safety Requirements". The revised standard updated the safety requirements for the use of scaffolds in construction and demolition operations. OSHA is required, under the provisions of Section 6 (b) (8) of the Act, to respond to the reason why the ANSI standard, "A10. 8-1988, Scaffolding Safety Requirements", will better protect workers than the OSHA standard. This, again, initiated a rulemaking session. The A10 Committee again revised A10.8, Scaffolding Safety Requirements in 2000. This is the current standard.

As a result of this session and its comments from employers, business, labor unions, trade associations, state governments, and other interested parties,

OSHA was able to develop a rulemaking record providing a sound basis for the promulgation of a revised Sub-Part L.

During this session many interesting facts were disclosed through the required economic analysis:

- There are 3.6 million construction workers currently covered by Sub-Part L.
- Of the 510,500 injuries occurring in the construction industry annually, 9,750 are related to scaffolds.
- Of the 924 occupational fatalities that occured, at least 79 are associated with working on scaffolding.

In OSHA's statistical estimates based on 4. 5 million construction workers then covered under Subpart L, and based on the BLS (Bureau of Labor Statistics) accident data, some of the findings show that:

- 72% of the scaffold accidents in which workers were injured were attributed either to the planking or support giving way, the employee slipping, or the employee being struck by a falling object. Planks' slipping was the most commonly cited cause.
- 70% of the workers learned about safety requirements for scaffolds through on-the-job training. Approximately 25% had no training at all.
- Only 33% of all scaffolding was equipped with a handrail.

OSHA, in making these kinds of findings, has determined that workers using scaffolds are exposed to significant risk of harm. Scaffold-related fatalities still account for approximately 9% of all fatalities in the construction workplace. Based on this kind of data, OSHA determined that revision of Subpart L was necessary. OSHA has also determined that the standard, as revised, clearly states the employer's duties and appropriate compliance measures. The new Subpart L was passed on August 30, 1996. Information for the above was taken from the Federal Register, Friday, August 30, 1996 issue, 29 CFR Part 1926, Safety Standards for Scaffolds Used in the Construction Industry, Final Rule. This publication is complete and fully covers the regulation and the comments taken into account for revising this regulation. The regulations as promulgated are included in various appendices.

Regulations of state plans may be more stringent. Contractors who work in the states that have their own regulations should obtain copies of state plans. Refer to the appendix where the state plans are listed.

The Structural Work Act

The Structural Work Act, also known as the "Scaffold Act", became law in Illinois on June 3, 1907. It was an important attempt to provide legal recourse for injury and was primarily meant to cover workers injured in falls from scaffolds who previously had little or no legal recourse.

Workers' Compensation did not come into effect in Illinois until 1913. It was intended to replace the Scaffold Act, covering all types of injuries incurred on the job. Workers' Compensation is designed as "no fault", that is, employers paid into the system and injured workers would receive benefits. The worker didn't have to go to court for compensation for his injury and the employer did not have any additional liability.

While the Workers' Compensation Act was intended to supersede the Scaffold Act, it was never repealed. An Illinois court in 1952 struck down the section of the Acts that prohibited third-party lawsuits. So, from then until 1995 when the Scaffold Act was repealed, any scaffold worker could collect from his Workers' Compensation and sue any contractor on the site, as well as the property owner and architect. The Illinois courts meanwhile expanded the legal meaning of "scaffold" to include almost anything on the construction site: scaffolds, trenches, the building itself, the floors of buildings, and ladders.

What was the impact of the Scaffold Act on the Illinois economy? The Watson-Wyatt group in 1988 gave a conservative estimate of the cost of the Structural Work Act from 1952 when it was re-instated. Their figures show:

- The Act cost employers $113 million in direct costs in insurance alone, plus as much as $170 million in indirect costs (court appearances, time lost from work for depositions, etc.). These costs were passed on to the consumer.
- Legal fees were hard to track, but it can cost as much as $10,000 for an uninvolved party to be removed from the suit. Involved parties could count on many times more than that defending the suit.

- Companies were relocating to neighboring states because of the additional insurance costs related to the Structural Work Act, costing the state of Illinois to lose revenue, jobs, and economic growth.

Supporters of the Structural Work Act say the act makes the job safer. Those against the Structural Work Act say no. The year the act was repealed showed the same number fatalities from falls as from the previous year. Since 1995, the number of fall-related deaths has been 20 or less each year. In Illinois specifically, injury rates dropped 32% from 1991 to 1997 and the rate of permanent partial disabilities dropped 43% from 1991 to 1997. OSHA has a good deal of influence in reducing these figures by their enforcement of the OSHA rules and regulations. Contractors cannot only be fined by OSHA for having unsafe worker conditions, but can go to prison. This has had more effect on reducing the accident fatality and injury rates than the Structural Work Act, according to those opposed to the SWA.

Supporters of the Structural Work Act contend that the Workers' Compensation Act does not provide adequate death benefits. Those against the Structural Work Act support reforming the Workers' Compensation system rather than adding the burden of the SWA on an industry already struggling with exorbitant insurance costs. Reinstatement of the SWA could result in substantial cost to the contractor in terms of:

- Increased general liability/umbrella insurance costs by 20 - 30%
- Insurance costs to general contractors and architects would have to be part of the job cost and passed on to the owner.
- Cost implications may very well have developers looking twice at potential projects, putting future Illinois jobs at risk.

These are very real concerns for the state of Illinois and its people. One 1998 estimate indicates that reinstatement of the SWA could add an additional $300 to $400 to the cost of building a new home in the Chicago area. A large company would be driven away by the prospect of hundreds of thousands of dollars in additional costs when building plants and offices in Illinois.

Both sides of this issue are passionate in their beliefs and the last word is that the Illinois legislature is looking at the possibility of reinstatement of the SWA. It will be up to the legislators now, however this is an issue worth tracking.

Chapter 2

Accidents, Injuries, and Fatality Statistics

Scaffolding, elevated work platforms, and aerial lifts are becoming ever more popular ways of providing access to elevated workstations in the construction industry. This increase is the result of several influences that have occurred in the construction industry.

HAZARDS OF ELEVATED WORK PLATFORMS

The recently published Subpart M, August 9, 1994, OSHA "regulations on fall protection," have increased the requirements for fall protection. These regulations have made the use of scaffolding and elevated work platforms for access to elevated workplaces more desirable and economical in lieu of allowing workers to work from unprotected portions of the building. The recent enactment of the new OSHA regulations "Steel Erection", Subpart R, have encouraged, if not mandated, the use of scissor lifts and aerial lifts in the process of steel erection and also, more predominately, the use of ladders, to obtain access to elevated work areas in the steel erection industry.

Another major factor influencing the increased use of scaffolding, aerial lifts, and other elevated work platforms, has been the dramatic increase in the variety and numbers of platforms available. New models of aerial lifts with larger capacities, longer reach, and most of all, availability, make them much more economical and expedient to erect, utilize, dismantle, and or move. One of the more convincing arguments, perhaps as a result of the usage, is the realization that many people are killed and/or injured because of falls from elevated areas. A simple phone call will have a lift with a reach of over one hundred feet delivered to the jobsite almost immediately.

Scaffolds, ladders, aerial lifts, and other elevated work platforms, by their nature, present severe exposures to workers on the scaffold and others on construction sites and also to adjacent properties. The primary hazard is one of falling from the height by the workers who are working on or erecting the scaffolds. In addition to falls, numerous electrocutions have occurred related to scaffolds, either coming into contact with high voltage lines as they are being relocated, scaffolding components coming into contact with high voltage lines as they are being erected, and/or workers handling materials on a scaffold coming into contact with adjacent power lines.

There have been numerous collapses of scaffolds which not only result in death and injury of the worker on the scaffold but also to adjacent persons on public property, particularly in large metropolitan areas where scaffolding is erected adjacent to or over a public right of way. In many cases the scaffolds are erected with some notoriety and in recent years have been erected in many public atmospheres. Recent scaffolding accomplishments include the Washington Monument, Statue of Liberty, the renovation of Grand Central Station in New York, and also currently the AOL/Time Warner Building. Scaffolding such as this, while monumental in nature and scope, as well as other scaffolding erected in less historical or monumental structures, still present severe exposure to adjacent structures, the public, or public facilities. Accidents relating to this kind of scaffolding involving the public or public property, although frequent, are not statistically documented in any of the literature. They do make the evening news, with resultant adverse publicity for involved parties.

Death and injuries to workers erecting, working from or around scaffolding, ladders, or elevated work platforms are reasonably well documented through the requirements of various agencies. The magnitude of the task of collecting injury information makes it difficult to classify accidents, to establish primary causal factors, and count the numbers of workers exposed and the time that they are exposed.

STATISTICS

It is impossible to count all of the injuries that occur. Many accidents go unreported because of the reporting requirements and the difficulties in collecting and compiling the data. A true count of the number of workers who are employed in any or all industries is no more accurate than the employment statistics published. The number of employees exposed to the hazards of falling or those who utilize scaffolds are nearly impossible to compute or approximate. The Bureau of Labor Statistics, The US Department of Labor, and other organizations collect and compile injury statistical data. While this data is subject to the above statistical errors, it is the best available.

These same government agencies also compile data relating to the number of fatalities that result from falls. While this data is subject to some of the same variables as those for injuries, it is somewhat more accurate and to some extent more valid. The National Traumatic Occupational Fatalities (NTOF) comprises the surveillance system of NIOSH, the National Institute of Occupational Safety and Health. Their figures are based on an analysis of death certificates from all 50 states and is, I believe, the most statistically valid evaluation available at this time. Based on the NTOF analysis of 15 years of data, the following tables may be used with reasonable confidence.

Studies performed by NIOSH and based on NTOF, as recently published, accumulate the fatalities relating to falls that occurred in the time period 1980-1994. Their statistics indicate that 8,102 fatal falls occurred in all industries in the United States. This is an average of 540 fatalities per year due to falls, which is 9.6% of the total work-related deaths in the United States (Table 2.1). Of these 8,102 fatal falls, 49.9% or 4,044 occurred in the construction industry.

Table 2.2 provides a breakdown of the construction fatalities by industry classification codes (SIC).

Table 2.3 indicates that 25% of all fatal falls are related to scaffolds and ladders. The increased use of this equipment will undoubtedly result in larger numbers.

Table 2.3 is one which we compiled using figures from the Bureau of Labor Statistics from 1980 to 1984 and extended with annual compilations through 2001. This table relates the total fatalities recorded by contractor SIC (Standard Industry Classification) codes to those caused by falls. As you can see, falls represent a significant contribution to the total deaths in most categories.

TABLE 2.1 Number and rate per 100,000 workers of fatal falls by industry division, United States, 1980-1984. *Source: National Traumatic Occupational Fatalities (NTOF) Surveillance system*

INDUSTRY DIVISION	N	%	RATE
Agriculture/Forestry/Fishing	507	6.3	0.99
Mining	211	2.6	1.69
Construction	4044	49.9	3.89
Manufacturing	943	11.6	0.30
Transportation/Communications/ Public Utilities	518	6.4	0.45
Wholesale Trade	145	1.8	0.22
Retail Trade	250	3.1	0.09
Finance/Insurance/Real Estate	106	1.3	0.10
Services	765	9.4	0.14
Public Administration	181	2.2	0.23

TABLE 2.2 Detailed SIC codes with 50 or more fatalities from elevations, United States, 1990-1994. *Source: National Traumatic Occupational Fatalities (NTOF) Surveillance System*

SIC CODE	DESCRIPTION	DEATHS
1542	General Contractors: Nonresidential Buildings, Other Than Industrial Buildings and Warehouses	205*
1611	Highway and Street Construction, Except Elevated Highways	177**
1761	Roofing, Siding, and Sheet Metal Work	147
1791	Structural Steel Erection	121
-17.%1	Carpentry Work	79
721	Painting and Paper Hanging	71
1521	General Contractors — Single Family Houses	63
1731	Electrical Work	59
7349	Building Cleaning and Maintenance Services, not elsewhere classified (n.e.c.)	56
783	Ornamental Shrub and Tree Services	56

* *Default category when death certificate specifies building construction*

TABLE 2.3 Fatal falls related to scaffolds and ladders *(continued on next page)*

	1980-1994	1994	1995	1996	1997	1998	1999	2000	2001
Fatal Injuries All Industries	8,102	6,588	6,210	6,112	6,218	6,026	6,023	5,915	5,900
Fatal Falls	3,969	658	645	684	715	699	716	733	808
Construction Fatalities	4,044	1,027	1,048	1,039	1,107	1,171	1,190	1,154	1,225
Construction Fatal Falls	1,573	329	335	336	377	382	378	373	421
General Building Fatalities		189	175	183	194	212	183	175	201
General Building Fatal Falls		79	78	81	77	87	79	75	94
General Building-Falls to Lower Level		72	76	79	77	87	77	76	
General Building Falls from Scaffold		17	13	9	10	26	17	17	
Residential Building Fatalities		82	152	82	89	114	98	97	99
Residential Building Fatal Falls		26	73	34	30	46	35	39	45
Residential Building Construction Falls to Lower Level		26	38	34	31	47	35	40	
Residential Building Falls from Scaffold			8			12	7	8	
Non-Residential Building Fatalities		96	83	88	92	79	76	67	83
Non-Residential Building Fatal Falls		45	40	42	43	36	40	33	41
Non-Residential Building -Falls to Lower Level		40	37	40	43	35	40	33	
Non-Residential Building Falls from Scaffold		13	5	5	6	13	10	9	
Heavy Construction -Except Building Fatalities		247	82	247	252	271	280	284	267
Heavy Construction - Except Building Fatal Falls		20	37	21	24	21	19	21	32
Heavy Construction-Except Bldg -Falls to Lower Level		18	20	21	25	21	19	21	
Highway and Street Construction Fatalities		76	90	91	82	104	86	82	98
Highway and Street Construction Fatal Falls	-	-		4	5				
Highway and Street Construction -Falls to Lower Level					5				
Heavy Construction -Except Highways Fatalities		166	154	155	163	166	189	199	168
Heavy Construction - Except Highways Fatal Falls		16	16	17	14	19	17	17	28
Heavy Construction-Except Hwys -Falls to Lower Level		17	16	17	16	19	17	18	
Special Trades Contractors Fatalities		591	613	599	648	679	709	672	735
Special Trade Contractors Fatal Falls		231	231	230	270	270	277	266	288
Special Trade Contractors -Falls to Lower Level		226	230	227	266	263	272	263	
Special Trade Contractors Falls from Scaffold		50	40	52	49	57	47	38	

TABLE 2.3 *(continued from previous page)* Fatal falls related to scaffolds and ladders

Plumbing, Heating, & Air-Conditioning Fatalities	71	62	62	61	65	66	73	79
Plumbing, Heating, & Air-Conditioning Fatal Falls	16	15	14	20	15	12	18	16
Painting and Paper Hanging Fatalities	40	47	41	45	42	36	45	50
Painting and Paper Hanging Fatal Falls	25	20	28	23	18	21	23	27
Painting and Paper Hanging Falls from Scaffold	9	8	11	8	6		8	
Electrical Work Fatalities	78	91	71	82	114	107	81	94
Electrical Work Fatal Falls	18	21	13	19	19	20	16	18
Masonry, Stonework, Tile Setting, and Plastering Fatalities	53	66	52	42	53	57	65	60
Masonry, Stonework, Tile Setting and Plastering Fatal Falls	27	31	24	21	27	27	34	25
Masonry, Stonework, Tile Setting, and Plastering Falls from Scaffold	16	16	15	11	20	12	15	
Carpentry and Floor Work Fatalities	31	38	36	45	42	46	58	66
Carpentry and Floor Work Fatal Falls	16	21	16	23	25	24	27	34
Roofing, Siding, and Sheet Metal Work Fatalities	89	95	99	104	91	96	98	116
Roofing, Siding, and Sheet Metal Work Fatal Falls	60	64	65	74	64	72	68	72
Roofing, Siding, and Sheet Metal Work Falls from Scaffolds	5		6	5		5		
Concrete Work Fatalities	34	34	29	33	28	42	28	31
Concrete Work Fatal Falls	4	7		6	7	6	5	7
Water Well Drilling Fatalities	8		11	10	10			15
Water Well Drilling Fatal Falls								
Miscellaneous Special Trade Contractors Fatalities	182	179	196	221	227	244	215	218
Miscellaneous Special Trade Contractor Fatal Falls	58	52	66	78	91	91	70	76
Miscellaneous Special Trade Contractor-								

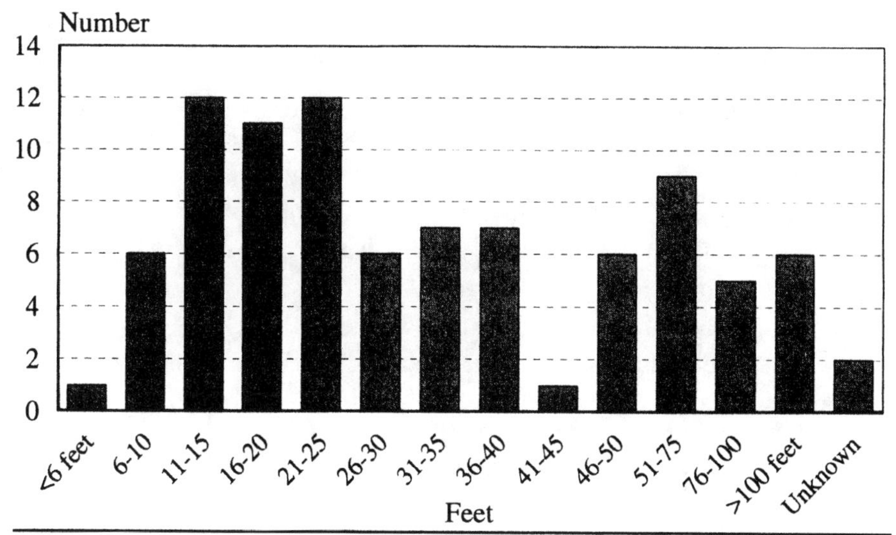

FIGURE 2.1 Distribution of Fatal Falls by Height of Fall, FACE Investigations, 1982-1997 (N=91)

These statistics that record fatalities occurring due to falls are significantly more reliable and accurate than attempts to collect data relating to the numbers of total injuries. These statistics, other than the number of fatalities, provide little information or insight as to the overall affect because the total numbers of workers exposed to scaffolding and the duration or percentage of their time spent on elevated work scaffolds or platforms is not accurately documented. Although it is impossible to ascertain a fatality or an injury rate from workers utilizing the scaffolding, these numbers are sufficient to show that a larger number of people are falling and dying than is acceptable. The rate of death, i.e., fatalities/man-hours or fatalities/worker, pales in comparison to the raw total numbers of fatalities.

In the same NIOSH study published in 2000, they tabulated 90 fatal falls investigated in great detail. These 90 fatal falls were classified from the accident report by fall distance (Figure 2.1).

It is significant to note that 42 of the 90 falls, or 46% of the fatalities, occurred with a fall distance of less than 25 feet. 19 of the 90, 21% or roughly 1 in 5, occurred at a fall distance of less than 15 feet. The predominant number of falls, but also statistically the number of injuries suffered in falls, are clustered in that range of less than 25 feet.

It can be argued that the predominance of the fall deaths being recorded as occurring at less than 25 feet is caused by innumerable reasons, i.e., more workers are subjected to that fall distance, or conversely, workers at greater heights are more careful. Either of these hypotheses could be argued and supported with rational means by studies. The end result, however, is certain. At fall distances between 6 and 25 feet there is a significant number of workers subjected to fatal falls and that (1) no fall is an acceptable fall and (2) no area can be excluded from consideration. The fall hazards associated must be addressed. Falls from heights result in significant deaths and injuries and the relative height is not significant; scaffolds, ladders, or elevated platforms cannot be eliminated from our concern and consideration on the basis of their working height.

These statistics demand that in the overall procedure of scaffold selection, erection, use, and dismantlement, we need to devote significant attention to the factors that impact on safety of scaffolds or elevated work platforms of any size. Pre-planning procedures prior to using the scaffold, the training of the workers who are going to erect the scaffold, as well as the training of employees who work on the scaffold are significant and necessary functions which must occur regardless of the height of the scaffold, ladder, or elevated work platform.

Why Do Scaffolds Fall?

Scaffold failures occur for a variety of reasons, and when analyzed in detail, there is more than one causal factor involved in the failure of a scaffold. These causal factors, primary, secondary, or other, which result in failures to scaffolds, can be analyzed and divided into four major categories.

The first category is an improper design of the scaffold. The large majority of scaffolds utilized are assembled from manufactured components of systems. These are designed and tested for the specific use as a component of a scaffold. Many times the entire system as an assembly is not considered.

It is a rare or unusual occasion when individual manufactured components fail structurally because of a design deficiency. The majority of the improper design failures in manufactured components result from a failure to properly analyze the supporting structure, the footings, the overhead suspension, or failure to properly analyze the loads that are going to be imposed on the scaffolding.

Engineering Design Analysis

A smaller number of scaffolds are designed for specific purposes, incorporating components other than standard manufactured components, and require a complete engineering of the analysis design . One of these scaffolds which was analyzed by the writer consisted of a large area work platform, approximately 400 square feet, which was to be suspended underneath a bridge for purposes of repair, maintenance, and painting of the bridge.

A detailed engineering analysis was performed of the loads on the scaffold during work operations. It was suspended by six cables from attachment points to the bridge girders that were all engineered in great detail. The suspension ropes were half-inch independent wire rope core, fitted to specially designed attachment clamps for the bridge girders and into special attachments to support the working platform. Although there were several discrepancies in the analysis, the overall design, as it was in place underneath the bridge, was adequate for its intended purpose.

However, the analysis did not consider lifting the scaffold into place and lowering the scaffold from the workplace. The plan to raise and lower the scaffold was to utilize four electric hoists, also attached to the bridge and to the scaffold. The load of the scaffold was transferred to four cables, one at each corner, as opposed to six. This fact increased the load on each individual cable by 50%. For example, 6 cables of 5k each = 30k total weight of scaffold divided by 4 cables during lowering = 7.5k per cable.

In addition, it was assumed that the entire load of the scaffold would be equally distributed among the four cables. The practical impossibility of simultaneously lowering four electrical hoists, independently operated, was not addressed. In this configuration it is reasonable to expect that transfer of the majority of the total load to two cables would occur. Cables on diagonal corners could support the entire platform in an almost balanced condition as the hoists are manipulated with the result that two $\frac{3}{8}$" cables from the electric hoists supported the entire weight of the scaffold, that is 7.5k x 2 or 15,000#/cable. The result is that the loads on the cable were increased from 5k/cable to 15k, or a 300% increase! This was a basic design flaw in the application of engineering principles and resulted in failure in the supporting cable from the hoist motor and the collapse of the scaffolding. This would be clearly classified in the category of improper design.

There were other deficiencies in the scaffolding that may or may not have contributed to the resulting injuries, but the primary cause of injuries and, indeed, the only cause of the scaffold collapse was the improper design of the suspension system. Whether the scaffold is a manufactured system or a special design, the design criteria has to evaluate the loading conditions of the scaffold throughout its life, not only during the erection phase, the work phase, and the dismantling phase. Each of these presents unique hazards in the utilization of a scaffold. Each type of scaffold has unique hazards that will be dealt with in the appropriate chapters of the text.

CASE STUDIES

The following insert illustrates improper use of equipment (not causally related to the accident) and improper procedure by the worker. Each example summarizes the incident, the investigation, and the recommendations. These are case studies from the NIOSH Publication: Worker Deaths by Falls. It was published in 2000. FACE is an acronym for Fatal Assessment and Control Evaluation. These case studies were taken from these FACE files.

FACE 9012 Case Study # 1

A journeyman painter died when the swing scaffold he was using to access the interior of a 68-foot-tall by 32-foot-diameter municipal water tank fell. The painter was working from a single point suspension scaffold near the top of the tank. The painter was wearing a safety belt and lanyard secured to a lifeline. When he finished painting the upper area of the tank, the painter disconnected his lanyard from the lifeline and moved to the other end of the scaffold to hand the spray paint gun he was using to his foreman. The foreman had just taken the spray paint gun from the victim when he heard a "pop" and saw the scaffold on which the victim was standing fall to the floor of the tank 65 feet below. Investigation after the incident revealed that the two "U" bolts on the cable which supported the block and tackle from which the scaffold was suspended had loosened enough to allow the cable to slip through them, causing both the scaffold and all of its supporting hardware to fall. The victim was pronounced dead at the local hospital approximately 1½ hours after the incident. NIOSH investigators concluded that, in order to prevent similar incidents in the future, employers must ensure that:

- Appropriate personal protective equipment be worn properly and consistently whenever the potential for a serious fall exists
- Suspension scaffold rigging be inspected periodically to ensure that all connections are tight and that no damage to the rigging has occurred since its last use

Introduction

On October 22, 1989, officials of the Indiana Occupational Safety and Health Administration notified the Division of Safety Research (DSR) of the death of a 37-year-old male painter who died on October 21, 1989, when the suspension scaffold he was working on fell 65 feet inside a municipal water tank. The Indiana Occupational Safety and Health Administration requested technical assistance, and on November 30, 1989, a DSR safety specialist conducted an investigation of this incident. The investigator discussed the case with state officials and emergency services personnel. The investigator reviewed the incident with company officials, and investigated and photographed the incident site.

The employer, a painting contractor with 20 employees, has been in business for seven years. The company has a designated safety officer and written safety rules and procedures, but no formal training program. The victim was hired as a journeyman painter, and had worked for the company for one month at the time of the incident. The victim had previously been employed as a painter by other contractors for approximately 10 years.

Investigation

The victim was a member of a three-man crew engaged in painting the interior and exterior of two 68-foot-tall by 32-foot-diameter municipal water tanks. The crew had been working on this project for two weeks prior to the incident, and had completed all work on one tank and most of the exterior work on the second.

On the day of the incident, the crew arrived at the worksite at approximately 11:30 a.m. The crew consisted of a foreman, the victim, and a groundman. The foreman was going to spray paint the interior of the water tank while the victim was to finish work on the exterior of the tank. The groundman was to work inside the tank handling the spray paint lines used in the operation. The victim, a journeyman painter, asked to paint the interior of the tank. The foreman agreed, and the victim proceeded to paint the interior of the tank while the foreman finished work on the exterior of the tank.

Access to the interior of the tank was provided through a manhole on the side of the tank at ground level, and a second manhole located on top of the tank. This second manhole was reached by climbing a fixed ladder on the exterior of the tank.

The interior sidewalls of the tank were reached via a swing scaffold rigged inside the tank. This scaffold consisted of an aluminum ladder secured to a steel "stirrup" (a steel bar bent into a box shape and installed perpendicular to the ladder) at each end. The ladder was thus subjected to loading while in a horizontal position, rather than in the vertical position for which it was designed. Cables from each stirrup ran to a common tie-off point. A cable from this common tie off point then passed through a block and tackle. By pulling on this cable the entire scaffold could be raised and lowered from the ground level of the interior of the tank (Figure 2.2). The block and tackle, which supported the scaffold, was secured by a single cable that looped around a vertical steel pipe on top of the tank and fastened back to itself by two "U" bolts.

The entire crew entered the tank through the lower manhole. The groundman and the supervisor then raised the scaffold with the victim on it to the top

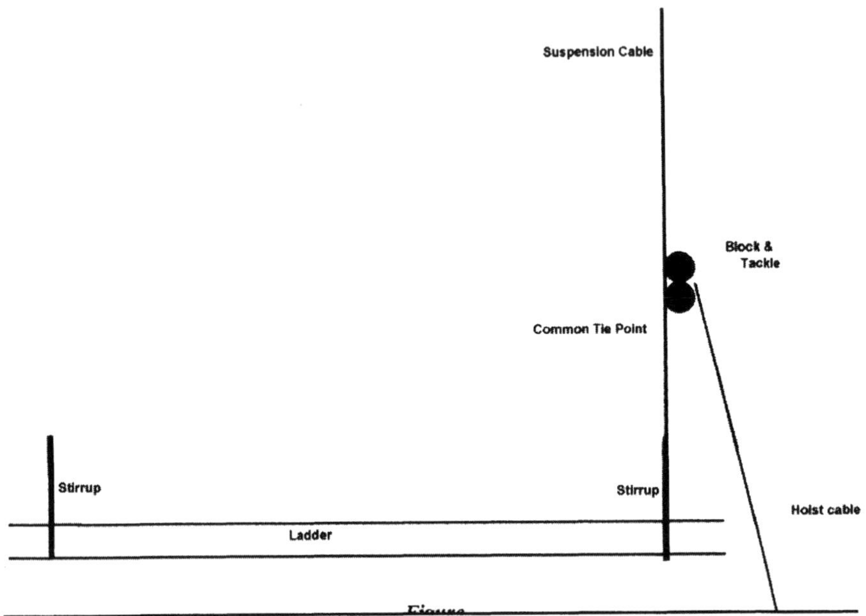

FIGURE 2.2 Raising and lowering a scaffold by use of a block and tackle

of the tank. The victim was wearing a safety belt and lanyard that was secured to a lifeline, with the lifeline secured to a steel railing on the top of the tank. The victim proceeded to paint the top few feet of the tank's interior. The foreman climbed the exterior ladder to the manhole on top of the tank to help complete work near the tank's top. At approximately 1:00 p.m., the victim completed painting at the upper level. He then disconnected his lanyard from his lifeline and moved over to where he could hand the paint spray gun to the foreman so the foreman could finish a small area at the top of the tank. The foreman had just taken the spray gun from the victim when he heard a "pop" and saw the victim and the scaffold on which he was standing, fall to the floor of the tank 65 feet below. The victim and the scaffold struck the floor of the tank, barely missing the groundman. The foreman called to the groundman and told him to go next door and call an ambulance. The foreman then descended the ladder on the exterior of the tank and went in to assist the victim. The Emergency Medical Service (EMS) unit arrived on the scene approximately five minutes after the incident, removed the victim from the tank via the lower manhole, and transported him to the local hospital. The victim was pronounced dead at the hospital at 2:29 p.m.

Investigation after the incident revealed that the two "U" bolts on the cable which supported the block and tackle had allowed the cable to slip through them, causing both the scaffold and all of its supporting hardware to fall. This particular rig had been used daily for two weeks preceding the incident with no problems.

Cause of Death

The cause of death was listed by the coroner as "hemorrhage from severe liver laceration and brain stem hematoma."

Recommendations/Discussion

Recommendation #1: Appropriate personal protective equipment should be worn at all times whenever the potential for a serious fall exists.

Discussion: In this case the victim was wearing a safety belt and lanyard, however, at the moment when the incident occurred, he was not hooked up to his lifeline. This failure to use PPE at all times during the job allowed the victim to experience a fatal fall when a scaffold failure occurred.

Recommendation #2: Suspension scaffold rigging should be inspected periodically to ensure that all connections are tight and that no damage to the rigging has occurred since its last use.

Discussion: The scaffold rigging in this case had been used daily for two weeks prior to the incident; however, no periodic inspection program was in place. It appears that the "U" bolts holding the scaffold had loosened over time, although this loosening had not been observed by workers at the site.

Recommendation #3: Equipment should only be used for the purpose for which it was designed.

Discussion: The "scaffold platform" in this incident was a simple aluminum ladder. This ladder was designed to support a load in a vertical position but was being utilized to support a load while in a horizontal position. While this did not directly contribute to the incident, the potential for a failure of the ladder while being used in this manner was certainly present.

Failure to Understand and/or Comply with Regulatory Requirements

The National Institute for Occupational Safety and Health (NIOSH), Division of Safety Research (DSR), performs Fatal Accident Circumstances and Epidemiology (FACE) investigations when a participating state reports an occupational fatality and requests technical assistance. The goal of these evaluations is to prevent fatal work injuries in the future by studying the working environment, the worker, the task the worker was performing, the tools the worker was using, the energy exchange resulting in fatal injury, and the role of management in controlling how these factors interact.

FACE 890 Case Study # 2

On November 15, 1988, a 53-year-old male foreman and a 28-year-old male painter died when the scaffold from which they were working collapsed, causing them to fall 48 feet to the ground below.

Contacts and Activities

State officials of the Occupational Safety and Health Program notified DSR of this fatality and requested technical assistance. On December 15, 1988, a DSR research industrial hygienist met with the state OSHA official who investigated the incident, representatives of various companies, and local police and fire departments that were involved in the incident, and photographed the site.

Overview of Employer's Safety Program

The employer is a painting company with 50 employees. The company consists of a painting division with 29 painters and a small construction division. Most of the company business involves painting buildings and other outdoor structures. The company's Hazard Communication Program consists of a brief verbal orientation to new employees concerning the potential hazards of various chemicals contained in paint. The company also has Material Safety Data Sheets (MSDS) available. However, the company has no written safety program, and did not have any safety meetings or training specifically addressing fall prevention or fall protection.

The foreman involved in this incident had a total of 20 years of experience as a painter, including 15 years with the company as a painter foreman. The other employee had 2 years of experience with the company as a painter.

It should be noted that two painters with the same company died in separate, previous work-related incidents. In 1987, a painter fell to his death from an aerial bucket, and in 1972, a painter suspended in a boatswain's chair came in contact with a power line and was electrocuted.

Synopsis of Events

The company was hired to paint the outside of several tanks at a petrochemical storage plant. The storage tanks are 48 feet high and 56 feet in diameter. Stairs that wind around the tanks provide access to the top. The tops of the tanks are smooth and have a slight downward slope that extends from the center to the outside edge.

The two workers began painting the tanks from the bucket compartment of an aerial bucket truck without wearing any type of fall protection equipment.

The painters used this painting method for several days and had completed one tank and were nearing completion of a second tank. However, gaining access to the unpainted side of the tank by using the bucket truck was not possible because other tanks were too close and some aboveground piping was in the way. Therefore, the foreman decided to finish painting the second tank using a two-point suspension scaffold.

The two workers arrived at the site in the morning on November 15, 1988 and set up the scaffold. The scaffold consisted of a worker platform of tubular steel, measuring 2 feet wide by 17 feet long, with two outside guardrails 24 inches and 48 inches above the platform. The platform was suspended by two wire suspension cables, each of which was 5/16 of an inch in diameter. The cables hung vertically from two tubular steel outriggers placed on top of the tank with the outboard ends extending 24 inches beyond the edge of the tank. The cables ran through an electrically operated hoist on each end of the scaffold platform. This allowed the workers to raise or lower the scaffold platform to the desired height.

Although there were no eyewitnesses to the incident, physical and circumstantial evidence suggests the following:

- The scaffold outriggers had been installed on top of the tank with only 200 pounds of counterweight. There were two 50-pound steel bars on each of the two outriggers. The outriggers had been set up to keep the suspension cables at a horizontal distance of 24 inches from the side of the tank. In order to maintain this horizontal distance, the scaffold manufacturer required a minimum of 600 pounds of counterweight for this type of scaffold (300 pounds on each outrigger) to counterbalance the workload.
- The outriggers were not tied off to prevent them from slipping.
- One end of a lifeline had been tied to a large vent pipe on the top center of the tank and the other end looped around the side of the scaffold guardrail.
- Two buckets, each containing approximately four gallons of paint, were placed on the scaffold platform.
- The two workers climbed onto the scaffold platform, raised the scaffold platform all the way to the top, got off on top of the tank, climbed down the tank stairs, and went to lunch.

- Presumably, some time during the afternoon while the workers were on the scaffold platform, the outriggers slid off the top edge of the tank and the entire scaffold along with the two workers fell approximately 48 feet to a hard-packed gravel surface below.

The two workers were not discovered until 4:56 p.m. At that time a truck driver at the petrochemical storage plant was on his way to lock up the plant premises when he noticed the bodies and scaffold wreckage. The truck driver immediately notified the local fire department emergency medical service. Paramedics arrived at the scene in approximately five minutes and upon examining the victims, could not detect any signs of life. The county coroner subsequently arrived and pronounced the two workers dead at the scene.

Cause of Death

The medical examiner reported the cause of death for both workers as multiple blunt force traumas.

Recommendations/Discussion

Recommendation #1: Employers should ensure that all employees required to work from elevated work platforms understand the potential danger of a fall, and the proper methods of erecting, placing, securing, and using scaffolds.

Discussion: Occupational Safety and Health Administration (OSHA) Safety and Health Standard 29 CFR 1926.451(g)(3) requires that the outriggers of this type of scaffold be securely anchored and that properly designed scaffolds, "... shall be constructed and erected in accordance with such design." For this type of scaffold and the way it was being used, the scaffold manufacturer recommends: (1) a minimum of 600 pounds of counterweight on the inboard end of the outrigger beams (300 pounds on each outrigger), and (2) that the outriggers also be securely tied back.

The fact that the workers only used 200 pounds of counterweight (100 pounds on each side) and that they did not tie back the outriggers indicates they did not fully understand the proper methods of erecting and securing this type of scaffold. The employer should ensure that all employees understand the danger of working on scaffolding. This includes the necessity of properly securing scaffold suspension points. Properly set up, the type of scaffold and anchoring system used in this incident would not have fallen.

Recommendation #2: Where the potential for a fall from an elevation exists, employers should ensure that fall protection equipment is provided and used by workers.

Discussion: Although a safety line had been tied to the top of the tank and the workers had safety belts with rope-grab devices at the site (and possibly on the scaffold) during the incident, they were not being worn by the workers. The use of a safety belt/lanyard combination is required by 29 CFR 1926.451(i)(8) for use on two-point suspension scaffolds. The use of the safety belt or body harness/lanyard with a rope grab device is appropriate for persons working from scaffolds at varying heights. Properly used, this type of fall protection would have prevented the workers in this incident from falling, even when the scaffolding fell.

Recommendation #3: Scaffolds should be erected under the supervision of persons who are competent in the use of scaffolds.

Discussion: OSHA Standard 1926.451(a))(3) states: "No scaffold shall be erected, moved, dismantled, or altered except under the supervision of Competent Persons. " The fact that the workers in this incident did not set up the scaffold according to the manufacturer's specifications points out that the workers did not understand the correct way to erect the scaffold under those circumstances. The scaffold erection should have been supervised by a worker experienced in erecting this type of scaffold.

Recommendation #4: When workers are assigned hazardous tasks, or must work at hazardous workstations (such as elevated scaffolds), a standby person should be assigned to continually observe, give assistance, and ensure timely response in the event of an emergency. Additionally, close supervisory contact should be maintained periodically throughout the duration of the work.

Discussion: On the day of the fatal incident, the two victims apparently worked alone, unobserved. They were not discovered until 4:56 p.m. when a truck driver was locking up the plant. No one was assigned to observe the work from the ground; additionally, the workers were apparently unsupervised from the time they installed the scaffold until the scaffold collapsed and they fell to the ground. Had the scaffold collapse and resultant fall been observed by someone standing by on the ground, help might have been summoned and emergency medical care administered promptly to the victims, improving their chances of surviving the traumatic injuries they received. In any workplace situation that involves the potential for traumatic injury, a "buddy system" and close, periodic supervision are essential to protect the lives of exposed workers.

Recommendation #5: The designers/owners of tanks of this type should design and install appropriate tank anchorage points for maintenance purposes.

Discussion: Permanent structures of this type are known to require extensive maintenance when they are designed. It is essential that designers/owners of these facilities incorporate anchorage points on tank roofs to which workers can adequately secure scaffolds and lifelines. Omission of designed anchor points causes workers to improvise anchors or not use them at all. This increases the possibility that a scaffold will be erected incorrectly. If scaffold anchor points had been available on the tank involved in this incident, the scaffold may not have been incorrectly erected, resulting in its failure. Also, if anchor points had been available, it's likely that the workers in this situation may have been tied off, thus preventing their fall when the scaffold fell.

Recommendation #6: All employers should develop and implement a safety program designed to help workers recognize, understand, and control hazards.

Discussion: Company management must ensure that employees are trained to recognize and avoid hazardous work conditions and that the work environment is safe. Employers should develop and implement a safety program to protect workers as required by OSHA Standard 1926.20. Additionally, OSHA Standard 1926.21(b)(2) requires employers to "...instruct each employee in the recognition and avoidance of unsafe conditions and the regulations applicable to his work environment to control or eliminate any hazards or other exposure to illness or injury." The company had no formal safety program, and there were no standard operating procedures for any of the tasks performed. Even after having two previous worker fatalities, the employer failed to provide written safety rules and training in safe work procedures. Although a relatively small company, the employer should immediately evaluate the tasks performed by workers, identify all potential hazards, and then develop and implement a safety program addressing these hazards. Prior to starting any job, the employer should conduct a job site survey, identify all hazards, and implement appropriate control measures.

Proper Assembly of Scaffolds

Scaffolds are assembled from manufactured components and, during assembly, great care must be exercised to assure that all of the components are properly assembled and that the components match. One of the most serious accidents involved the movement of a mobile tower. The tower was assembled from frame scaffold sections mounted on rollers. The scaffold was being rolled forward and the frames shifted, one side moved ahead and the effective base width narrowed, because there was no diagonal brace to prevent the occurrence. The erection of scaffolding must assure that all components are present and installed. Manufacturers and/or distributors do not always supply the necessary components. Although it is expected that the manufacturer/distributor would supply the necessary components to safely erect the scaffold.

FACE 9013 Case Study #3

The following case study demonstrates the improper erection and maintenance of the scaffold.

Summary

A 21-year-old asbestos worker died as a result of injuries sustained in a 12-foot fall from a scaffold. The victim was a member of a six-man crew engaged in the removal of asbestos-contaminated insulation from a series of large ducts on the exterior of an electric power generation plant. The victim was removing asbestos insulation from a large outdoor metal duct approximately 14 feet above the ground. The worksite was accessed by tubular metal scaffolding. The victim was working at the 12-foot level of the scaffold. The scaffold was not decked at this level. Instead, the crew had installed a single 2-inch by 12-inch plank across the tubing. The plank extended beyond the tubing on both sides and was not fastened in position to the tubing. Instead, the crew had driven two nails into each end of the plank at 45-degree angles to hold the plank against the tubing while allowing them to slide the plank along the tubing to various areas where they were working. The nails on one end of the plank had loosened sufficiently to slip free from the scaffold. The weight of the victim on the opposite end of the plank caused the plank to rise up in the air, dropping the victim to the ground below. NIOSH investigators concluded that, in order to prevent similar occurrences in the future, employers and employees must:

- Fully deck all scaffolds and secure decking material in accordance with existing OSHA regulations
- Provide appropriate fall protection equipment to all employees whenever the potential for a serious or fatal fall exists
- Provide safety training to all employees that addresses all potential hazards to which the employee may be exposed, especially the proper use of scaffolding and fall protection equipment.

Case Study # 4

Introduction

On November 2, 1989, officials of the Indiana Occupational Safety and Health Administration notified DSR of the death of a 21-year-old male asbestos worker who died as a result of a 12-foot fall from a scaffold on August 18, 1989 and requested technical assistance. On November 29, 1989 a DSR safety specialist conducted an investigation of this incident. The case was discussed with state officials and emergency services personnel, and the incident was reviewed with company officials.

The employer is a large, multistate insulation contractor. The company employs 500 individuals, including 100 asbestos workers who remove asbestos-contaminated insulation. The company has a designated safety officer and written safety policy and procedure manuals. The victim had been employed by the company for one month at the time of the incident. Although the victim had received safety training from the company, the primary focus of this training was asbestos removal procedures. (Note: The company had no policy in place requiring the use of fall protection equipment at the time this incident occurred. Since the incident, a policy has been implemented requiring the use of safety belts/lifelines whenever employees are working on any elevated surface.)

Investigation

On the day of the incident, a six-man crew was removing asbestos-contaminated insulation from a series of large ducts on the exterior of an electric power generation plant. The crew had been working intermittently at the plant (as environmental conditions permitted) for several days prior to the incident.

On the morning of the incident, the crew started work at 7:00 a.m. The victim was removing asbestos insulation from a large outdoor metal duct approximately 14 feet above the ground. The worksite was accessed via metal tubular scaffolding.

Each section of the scaffolding formed a 10-foot by 6-foot rectangle. The victim was working at the 12-foot level where the scaffold was not decked. Instead, the work crew had installed a single 8-foot-long, 2-inch by 12-inch plank across the tubing. This plank extended approximately 14 inches past the end of the scaffold tubing on one side, and approximately 10 inches past the tubing on the other side. This plank was not fastened in position on the scaffold tubing; rather, the crew had driven two nails into each end of the plank at 45-degree angles, to hold the plank against the tubing (Figure 2.3). This procedure allowed the workers to slide the plank along the tubing (along the 10-foot side) to various areas where they were working.

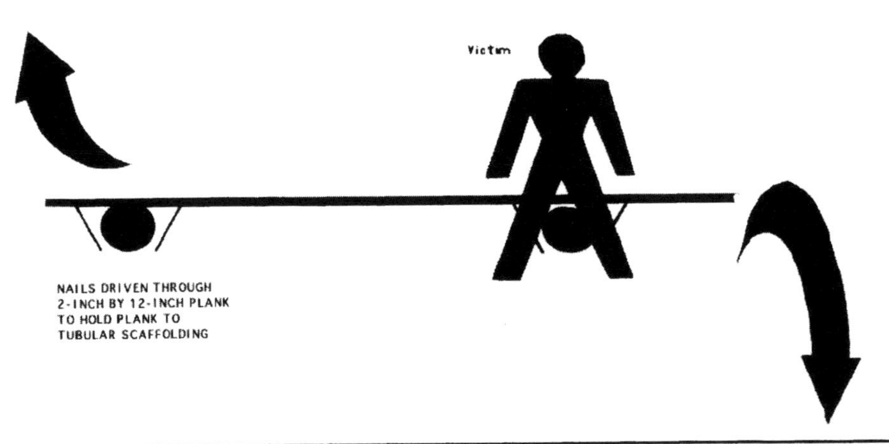

FIGURE 2.3 Nails driven through 2-inch by 12-inch plank to hold plank to tubular scaffold

The victim was sitting astride the tubing, on the end of the plank with the 14-inch overhang, in order to remove asbestos from the duct. Two co-workers had stepped off of the same plank about five minutes earlier.

Although no one witnessed the incident, it appears that the nails on one end of the plank had loosened sufficiently to allow the plank to slip free from the scaffold. The weight of the victim on the opposite end of the scaffold caused the plank to rise up in the air, dropping the victim to the ground below where he was struck by the falling plank. The two co-workers heard the victim and the plank strike the ground. The co-workers immediately called for help and went to the victim. The victim was conscious but told the co-workers that he "couldn't feel anything." He asked the co-workers to "put my hands on my chest," which they did.

Local Emergency Medical Service (EMS) personnel arrived on the scene approximately eight minutes after the incident, and promptly transported the victim to a local hospital. The victim died in the hospital 65 hours after the incident.

Cause of Death

The Coroner gave the cause of death as bronchopneumonia and sepsis complicating blunt force injury of the neck.

Recommendations/Discussion

Recommendation #1: All scaffolding should be fully decked and all decking material secured in accordance with 29 CFR 1910.28(1) and 1926.451(2) (Code of Federal Regulations, Washington, D.C.: U.S. Government Printing Office, Office of the Federal Register.)

Discussion: The scaffold in this incident was not properly decked, and the planking used for decking was not properly secured. These two conditions were major contributors to this incident.

Recommendation #2: Appropriate fall protection equipment should be employed wherever the potential for a serious or fatal fall exists.

Discussion: The victim was not using any type of fall protection equipment when this incident occurred. A safety belt and lanyard could have prevented this fatality had they been utilized.

Recommendation #3: Employee safety training should address all potential hazards to which an employee may be exposed.

Discussion: While the employer in this case did have a safety training program, this program dealt specifically with the hazards of asbestos removal work. The employer's program failed to address other hazards to which employees may be exposed, such as falls and the proper installation and use of scaffolding. A comprehensive safety training program emphasizing the hazards posed by falls and stressing the use of appropriate personal fall protection equipment, might have prevented this fatality.

FACE 8827 Case Study #5

Dry Wall Finisher Dies in Fall from Ladder on Scaffold

On June 23, 1988, a 55-year-old male dry wall finisher was fatally injured when he fell 22 feet from a portable wooden stepladder that was on top of a 17-foot-high mobile scaffold.

Contacts/Activities

On June 27, 1988, a state Occupational Safety and Health official notified DSR of this fatality and requested technical assistance. On July 12, 1988, NIOSH met with a company representative, discussed the incident with the OSHA compliance officer, photographed the site, interviewed a co-worker who witnessed the incident, and obtained a report from the local fire department's emergency medical service (EMS) rescue squad that responded.

Overview of Employers' Safety Program

The victim was a dry wall finisher working for a general contracting construction company. The company has been in business for approximately four years and currently employs 90 employees, including four dry wall finishers. The company uses written safety rules and procedures and provides on-the-job training to employees. The construction job site superintendent is responsible for administering the safety program that includes conducting weekly job site safety meetings with all the employees. The victim had almost four years of experience as a dry wall finisher. He had never received a reprimand for violating safety rules or procedures.

Synopsis of Events

The construction company had been contracted to build a multilevel brick high school. Construction started in October 1986, with completion scheduled for September 1988. At the time of the incident, most of the exterior work had been completed and the interior finishing work was in progress.

On June 23, 1988, two dry wall finishers were putting filler compound over the heads of the screws that secured sheetrock panels to the interior walls. They were working in the same room from separate scaffolds. The scaffolds were mobile metal scaffolds, 17 feet high, 7 feet long, and 5 feet wide, which were equipped with 8-inch rubber tires with locking casters. The victim's work platform was made up of two 2-inch by 10-inch, 7-foot-long wooden boards and one 2-foot-wide by 7-foot-long standard aluminum plank mounted across the top railing of the scaffold. Additionally, the victim placed an 8-foot wooden stepladder on top of the work platform to reach the upper sections of the wall, which was 25 feet high.

Prior to the incident a co-worker told the victim that the casters on the scaffold were not locked. The victim replied, "I want them that way." The victim positioned the stepladder on the scaffold platform and leaned the top of the ladder against the wall. When the victim climbed the ladder, the force exerted at the ladder's foot caused the scaffold to roll. The victim fell headfirst onto a concrete floor 22 feet below.

The construction superintendent, who was in an adjacent room, heard a disturbance and ran to the incident site. He immediately called the local EMS squad using a two-way walkie-talkie. An ambulance arrived four minutes later, and EMS personnel provided advanced life support. The victim was transported to a local hospital where he was pronounced dead on arrival.

Cause of Death

The coroner reported the cause of death as traumatic injuries to the head and chest.

Recommendations/Discussion

Recommendation #1: Employers should ensure that all employees required to work from elevated work platforms understand the potential danger of a fall, and the proper methods of erecting, placing, securing, and using scaffolds and ladders.

Discussion: Occupational Safety and Health Administration (OSHA) Safety and Health Standard 29 CFR 1926.451(e)(8) states that, "Scaffolds in use by any persons shall rest upon a suitable footing and shall stand plumb, also the casters or wheels be locked to prevent any movement." The employer should ensure that all employees understand the danger of working on scaffolding; this includes the necessity of locking casters or wheels. Employers should also instruct all employees to report all unsafe working conditions (e.g., the unlocked casters observed by the co-worker) to the supervisor. If the victim had locked the casters or the co-worker had reported this unsafe working condition, this fatality may have been prevented.

Recommendation #2: Employers should ensure that appropriate guardrails and toeboards are installed on mobile scaffolding used for work at levels exceeding 10 feet above the ground or floor.

Discussion: OSHA Safety and Health Standard 29 CFR 1926.451(a)(4) requires that guardrails and toeboards be installed on all open sides and ends of platforms more than 10 feet above the ground or floor. The work platform of the mobile scaffolding was 17 feet above the floor, and all four sides surrounding the platform were open. The employer should have equipped the mobile scaffolding with guardrails and toeboards before the platform was used.

Recommendation #3: Employers should ensure that mobile scaffolding platforms are tightly planked.

Discussion: OSHA Safety and Health Standard 29 CFR 1926.451(e)(4) requires that mobile scaffolding platforms be tightly planked for the full width of the scaffold. In addition to the hazard created by leaning an 8-foot wooden stepladder against the wall, the platform was only partially planked, creating an opening approximately 17 inches wide by 7 feet long. The employer should regularly inspect to ensure that all scaffolding meets the requirements established by the OSHA Safety and Health Standards (e.g., locked casters, installed guardrails, and tightly planked platforms, etc.).

Recommendation #4: In the event an employee is injured on the job, the employer should review, and revise if necessary, the safety rules and procedures, inspect the work site for unsafe working conditions, and initiate actions to ensure safe working conditions before work activities continue.

Discussion: This fall is one of four falls experienced by employees of the contractor or sub-contractor at this specific job site (initiated October, 1986). Although the previous three falls did not result in death, the workers involved received severe injuries including fractures and lacerations. One of these workers is permanently paralyzed as a result of a fall. It is evident that safety conditions are poor at this specific work site; the employer should initiate immediate action to correct these unsafe working conditions.

Listing of Citations

The regulations, as promulgated by the Occupational Safety and Health Act and as outlined in the previous chapter, also have associated penalties for lack of compliance. The Occupational Safety and Health Administration regularly publishes a listing of citations issued for a 12-month period. In the period of October 2001 through September 2002, the listing of citations (See Appendix F) indicates that 44,409 citations for violations of the Occupational Safety and Health Act were issued. These citations accrued $38,645,543 in assessed fines and penalties. The number one citation is a violation of 29CFR1926.451, General Requirements for All Types of Scaffolding. There were 8,377 citations issued, with total assessed penalties of $7,631,205.

The twelfth most issued citation was for violation of 29CFR1926.453, with 1,103 citations issued, resulting in penalties of $962,356 for violations of Manually Propelled Mobile Ladder Stands and Scaffolds. The fifteenth most issued citation was for training violations, 29CFR1926.454, with a total of 919 citations with $376,312 in penalties. The twentieth most issued citation was for additional scaffold violations under 29CFR1926.452, with 669 citations and $415,924 in assessed penalties. Scaffolds accounted for 4 of the 20 most frequently issued citations during the report period. This total of 11,068 citations out of 44,409, or 24.9% of all citations OSHA issued in the past fiscal year, was for violations of some form of scaffold standard. The total penalties assessed for these four scaffold violations categories is $9,385,592, a total of 24.6% of the total penalties assessed by OSHA.

These numbers, the fatalities which are occurring, the injuries, the numerous accidents, and the numbers of citations and penalties represent many, many fatalities and millions of dollars expended not only on fines, but damage and delay to construction sites. These items make a significant impact on the safety of the construction site and also the dollars expended on construction or expended by the contractors.

In addition, there are also numerous citations for fall protection and guardrails, which may or may not be related to scaffolds. There are also numerous citations and penalties assessed for ladders.

Chapter 3

Preplanning — Selection & Use

Preplanning is Essential!

No plan is a plan, a plan for failure. In this chapter we will discuss preplanning in general and preplanning in the selection and use of work platforms, scaffolds, aerial lifts, stairways, and ladders.

Pre-planning is identifying opportunities and hazards, eliminating those hazards that you can and managing those that you cannot. Make use of the best opportunities that you can identify. Choose which work platforms, scaffolds, aerial lifts, stairways, and ladders can help you achieve your goal of a productive and safe workplace.

Beginning

When should preplanning begin? As soon as possible! Preplanning should begin during the pre-bid phase and continue through the bid, award, prior to signing the contract, prior to starting the job, during the job, making change orders, and up to the completion of the project. Any time a change can be made in your means and methods that will enhance the performance or safety of the work, it should be recognized and implemented. It is just good management.

The Pre-Bid Process

During the pre-bid/bid process you need to determine the scope of the project. Review the contract documents for any specific requirements that you need to address. Determine the "who, what, when, and where." Who is responsible for determining what work platforms, scaffolds, aerial lifts, stairways, and ladders will be used, when they will be used and where? Each type of equipment has its positive and negatives. The selection and use of the most appropriate equipment is paramount to a well-managed, productive, and safe project.

Job Hazard Analysis and Other Planning Instruments

When you become the low bidder, and prior to the award and the signing of the contract, you need to do a Job Hazard Analysis (Figure 3.1) and determine what specific means and methods you will employ. This is the opportunity to determine what job hazards there are and what and how you will eliminate or manage them. Once you have determined what the project encompasses, it is time to pick your team to manage your jobsite and safety responsibilities. This team needs to review in detail the contract documents. Now is your best and perhaps last chance to make changes. Prepare a site specific plan, including, in as much detail as possible, how you plan to erect, use, and dismantle work platforms, scaffolds, aerial lifts, stairways, and ladders and how they are an integral part of your means and methods. Determine and describe who will be your engineer of record, Competent Persons, and qualified persons for each equipment and task as necessary. Remember, training must be accomplished under the direction of a Competent Person and, in some cases, licensing or certification may be mandatory. Remember only trained persons are allowed to use the equipment, and the responsible owner or employer is liable for misuse or inappropriate use of the equipment by his employees or others.

Prior to starting the job, you need to ensure that your site specific plan is completed and approved by the contract-appointed approval authority. This is most important because the approval endorses your means and methods and specifies what personnel will use the equipment. It also ensures that you will be in compliance with the contract documents and any subsequent changes that impact your means and methods may be a chargeable change order.

The Checklist below is minimal and should be added to and adapted to individual company and manufacturer's requirements. Checklists for various activities are effective when corrective action is taken. These corrective actions must be verified in order to render these inspections effective.

SAFETY PROGRAM/PLAN CHECKLIST

YES	NO		CORRECTIVE ACTION	DATE	BY WHO?
		Is there a written site-specific safety plan on site and available for viewing?			
		Is there a written safety program on site and available for viewing?			
		Has the site-specific safety plan been implemented by all contractors on site?			
		Have appropriate safety analyses been performed as required?			
		Are the results of these analyses available to affected contractors and their employees?			
		Are appropriate controls in place as specified by the safety analyses?			
		Are required competent persons designated and available on site for this project?			
		Is there a Professional Engineer on site where required?			
		Have competent persons conducted work site walk throughs as required by OSHA standards?			
		Do all contractors and subcontractors have the appropriate competent persons on site?			
		Does the site-specific safety plan reflect the actual site conditions?			
		If any scheduled activities create special hazards, are proper controls in effect to control those hazards?			

The above checklist is meant for use by persons who conduct daily work site inspections of safety programs and plans.

This checklist is not intended for the setup or initial implementation of the safety programs or plans, nor is it intended to provide a comprehensive evaluation of the safety and health plans.

Additional Comments:

Inspected by: ______________________________ Date: ____________________

Verification of All Corrective Action Taken:

Name: ______________________________ Date____________________

FIGURE 3.1 Job hazard analysis

Equipment and Training

Once you have established what equipment you are going to use, where it will be used, and who will be using it, you need to reduce your plan to a task specific schedule and plan. You need to insure that all personnel that are assigned specific tasks are qualified and have been trained. For example, the Competent Person who is going to erect tube & coupler scaffolding needs to be identified, as do other Competent Persons for the specific equipment that you are going to erect or use. Once the equipment and personnel have been chosen and trained, appropriate fall protection equipment needs to be chosen for those erecting the equipment, using the equipment, and removing or dismantling the equipment. Different fall protection equipment may be chosen for each task and different Competent Persons may be needed for erecting and dismantling different types of equipment and for training.

Some of the factors you will want to consider in choosing equipment will be the work to be performed, ease of access to the work area, number of people needing access, materials and tools to be stored as needed, maximum intended loads, methods of fall protection, and how the equipment works with your means and methods. Obviously, you will need to meet or exceed all OSHA, federal, state, and local laws and regulations as well as the designers or manufacturer's instructions, any consensus standards such as ANSI (American National Standards Institute), and the requirements of the contract documents.

As you make your selections, you will evaluate portable ladders, for instance, taking into consideration that the maximum load is 300 pounds while stairways need to be designed taking into consideration what the maximum intended load of both personnel and equipment will be.

In selecting aerial lifts, you will want to consider what the ground and weather conditions will be, and the heights and angles that you will be working. You will need to preplan your lifts taking into consideration the number and weight of personnel, materials, and tools and determine from a load chart the safe weight and extension limitations of the appropriate equipment. Additionally a critical consideration will be what, if any, of your means and methods will change the wind loads or center of gravity of the equipment.

When evaluating what scaffolds or work platforms you will want to use, you will need to consider the work to be performed, and the accessibility needs to the work area. This book deals with all of the commonly used scaffolds and work platforms. Specifically, there are chapters outlining:

- Mason's frame scaffolds
- Tube and coupler scaffolds
- Wood pole scaffolds
- Pump jack and ladder jack scaffolds

- Job manufactured scaffolds and formwork scaffolds
- Suspended scaffolds
- Two point-suspended and large area scaffolds
- Adjustable suspended platforms
- Mobile scaffolds, including manually propelled and rolling towers, and special engineered scaffolds
- Working platforms including scissor lifts, boom-type lifts, and vehicle-mounted platforms

Each chapter includes the OSHA regulations and industry standards that apply to the specific type of equipment. There are illustrations of the specific types of equipment and the common and appropriate uses of each scaffold are discussed. General instructions for installation and removal of equipment are outlined and need to be supplemented by the manufacturer's instructions under the supervision of a Competent Person. Also outlined in each chapter are the training and inspection requirements together with inspection checklists (Figure 3.2-3.7).

Some of the typical means of using appropriate fall protection on specific equipment are addressed together with common hazards, typical malfunctions and accidents, and how to eliminate them.

Special Equipment

As mentioned earlier, it should be noted that in the selection and use of any equipment, careful consideration needs to be paid to the limitations that are imposed by laws, regulations, and the manufacturer's instructions. Further, site specific plans need to be developed as to exactly how and what the equipment will be incorporated into accomplishing the tasks at hand. For example, in special situations where materials are to be lifted along with personnel, this may encompass a new or unusual use of the equipment. These applications will need to be designed by a qualified engineer and in some cases approved by the manufacturer. Certainly specialized applications need close supervision by a Competent Person and training for any additional risks or hazards needs to be site specific.

The Checklist below is minimal and should be added to and adapted to individual company and manufacturer's requirements. Checklists for various activities are effective when corrective action is taken. These corrective actions must be verified in order to render these inspections effective.

SCAFFOLDING CHECKLIST

YES	NO		CORRECTIVE ACTION	DATE	BY WHO?
		Are the scaffolds inspected before each work shift by a competent person?			
		Are all scaffold components inspected before each work shift by a competent person?			
		Has the competent person trained all employees who erect, disassemble, move, operate, repair, maintain, or inspect the scaffolds?			
		Have persons working on scaffolding or aerial lifts been trained by a qualified person?			
		Are lean-to jack scaffolds prohibited on the site and removed from use?			
		Are shore scaffolds prohibited on the site and removed from use?			
		Are nailed brackets prohibited on the site and removed from use?			
		Are loose tile, brick, blocks used to level scaffolds prohibited from on the site and removed from use?			
		Is the capacity of the scaffold known ?			
		Are ladders or stairways fixed or built-in and located so as not to make the scaffold unstable?			
		If a ladder is used for access, does it extend 3 feet above platform level?			

The above is a minimal check meant for use by persons who conduct daily work site safety inspections of scaffolding.

This checklist is not intended for setup or placement of scaffolds, nor is it intended to provide a comprehensive safety evaluation of all scaffold operations.

Additional Comments:

Inspected by: ______________________ Date ______________

FIGURE 3.2 Scaffolding checklist

OSHA STANDARDS

As outlined in "A Guide to Scaffold Use In the Construction Industry" OSHA 3150 2001(Revised), employers and employees should be familiar with the seven key provisions of the revised scaffolding standard:

- The standard requires fall protection at a 10-foot height above a lower level for employees. 1926.451(g)(1)

The Checklist below is minimal and should be added to and adapted to individual company and manufacturer's requirements. Checklists for various activities are effective when corrective action is taken. These corrective actions must be verified in order to render these inspections effective.

CRANE CHECKLIST

YES	NO		CORRECTIVE ACTION	DATE	BY WHO
		Is there a designated competent person?			
		Is the crane clear of any overhead electrical lines?			
		Is the crane blocked and secured?			
		Is the crane area barricaded to keep unauthorized personnel away from the area?			
		Are personnel prohibited from riding the load?			
		Is the load rating chart and other warnings and instructions secured to the crane and are they accessible to the operator?			
		Are all lifts performed within the stated capacity of the crane and rigging?			
		Is there a functioning audible warning device on board?			
		Is there a fire extinguisher in the cab? Is it current?			
		Are the foot walks capable of supporting a worker?			
		Are there standard handrails and toeboards present?			
		Is the cab equipped with locks and seatbelts?			
		Is the crane equipped with boom stops and jib boom stops?			
		Are exposed moving parts guarded? This includes gears, shafts and sprockets.			
		Is the crane operator receiving signals from only one designated signal person?			
		Is this signal person in sight of the operator at all times during lifts?			
		Is the main boom free of any evidence of cracks, corrosion or other degradation?			
		Is the jib boom free of any evidence of cracks, corrosion or other degradation?			
		Are the boom extensions free of any evidence of cracks, corrosion or other degradation?			
		Is the rigging equipment in good condition?			
		Are the rung ropes free of distortion, birdcaging, corrosion or broken or cut strands?			
		Has this crane been inspected within the last year?			

The above checklist is meant for use by persons who conduct daily work site safety inspections of crane operations.

This checklist is not intended for setup or placement of cranes, nor is it intended to provide a comprehensive safety evaluation of all crane operations.

Additional Comments:

Inspected by: ______________________________ Date______________________

Verification All Corrective Action Taken:

Name: ______________________________ Date______________________

FIGURE 3.3 Crane checklist

The Checklist below is minimal and should be added to and adapted to individual company and manufacturer's requirements. Checklists for various activities are effective when corrective action is taken. These corrective actions must be verified in order to render these inspections effective.

LADDER CHECKLIST

YES	NO		CORRECTIVE ACTION	DATE	BY WHO?
		Is there a designated competent person?			
		Have employees been properly trained in the hazards associated with ladders?			
		Is the ladder in good condition and free from defects, defective rungs or side rails, protruding nails, screws, splinters, oil or grease on the rungs, defective footings and bases corrosion on open end of hollow rungs, frayed ropes, other indications of hazards?			
		Are ladders secured at the top?			
		Are ladders placed on firm dry ground?			
		Are ladders clear of doorways, driveways, passageways?			
		Are ladders being used as intended? NO use as horizontal platform, runways or scaffolds?			
		Do ladders extend at least 3 feet above the place of landing?			
		Are only non-conductive ladders used near electrical lines or energized parts?			
		When using job built ladders are the rails made of Douglas Fir without knots?			
		When using job built ladders are the rungs made from clear straight grained lumber?			
		When using job built ladders are the rungs spaced 12 inches apart vertically?			
		When using job built ladders are the rungs blocked or notched into the side rails?			

The above is a minimal check meant for use by persons who conduct daily work site safety inspections of ladders.

This checklist is not intended for setup or placement of ladders, nor is it intended to provide a comprehensive safety evaluation of all ladder operations.

Additional Comments:

Inspected by: ______________________ Date ______________________

Verification All Corrective Action Taken:

Name: ______________________ Date ______________________

FIGURE 3.4 Ladder checklist

- Guardrail height—the height of the toprail for scaffolds manufactured and placed in service before January 1, 2000 can be between 36 inches (0.9m) and 45 inches (1.2m). The height of the toprail for scaffolds manufactured and placed in service after January 1, 2000 must be between 38 inches (0.97m) and 45 inches (1.2m). 1926.451(g)(4)(ii) When the crosspoint of crossbracing is used as a toprail, it must be between 38 inches (0.97m) and 48 inches (1.3m)

The Checklist below is minimal and should be added to and adapted to individual company and manufacturer's requirements. Checklists for various activities are effective when corrective action is taken. These corrective actions must be verified in order to render these inspections effective.

FALL PROTECTION CHECKLIST

YES	NO		CORRECTIVE ACTION	DATE	BY WHO?
		Is there a designated competent person?			
		Are employees exposed to fall hazards properly trained?			
		Are safety nets provided where there is no other means of fall protection?			
		Toe boards, and/or catch nets are provided on walkways and platforms where tools, equipment or other objects could fall to prevent a falling object hazard to workers below?			
		Stairs and stairwells are guarded?			
		Are holes or other floor openings covered or guarded?			
		Are trenches and excavations guarded?			
		Are catwalks and crossovers guarded?			
		Are guardrails made of select lumber free of defects or made of equally substantial material?			
		If pipe or wire line is used as a guardrail, is it free of defects and capable of supporting required loads?			
		If lifelines are used as fall protection, are they in good repair?			

The above checklist is meant for use by persons who conduct daily work site safety inspections of fall protection.

This checklist is not intended for setup or placement of fall protection, nor is it intended to provide a comprehensive safety evaluation of all fall protection operations.

Additional Comments:

Inspected by: ______________________ Date ______________

Verification All Corrective Action Taken:

Name: ______________________ Date ______________

FIGURE 3.5 Fall protection checklist

above the work platform. 1926.451(g) (4) (xv) Midrails must be installed approximately halfway between the toprail and the platform surface. When a crosspoint of crossbracing is used as a midrail, it must be between 20 inches (0.5m) and 30 inches (0.8m) above the work platform. 1926.451 (g) (4)

- Erecting and dismantling—after September 2, 1997, when erecting and dismantling supported scaffolds, a Competent Person must determine the feasibility of providing a safe means of access and fall protection for these operations. 1926.451(e) (9) & (g) (2)

The Checklist below is minimal and should be added to and adapted to individual company and manufacturer's requirements. Checklists for various activities are effective when corrective action is taken. These corrective actions must be verified in order to render these inspections effective.

SUPPORT SCAFFOLD SYSTEMS

YES	NO		CORRECTIVE ACTION	DATE	BY WHO?
		Are the scaffolds square and plumb?			
		Are planks of a scaffold grade material?			
		Are scaffolds fully planked?			
		Do planks extend between 6 inches and 12 inches over supports?			
		Are these planks in good condition?			
		Are these planks free from obvious defects?			
		Are heavy items placed on planks near load bearing members?			
		Is the scaffold footing on base plates and mud sills level, sound and rigid?			
		Are 4:1 (height to width) fabricated frame scaffolds secured to building/structure per the manufacturer or at every 30 feet horizontally plus each end and every 26 feet or less vertically when the scaffold is wider than 3 feet?			
		Does the scaffold have guardrails?			
		Does the scaffold have toe boards?			
		Are vertical lifelines secured to a structurally sound anchor?			
		Are tie backs secured to a structurally sound anchorage? This must NOT be the same anchorage used for a lifeline.			

The above is a minimal checklist meant for use by persons who conduct daily work site safety inspections of scaffolding.

This checklist is not intended for setup or placement of scaffolds, nor is it intended to provide a comprehensive safety evaluation of all scaffold operations.

Additional Comments:

Inspected by: ______________________ Date______________

Verification All Corrective Action Taken:

Name: ______________________ Date______________

FIGURE 3.6 Support scaffold systems

- Training—employers must train each employee who works on a scaffold on the hazards and the procedures to control the hazards. 1926.454
- Inspections—before each work shift and after any occurrence that could affect the structural integrity, a Competent Person must inspect the scaffold and scaffold components for visible defects. 1926.451(f)(3)

PROJECT SAFETY ANALYSIS

Activity/Operation

Unsafe Condition, Action or Hazard

Preventative or Corrective Action

Discussion:

ASA Required: Yes No
JSA Required: Yes No
Responsible Supervisor: ______________________
Competent Person: ______________________
Inspection Required: Yes No Frequency: __________

Approved: ______________________
Signature Date

FIGURE 3.7 Project safety analysis *(continued on next page)*

PROJECT SAFETY AND HEALTH DOCUMENTATION

The Project Supervisor shall maintain on the project work site, a log of safety and health information. The log shall be accessible to project employees or their representatives upon request and shall contain documentation of:

Job site injuries, illnesses or near misses, including the dates of occurrence, names of those involved, a description of the related circumstances and the results of related incident investigations.

Safety inspections, including the name of the inspector, date of inspection, description of findings and corrective action resulting from the inspection.

Project safety training (i.e., project safety orientation, safety task assignment meetings and "tool box" safety meetings), including names and addresses of attendees, name of the instructor, date and time of training, type of training and a brief description of the content of the training (a sign-in sheet that includes the required information shall serve as sufficient documentation).

Any written project or activity safety analyses performed for the project.

Results of any exposure monitoring or medical surveillance performed in support of the project (employees or their representatives shall have access only to their own medical surveillance and exposure monitoring records).

FIGURE 3.7 Project safety analysis *(continued on next page)*

- Overhand bricklaying—a guardrail or personal fall arrest system on all sides, except the side where the work is being done, must protect employees doing overhand bricklaying from supported scaffolds. 1926.451(g)(1)(vi)
- The standards for aerial lifts have been relocated from 1926.556 to 1926.453.

OSHA's scaffolding standard defines a Competent Person as "one who is capable of identifying existing and predictable hazards in the surroundings or working conditions, which are unsanitary, hazardous to employees, and who has authorization to take prompt corrective measures to eliminate them." The

JOB INJURIES, ILLNESSES OR NEAR MISSES

INCIDENT, ILLNESS, ETC.	DATE OF OCCUR-RENCE	DESCRIPTION	RESULTS

Reported By:________________________ Date:______________

Inspected By: ________________________ Date:______________

Need for Further Follow-Up? Yes No

FIGURE 3.7 Project safety analysis *(continued on next page)*

standard requires a Competent Person to perform the following duties under these circumstances:

In General:

- To select and direct employees who erect, dismantle, move, or alter scaffolds 1926.451(f)(7)

SAFETY INSPECTIONS

DESCRIPTION OF FINDINGS	CORRECTIVE ACTION

Inspector:____

Date of Inspection:____________________________________

FIGURE 3.7 Project safety analysis *(continued on next page)*

- To determine if it is safe for employees to work on or from a scaffold during storms or high winds and to ensure that a personal fall arrest system or wind screens protect these employees (Note: Windscreens should not be used unless the scaffold is secured against the anticipated wind forces imposed.) 1926.451(f)(12)

PROJECT SPECIFIC TRAINING SIGN-IN SHEET

NAME OF ATTENDEE	COMPANY & ADDRESS

Type of Project Specific Training:________________________________

Date and Time ____________ Name of Instructor____________________

Description of content of training:________________________________

____

FIGURE 3.7 Project safety analysis *(continued on next page)*

For Training:

- To train employees involved in erecting, disassembling, moving, operating, repairing, maintaining, or inspecting scaffolds to recognize associated work hazards 1926.454(b)

EXPOSURE MONITORING/MEDICAL SURVEILLANCE

Date:____________________________

Condition requiring monitoring or medical surveillance: ____________________

____

____

Individuals Monitored/ Under Surveillance	Results of Monitoring/Surveillance

Monitored/Surveyed by: ____

FIGURE 3.7 Project safety analysis *(continued on next page)*

WEEKLY SAFETY INSPECTION SUMMARY

DESCRIPTION OF FINDINGS	CORRECTIVE ACTION

Inspector:______________________________________

Date of Inspection:________________________________

FIGURE 3.7 Project safety analysis

For Inspections:

- To inspect scaffolds and scaffold components for visible defects before each work shift and after any occurrence which could affect the structural integrity and to authorize prompt corrective actions 1926.451(f)(3)
- To inspect ropes on suspended scaffolds prior to each workshift and after every occurrence which could affect the structural integrity and to authorize prompt correct actions 1926.451(d)(10)
- To inspect manila or plastic (or other synthetic) rope being used for toprails or midrails 1926.451(g)(4)(xiv)

For Suspension Scaffolds:

- To evaluate direct connections to support the load 1926.451(d)(3)(i)
- To evaluate the need to secure two-point and multi-point scaffolds to prevent swaying 1926.451(d)(18)

For Erectors and Dismantlers:

- To determine the feasibility and safety of providing fall protection and access 1926.451(e)(9) and 1926.451(g)(2)
- To train erectors and dismantlers (effective September 2, 1997) to recognize associated work hazards 1926.454f(b)

For Scaffold Components:

- To determine if a scaffold will be structurally sound when intermixing components from different manufacturers 1926.451f(b)(10)
- To determine if galvanic action has affected the capacity when using components of dissimilar metals 1926.451(b)(11)

The standard defines a qualified person as "one who—by possession of a recognized degree certificate, or professional standing, or who by extensive knowledge, training, and experience—has successfully demonstrated his/her ability to solve or resolve problems related to the subject matter, the work, or the project." The qualified person must perform the following duties in these circumstances:

In General:

- To design and load scaffolds in accordance with that design 1926.451(a)(6)

For Training:

- To train employees working on the scaffolds to recognize the associated hazards and understand procedures to control or minimize those hazards 1926.454(a)

For Suspension Scaffolds:

- To design the rigging for single-point adjustable suspension scaffolds 1926.452(o)(2)(i)
- To design platforms on two-point adjustable suspension types that are less than 36 inches (0.9m) wide to prevent instability 1926.452(p)(1)
- To make swaged attachments or spliced eyes on wire suspension ropes 1926.451(d)(11)

For Components and Design:

- To design scaffold components construction in accordance with the design 1926.451(a)(6)

The standard requires a registered professional engineer to perform the following duties in these circumstances:

For Suspension Scaffolds:

- To design the direct connections of masons' multi-point adjustable suspension scaffolds 1926.451(d)(3)(i)

For Design:

- To design scaffolds that are to be moved when employees are on them 1926.451(f)(5)
- To design pole scaffolds over 60 feet (18.3m) in height 1926.452(a)(10)
- To design tube and coupler scaffolds over 125 feet (38m) in height 1926.452(b)(10)

- To design fabricated frame scaffolds over 125 feet (38m) in height above their base plates 1926.452(c)(6)
- To design brackets on fabricated frame scaffolds used to support cantilevered loads in addition to workers 1926.452(i)(8)
- To design outrigger scaffolds and scaffold components 1926.452(i)(8).

Capacity

- Each scaffold and scaffold component must support without failure its own weight and at least four times the maximum intended load applied or transmitted to it. 1926.451(a)(1)
- A qualified person must design the scaffolds, which are loaded in accordance with that design. 1926.451(a)(6)
- Scaffolds and scaffold components must not be loaded in excess of their maximum intended loads or rated capacities, whichever is less. 1926.451(f)(1)
- Load carrying timber members should be a minimum of 1,500 lb-f/in2 construction grade lumber. Appendix A(1)(a)

Scaffold Platform Construction

- Each platform must be planked and decked as fully as possible with space between the platform and uprights not more than one inch (2.5cm) wide. The space must not exceed nine inches (24.1cm) when side brackets or odd-shaped structures result in a wider opening between the platform and the uprights. 1926.451(b)(1)
- Scaffold planking must be able to support, without failure, its own weight and at least four times the intended load. 1926.451(a)(1)
- Solid sawn wood, fabricated planks and fabricated platforms may be used as scaffold planks following the recommendations by the manufacturer or a lumber grading association or inspection agency. Appendix A(1)(b)&(c)

- Tables showing maximum permissible spans, rated load capacity, and nominal thickness are in Appendix A(1)(b) & (c) of the standard.
- The platform must not deflect more than 1/60 of the span when loaded. 1926.451(f)(16)
- The standard prohibits work on platforms cluttered with debris. 1926.451(f)(13)
- Each scaffold platform and walkway must be at least 18 inches (46 cm) wide. When the work area is less than 18 inches (46 cm) wide, guardrails and/or personal fall arrest systems must be used. 1926.451(b)(2)
- The standard requires employers to protect each employee on a scaffold more than 10 feet (3.1m) above a lower level from falling to that lower level. 1926.451(g)(1)

To ensure adequate protection, install guardrails along all open sides and ends before releasing the scaffold for use by employees, other than the erection and dismantling crews. 1926.451(g)(4) Guardrails are not required, however:

- When the front end of all platforms are less than 14 inches (36 cm from the face of the work 1926.451(b)(3)
- When outrigger scaffolds are three inches (8cm) or less from the front edge 1926.451(b)(3)(1)
- When employees are plastering and lathing 18 inches (46cm) or less from the front edge 1926.451(b)(3)(ii)
- Steel or plastic banding must not be used as a toprail or midrails 1926.451(g)(4)(xiii)

Criteria for Supported Scaffolds

Supported scaffolds are platforms supported by legs, outrigger beams, brackets, poles, uprights, posts, frames, or similar rigid support. 1926.451(b) The structural members, poles, legs, posts, frames, and uprights must be plumb and braced to prevent swaying and displacement. 1926.451(c)(3)

All employees must be trained by a qualified person to recognize the hazards associated with the type of scaffold being used and how to control or minimize those hazards. The training must include fall hazards, falling object hazards, electrical hazards, proper use of the scaffold, and handling of materials. 1926.454(a)

Supported scaffolds with a height to base width ratio of more than 4:1 must be restrained by guying, tying, bracing, or an equivalent means. 1926.451(c)(1) Either the manufacturers' recommendation or the following placements must be used for guys, ties, and braces:

- Install guys, ties, or braces at the closest horizontal member to the 4:1 height ad repeat vertically with the top restraint no further than the 4:1 height from the top.
- Vertically—every 20 feet (6.1m) or less for scaffolds less than three feet (0.91m) wide; every 26 feet (7.9m) or less for scaffolds more than three feet (0.91m) wide
- Horizontally—at each end, at intervals not to exceed 30 feet (9.1m) from one end 1926.451(c)(1)
- Supported scaffolds' poles, legs posts, frames and uprights must bear on base plates and mud sills, or other adequate firm foundation. 1926.451(c)(2)(i)&(ii)
- Forklifts can support platforms only when the entire platform is attached to the fork and the forklift does not move horizontally when workers are on the platform. 1926.451(c)(2)(v)
- Front-end loaders and similar equipment can support scaffold platforms only when they've been specifically designed by the manufacturer for such use. 1926.451(c)(2)(iv)
- Stilts may be used on a large area scaffold. When a guardrail system is used, the guardrail height must be increased in height equal to the height of the stilts. The manufacturer must approve any alterations to the stilts. 1926.452(v)

Note: A large area scaffold consists of a pole, tube and coupler systems, or a fabricated frame scaffold erected over substantially the entire work area. 1926.451(b)

Criteria for Suspended Scaffolds

A suspension scaffold contains one or more platforms suspended by ropes or other non-rigid means from an overhead structure, 1926.450(b), such as the following scaffolds:

- Single-point
- Multi-point
- Multi-level
- Two-point
- Adjustable
- Boatswains' chair
- Catenary
- Chimney hoist
- Continuous run
- Elevator false car
- Go-devils
- Interior hung
- Masons'
- Stone setters'

Some of the requirements for all types of suspension scaffolds include:

- Employers must ensure that all employees are trained to recognize the hazards associated with the type of scaffold being used. 1926.451(d)(1)
- All support devices must rest on surfaces capable of supporting at least four times the load imposed on them by the scaffold when operating at the rated load of the hoist, or at least one-and-a-half times the load imposed on them by the scaffold at the stall capacity of the hoist, whichever is greater. 1926.451(d)(1)

- A Competent Person must evaluate all direct connections prior to use to confirm that the supporting surfaces are able to support the imposed load. 1926.451(d)(1)
- All suspension scaffolds must be tied or otherwise secured to prevent them from swaying, as determined by a Competent Person. 1926.451(d)
- Guardrails, a personal fall arrest system, or both must protect each employee more than 10 feet (3.1m) above a lower level from falling. 1926.451(g0
- A Competent Person must inspect ropes for defects prior to each workshift and after every occurrence that could affect a rope's integrity. 1926.451(d)(10)
- When scaffold platforms are more than 24 inches (61cm) above or below a point of access, ladders, ramps, walkways, or similar surfaces must be used. 1926.451(e)(1)
- When using direct access, the surface must not be more than 24 inches (61cm) above or 14 inches (36cm) horizontally from the surface. 1926.451(e)(8)
- When lanyards are connected to horizontal lifelines or structural members on single-point or two-point adjustable scaffolds, the scaffold must have additional independent support lines equal in number and strength to the suspension lines and have automatic locking devices. 1926.451(g)(3)(iii)
- Emergency escape and rescue devices must not be used as working platforms, unless designed to function as suspension scaffolds and emergency systems. 1926.451(d)(19)

Counterweights used to balance adjustable suspension scaffolds must be able to resist at least four times the tipping moment imposed by the scaffold operating at either the rated load of the hoist, or one-and-a-half (minimum) times the tipping moment imposed by the scaffold operating at the stall load of the hoist, whichever is greater. 1926.451(a)(2)

Only those items specifically designed as counterweights must be used: 1926.451(d)(3)(iii):

- Counterweights used for suspended scaffolds must be made of materials that cannot be easily dislocated. Flowable material, such as sand or water, cannot be used. 1926.451(f)(3)(ii)
- Counterweights must be secured by mechanical means to the outrigger beams. 1926.451(d)(3)(iv)
- Vertical lifelines must not be fastened to counterweights. 1926.451(g)(3)(i)
- Sand, masonry units, or rolls of roofing felt cannot be used as counterweights. 1926.451(d)(3)(ii)&(iii)

Outrigger beams (thrustouts) are the structural members of a suspension or outrigger scaffold that provide support. 1926.450(b) They must be placed perpendicular to their bearing support. 1926.451(d)(3)(viii)

Tiebacks must be secured to a structurally sound anchorage on the building or structure. Sound anchorages do not include standpipes, vents, or other piping systems, or electrical conduit. 1926.451(d)(3)(ix)&(d)(5)

A single tieback must be installed perpendicular to the face of the building or structure. Two tiebacks installed at opposing angles are required when a perpendicular tieback cannot be installed. 1926.451(d)(3)(x)

The suspension ropes must be long enough to allow the scaffold to be lowered to the level below without the rope passing through the hoist, or the end of the rope configured to prevent the end from passing through the hoist. 1926.451(d)(6)

The standard prohibits using repaired wire. 1926.451(d)(7)

Drum hoists must contain no less than four wraps of the rope at the lowest point. 1926.451(d)(6)

Employers must replace wire rope when the following conditions exist: kinks, six randomly broken wires in one rope lay or three broken wires in one strand in one lay, one third of the original diameter of the outside wires is lost, heat damage, evidence that the secondary brake has engaged the rope, and any other physical damage that impairs the function and strength of the rope. 1926.451(d)(10)

Suspension ropes supporting adjustable suspension scaffolds must be a diameter large enough to provide sufficient surface area for the functioning of brake and hoist mechanisms. 1926.451(f)(10) They must be shielded from heat-producing processes. 1926.451(f)(11)

Hoists

- Power-operated hoists used to raise or lower a suspended scaffold must be tested and listed by a qualified testing laboratory. 1926.451(d)(13)
- The stall load of any scaffold hoist must not exceed three times its rated load. 1926.451(a)(5) The stall load is the load at which the prime-mover (motor or engine) of a power-operated hoist stalls or the power to the prime-mover is automatically disconnected. 1926.451(b)
- Gasoline power-operated hoists or equipment are not permitted. 1926.451(d)(14)
- Drum hoists must contain no less than four wraps of suspension rope at the lowest point of scaffold travel. 1926.451(d)(6)
- Gears and brakes must be enclosed. 1926.451(d)(15)
- An automatic braking and locking device, in addition to the operating brake, must engage when a hoist makes an instantaneous change in momentum or an accelerated overspeed. 1926.451(d)(16)
- Manually operated hoists used to raise or lower a suspended scaffold must be tested and listed by a qualified testing laboratory. 1926.451(d)(13)
- These hoists require a positive crank force to descend. 1926.451(d)(17)

Welding can be done from suspended scaffolds when:

- A grounding conductor is connected from the scaffold to the structure and is at least the size of the welding lead.

- The grounding conductor is not attached in series with the welding process or the work piece.
- An insulating material covers the suspension wire rope and extends a least four feet (1.2m) above the hoist.
- Insulated protective covers cover the hoist.
- The tail line is guided, retained, or both, so that it does not become grounded.
- Each suspension rope is attached to an insulated thimble.
- Each suspension rope and any other independent lines are insulated from grounding. 1926.451(f)(17)

No materials or devices may be used to increase the working height on a suspension scaffold. This includes ladders, boxes, and barrels. 1926.451(f)(14)&(15)

Access Requirements

Employers must provide access when the scaffold platforms are more than two feet (0.6m) above or below a point of access. 1926.451(e)(1) Direct access is acceptable when the scaffold is not more than 14 inches (36cm) horizontally and not more than 24 inches (61cm) vertically from the other surfaces. 1926.451(e)(8) The standard prohibits the use of crossbraces as a means of access. 1926.451(e)(1) Several types of access are permitted:

- Ladders, such as portable, hook-on, attachable, and stairway (1926.451(e)())
- Stair towers (1926.451(e)(4))
- Ramps and walkways (1926.451(e)(5))
- Integral prefabricated frames (1926.451(e)(6))

Effective September 2, 1997, employees erecting and dismantling supported scaffolding must have a safe means of access provided when a Competent Person has determined the feasibility and analyzed the site conditions. 1926.451(e)

Use Requirements

Shore and lean-to scaffolds are strictly prohibited. 1926.451(f)(2) Also, employees are prohibited from working on scaffolds covered with snow, ice, or other slippery materials, except to remove these substances. 1926.451(f)(8) The standard does require specific clearance distances. 1926.451(f)(6)

Fall Protection Requirements

Employers must provide fall protection for each employee on a scaffold more than 10 feet (3.1m) above a lower level. 1926.451(g)(1) After September 2, 1997, a Competent Person must determine the feasibility and safety of providing fall protection for employees erecting or dismantling support scaffolds. 1926.451(g)(2) Fall protection includes guardrail systems and personal fall arrest systems. Personal fall arrest systems include body belts (Note that body belts will not be acceptable after January 1, 1998), harnesses, components of the harness/belt such as Dee-rings and snap hooks, lifelines, and anchorage point. 1926.451(g)(3)

Vertical or horizontal lifelines may be used. 1926.451(g)(3)(ii)-(iv) Lifelines must be independent of support lines and suspension ropes and not attached to the same anchorage point as the support or suspension ropes. 1926.451(g)(3)(iii)&(iv)

When working from an aerial lift, attach the fall arrest system to the boom or basket. 1926.453(b)(2)(v). Note to paragraph (b)(2)(v): As of January 1, 1998 subpart M of this part (1926.502(d)) provides that body belts are not acceptable as part of a personal fall arrest system. The use of a body belt in a tethering system or in a restraint system is acceptable and is regulated under 1926.502(e)

Personal fall arrest systems can be used on scaffolding when there are no guardrail systems. 1926.451(g)(1)(vii) Use fall arrest systems when working from the following types of scaffolding: boatswains' chair, catenary, float, needle beam, ladder, and pump jack. 1926.451(g)(1) Use fall arrest systems also when working from the boom/basket of an aerial lift. 1926.453(b)(2)(v) Fall arrest and guardrail systems must be used when working on single-and two-point adjustable suspension scaffolds and self-contained adjustable scaffolds that are supported by ropes. 1926.451(g)(1)

Falling Object Protection

To protect employees from falling hand tools, debris, and other small objects, install toeboards, screens, guardrail systems, debris nets, catch platforms, canopy structures, or barricades. In addition, employees must wear hard hats. 1926.451(h)(1)&(2)&(3)

Aerial Lift Requirements 1926.453

Vehicle-mounted aerial devices used to elevate employees, such as extensible boom platforms, aerial lifts, articulating boom platforms, and vertical towers, are considered "aerial lifts." 1926.453(a)(1) The 1926.453 and 1926.454 standards apply to aerial lifts and the 1926.454 standards apply to mobile scaffolds. Some specific requirements for aerial lifts include:

- Only authorized personnel can operate aerial lifts.
- The manufacturer or equivalent must certify any modification.
- The insulated portion must not be altered to reduce its insulating value.
- Lift controls must be tested daily.
- Controls must be clearly marked.
- Brakes must be set and outriggers used.
- Boom and basket load limits must not be exceeded.
- Employees must wear personal fall arrest systems, with the lanyard attached to the boom or basket.
- No devices to raise the employee above the basket floor can be used. 1926.453(b)
- All employees who work on a scaffold must be trained by a person qualified to recognize the hazards associated with the type of scaffold used and to understand the procedures to control and minimize those hazards. 1926.454(a)
- A Competent Person must train all employees who erect, disassemble, move, operate, repair, maintain, or inspect scaffolds. Training

must cover the nature of the hazards, the correct procedures for erecting, disassembling, moving, operating, repairing, inspecting, and maintaining the type of scaffold in use. 1926.454(b)

- Other recommended training topics include erection and dismantling, planning, personal protective equipment, access, guys and braces, and parts inspection.
- The standard requires retraining when (1) no employee training has taken place for the worksite changes, scaffold changes, or falling object protection changes; or (2) where the employer believes the employee lacks the necessary skill, understanding, or proficiency to work safely. 1926.454(c)1

Chapter 4

Training

SPECIFIC REQUIREMENTS

One of the most often repeated regulations published in the Occupational Safety and Health Regulations is that of training and training requirements. In 1926.21(b) (2), OSHA has general overall requirements for training. Throughout the Occupational Safety and Health Act in at least 264, plus or minus, locations, there are additional specific requirements to provide training to the workers. As they address scaffolds, the specific requirements are published in 1926.454 entitled "Training Requirements". OSHA frequently cites, when they perform job-site inspections, "lack of training" or "deficient training" as one of the more frequently cited violations.

TRAINING MATERIALS

OSHA has a policy in which it directs its attention to a few major items it believes causes the majority of the accidents. In the Focused Inspection Guideline form (Figure 4.1) that the inspector utilizes to determine if the site qualifies for a Focused Inspection, there are eight check boxes that give the compliance officer guidance. The 4th item, "Supervisor and Employee Training", includes recognition, reporting, and avoidance of hazards and applicable standards. Item 6, "Documentation of Training Permits and Inspections" is an area requir-

Construction Focused Inspection Guidelines 25

This guideline is to assist the compliance officer
to determine if there is an effective project plan to qualify for a Focused Inspection.

	YES/NO
PROJECT SAFETY AND HEALTH COORDINATION: Are there procedures in place by the general contractor, prime contractor, or other such entity to ensure that all employers provide adequate protection for their employees?	
Is there a DESIGNATED COMPETENT PERSON responsible for the implementation and monitoring of the project safety and health plan who is capable of identifying existing and predictable hazards and has authority to take prompt corrective measures?	
PROJECT SAFETY AND HEALTH PROGRAM/PLAN* that complies with 1926 Subpart C and addresses, based upon the size and complexity of the project, the following:	

_______ Project Safety Analysis at initiation and at critical stages that describes the sequence, procedures, and responsible individuals for safe construction.

_______ Identification of work/activities requiring planning, design, inspection, or supervision by an engineer, competent person, or other professional.

_______ Evaluation monitoring of subcontractors to determine conformance with the Project Plan.(The Project Plan may include, or be utilized by subcontractors.)

_______ Supervisor and employee training according to the Project Plan including recognition, reporting, and avoidance of hazards, and applicable standards.

_______ Procedures for controlling hazardous operations such as: cranes, scaffolding, trenches, confined spaces, hot work, explosives, hazardous materials, leading edges, etc.

_______ Documentation of: training, permits, hazard reports, inspections, uncorrected hazards, incidents, and near misses.

_______ Employee involvement in the hazard: analysis, prevention, avoidance, correction, and reporting.

_______ Project emergency response plan.

* For examples, see owner and contractor association model programs, ANSI A10.33, A10.38, etc.

The walkaround and interviews confirmed that the Plan has been implemented, including:

_______ The four leading hazards are addressed: falls, struck by, caught in\between, electrical.

_______ Hazards are identified and corrected with preventative measures instituted in a timely manner.

_______ Employees and supervisors are knowledgeable of the project safety and health plan, avoidance of hazards, applicable standards, and their rights and responsibilities.

THE PROJECT QUALIFIED FOR A FOCUSED INSPECTION. ☐

Construction Focused Inspection Guidelines

FIGURE 4.1 Construction Focused Inspection Guidelines

ing physical records to document these activities. All of these are required items that will key the inspector as to whether he should issue a citation.

Also in their compliance document CPL2-1.23, paragraph 11(b)(2), OSHA requires that compliance officers show they interviewed those employees engaged in erecting and dismantling scaffold operations to ascertain whether they have received the necessary training required under 1926.454(b)(1)-(4). The Occupational Safety and Health Act and the enforcement arm places great emphasis on training and, particularly, training of those workers who are on scaffolding. This is because of the recognition that most scaffolding can present a hazardous workplace and because conditions on the scaffolding are constantly changing.

There is a need for the employees utilizing the scaffold to be alert to these changing conditions and the hazards that they might create so that they can either correct the conditions or avoid the hazards. The Occupational Safety and Health Act in 1926.454 defines a minimum of two levels of training, which is required, as well as a third level of retraining. In the first section 1926.454(a) the requirements are as defined below:

- **1926.454(a)** The employer shall have each employee who performs work while on a scaffold trained by a person qualified in the subject matter to recognize the hazards associated with the type of scaffold being used and to understand the procedures to control or minimize those hazards. The training shall include the following areas, as applicable:
  - **1926.454(a)(1)** The nature of any electrical hazards, fall hazards and falling object hazards in the work area
  - **1926.454(a)(2)** The correct procedures for dealing with electrical hazards and for erecting, maintaining, and disassembling the fall protection systems and falling object protection systems being used
  - **1926.454(a)(3)** The proper use of the scaffold, and the proper handling of materials on the scaffold
  - **1926.454(a)(4)** The maximum intended load and the load-carrying capacities of the scaffolds used
  - **1926.454(a)(5)** Any other pertinent requirements of this subpart

What is the best way for a contractor to accomplish this type of training and to document it as is required? First of all, one should note that the training should be done by a person qualified in the subject matter. This is very similar to the requirements for a Competent Person so that, in the general context, the training for workers should be done by the person designated by the contractor as a Competent Person.

Secondly, a review of the plan which identifies the hazards associated with the scaffold and, as we pointed out in the selection program, the hazards associated with each type of scaffold is a major deciding factor in which type of scaffolding should be used. The review of the plans should identify the hazards and also identify the protective measures that are to be utilized in preventing or eliminating the hazard.

There are numerous training materials available from the manufacturers of scaffolds, as well as from contractors associations, such as the ABC (Associated Building Contractors), the AGC (Associated General Contractors), and Scaffold Industry Association. Training for employees should consist of a complete indoctrination prior to the start of work on the scaffold. This can take the form of a Toolbox Talk, which is presented to the employees prior to starting work on the scaffold and/or during a short period of instruction in the office. Employees should be given handout materials advising them of the hazards associated with scaffolding in general, as well as specific materials for the type of scaffold that is being utilized. In addition to the initial training, Toolbox Talks should be relatively frequent, at least weekly. These should deal with specific hazards detected as a result of the inspections of the scaffold and job observation reports. Both the initial training as well as the Toolbox Talks should be well documented as to their relevance to the scaffold in use, as well as to the attendance of the employees.

- **1926.454(b)** The employer shall have each employee who is involved in erecting, disassembling, moving, operating, repairing, maintaining, or inspecting a scaffold trained by a Competent Person to recognize any hazards associated with the work in question. The training shall include the following topics, as applicable:
  - **1926.454(b)(1)** The nature of scaffold hazards
  - **1926.454(b)(2)** The correct procedures for erecting, disassembling, moving, operating, repairing, inspecting, and maintaining the type of scaffold in question

- **1926.454(b)(3)** The design criteria, maximum intended load-carrying capacity, and intended use of the scaffold
- **1926.454(b)(4)** Any other pertinent requirements of this subpart

The second major category listed in OSHA is in 1926.454(b) and relates to the erection and dismantling of the scaffold. It is presumed that this would be a different crew and the erection portion of the scaffolding would occur before the actual work on the scaffold. It should be noted that this training is specifically to be performed by a Competent Person. This requires that a Competent Person, meeting the requirements as defined in OSHA and outlined previously, has been designated by the contractor. The nature of scaffold hazards in general is widely recognized, however, specific hazards associated with the scaffold in question need to be identified from the plan and included in the training of the erection crew. This specifically should include areas dealing with electrical hazards, fall protection hazards associated with scaffolding, and the means and methods selected by the contractor to provide fall protection.

Paragraph 1926.454(b)(2), defining the correct procedures for the erection of the scaffold, can easily be obtained from the manufacturer of the scaffold and these specific manufacturers' procedures should be on the jobsite and in the possession of the Competent Person prior to the beginning of the erection of the scaffold. Each crewmember who is erecting scaffolds should be instructed in detail as to the procedures for erecting the scaffold. Each scaffold should be treated as an individual scaffolding and, no matter how many times similar scaffolds have been erected, the procedures for erecting and operating the scaffolding should be examined in detail and reviewed with the crew. Of course this review procedure should be well documented. It is suggested that the manufacturers' checklist be utilized or a checklist similar to the one provided (Figure 4.2). The third area of OSHA training is when an employer has reason to believe that an employee lacks the skill or understanding needed for safe work involving erection use or dismantling the scaffolds. Retraining is required when changes in the worksite present a hazard about which an employee has not previously been trained. These items require that the contractor carefully document the inspection program on the scaffold on a daily basis so that the condition of the scaffold can be maintained and monitored. Some form of formal job observation report must be performed and documented so that the performance of each employee can be properly evaluated.

Example of a General Scaffold CheckList

Job Location______________________________

Date to be Erected______________________________

Date to be Disassembled______________________________

Competent Person______________________________

Scaffold Designation______________________________

	Yes	No
1. Was the Competent Person on site for:		
a. Erection of the Scaffold	___	___
b. Modification of the Scaffold	___	___
c. Dismantling of the Scaffold	___	___
2. Is this scaffold inspected daily by a Competent Person?	___	___
3. Is there written documentation for 1 and 2 above?	___	___
4. Is scaffold grade planking being used?	___	___
5. Are poles, legs, uprights for the scaffolding plumb?	___	___
6. Is the scaffold securely and rigidly braced to avoid swaying and displacement and assure the scaffold is plumb, square and rigid?	___	___
7. Are scaffold legs sound and capable of carrying the maximum intended load?	___	___
8. Are damaged scaffold members immediately repaired or taken out of service?	___	___
9. If taken out of service, or put aside for repair, are these members clearly marked "not usable"?	___	___
10. Are guardrails and midrails on open sides provided?	___	___
11. Are toe boards installed?	___	___
12. Is an access ladder or equivalent safe access provided?	___	___

FIGURE 4.2 Example of a general scaffold checklist *(continued on next page)*

13. Have weather conditions occurred which may impact on the condition of the scaffolding? ___ ___

14. Has anyone else used the scaffolding? ___ ___

15. Does the Competent Person require any additional users to have him/her present if any changes to the scaffold are made? ___ ___

16. Is a tag line provided for the hoisting of materials onto the scaffold? ___ ___

17. Has this tag line been rated for capacity and is that capacity clearly marked for users to assure they are not trying to hoist materials heavier than the tag line can handle? ___ ___

Comments:

Supervisor Signature Date:

_______________________ _______________________

This form is not intended as a final product; rather, it is provided as a beginning checklist which must be modified each and every time the checklist is used to include any specific specifications necessary for each individual use and also to reflect company policy as to scaffold use.

FIGURE 4.2 *(continued from previous page)* Example of a general scaffold checklist

Competent Person

An additional area, which I believe should be evaluated, is the training of the "Competent Person". The Competent Person requires specific knowledge about the hazards and the specific type of scaffolding. He/she needs to be familiar with the detailed plan for utilizing the scaffold and the various components of the scaffold. He/she also needs to be able to perform a detailed inspection of the components as well as the erected scaffolding and be capable of providing such documentation.

The training of the Competent Person, as well as the training of the erection crew, is critical to erecting and maintaining a safe scaffold. Without the proper erection and maintenance, hazards created by the faulty erection of the scaffold are present throughout the jobsite and create hazards for all workers on the scaffold. Although most construction scaffolds are in place for relatively short periods of time, they are frequently changed, modified, and extended, so there is probably a great deal of difficulty in having a crew for the erection and a crew for use of the scaffold.

The difficulties associated with training an erection crew can be avoided by the selection of an independent contractor who can provide trained crews to erect the scaffolding. Most distributors or manufacturers are capable of providing contractor erection services. Certainly the cost of erection must be weighed against the cost for training an unfamiliar crew, particularly as larger and more complex scaffolds are erected.

Competent, Qualified, and Authorized Designations

These are important designations throughout OSHA. The Competent Person, qualified person, and authorized person designations are very specific and are used consistently throughout 29CFR1926. Each of these designations has a specific meaning and a jobsite may be closed down if the proper person is not present to supervise or monitor the work.

For instance, 1926.32(d) "Authorized person" means a person approved or assigned by the employer to perform a specific type of duty or duties or to be at a specific location or locations at the jobsite. The term "authorized" may be applied to a crane operator, for instance. In that instance, "authorized" means that the crane operator has some specific type of duty for which he/she has received certification and becomes "authorized", or through employment with a crane firm as an operator, has been "authorized" to operate the crane.

In 1926.32(f) "Competent Person" means one who is capable of identifying existing and predictable hazards in the surroundings or working conditions which are unsanitary, hazardous, or dangerous to employees, and who has authorization to take prompt corrective measures to eliminate them. The term "competent" refers to a person who has been designated by his/her employer as being competent through experience, training or other means to recognize and identify existing and predictable hazards and do something to correct the condition. The operative word is designated. An employer must designate one or more persons to perform the duties of a Competent Person. A contractor may utilize the services of a consultant to perform some or all of the duties of a Competent Person or may use the services of an employee of another contractor. The critical requirement is that the Competent Person be designated for each of the required functions and that a list of the Competent Person assignments be maintained.

On many jobsites, more than one contractor may utilize the scaffold and a record must be maintained of who is the Competent Person and assure that he/she is available to perform the duties required of a Competent Person. When more than one contractor is utilizing a scaffold, OSHA will use this documentation to determine the various contractors' responsibilities under OSHA's multi-employer citation policy.

These decisions must be made during the preplanning phase of the project. Many contractors include specific requirements delineating who is to supply the scaffold, be responsible for erection and inspection on a daily basis, and supply the Competent Person as required. A procedure must be established for the site, defining these responsibilities. For these reasons many site safety programs require that a scaffold be tagged with this information and the tag utilized to record this information.

Scaffold Tags

While scaffold tags are not required by OSHA, they are an excellent idea and many, many companies tag their scaffolds as a matter of good safety policy.

What do we mean by "tagging scaffolds"? Well, tagging a scaffold (and we will view samples later in this section) is a means of indicating whether or not a scaffold is safe for use. The tag should be affixed to the scaffold AT EACH MEANS OF ENTERING THE SCAFFOLD in a manner that will allow easy changing of the tag, but will withstand wind, weather, and normal construction activities. The tag can contain all kinds of information as to when the scaffold was last inspected, who (Competent Person) inspected it, if it is ready for use, or if it is in a partially constructed state and should not be used. Since a set of scaffolds is sometimes used by many trades on a construction site, it is important to know just what condition the scaffold is in. It is not smart to hop on a scaffold and assume anything about its construction or condition.

Commonly, the red scaffold tag indicates that the scaffold has not been inspected or has been inspected and is not safe for use. The color red, just like a stoplight, indicates "STOP". Do not proceed further.

A green scaffold tag indicates that the scaffold has been inspected and is safe for use. Accompanying this tag should be information as to when the scaffold was inspected, who inspected it, the date of the next inspection, as well as any other information deemed necessary by the scaffold owner/erector. Depending upon the configuration or size of the scaffolding, information regarding load ratings, number of planked levels, and other such information may be required.

The yellow tag indicates that the scaffold has been modified to meet work requirements. Because this is a new condition, workers should be alerted as to what may have been changed and how it can, and will, affect what they are doing. The yellow tag might indicate a worker is to wear a harness or other type of fall protection equipment, or that certain other unusual conditions exist requiring a change of procedures. Whatever the condition, the yellow color, like the caution light, means just that — CAUTION. PROCEED CAREFULLY. The tag should indicate what precautionary measures should be taken to use the scaffold in a safe manner.

Many companies design their own tags allowing information as listed below to be displayed:

- Scaffold identification tag number for tracking purposes.
- Date erected

- Inspection information
- Company responsible for erection, maintaining, dismantling
- Modifications made to the scaffold, with dates of modification
- Competent Person erecting the scaffold
- Competent Person inspecting the scaffold

It is entirely possible that the person designated as Competent Person for erection may not be the person designated as Competent Person for inspection, and if the Competent Person for inspection is not available on the site, the scaffold should not be used, no matter what color the tag is. The Competent Person for inspection of the scaffold is the person you would report any problems to and is the one who should sign and date the tag each day.

Scaffold tags are often encased in a clear plastic envelope so that weather does not affect the tag condition. Sometimes the tag itself is plastic, and information to be filled in is done permanent ink. There is much available on the market and tags can be made for the individual user to accommodate a company's own particular specifications for documented safe use of the scaffold. Tagging, while not required by OSHA, is an excellent way to communicate the condition of the scaffolding to all those using it. Tagging also provides a way to verify inspection of the scaffold.

An OSHA interpretation regarding scaffold tagging, which is shown at the in Figure 4.3, is interesting to note and leaves open the question of being cited for not advising an unsafe condition. Certainly, should an accident occur on an untagged scaffold, this may very well become an important issue. By not, in some way, alerting your employees or the employees of others on the construction site regarding the condition of your scaffolding, you leave yourself open for some pretty dire consequences.

Competent and Qualified Persons on the Jobsite

Now that we have discussed "tagging", let's continue our discussion of competent and qualified persons on the jobsite. The authority invested in the "competent" person to stop work to correct hazards or dangerous conditions is what

April 6, 1992

Mr. Roy Gurnham,
Office of Construction & Maritime
Compliance OSHA
200 Constitution Avenue N.W.
Room N 3610
Washington, D.C. 20210

Dear Mr. Gurnham,

I am writing to introduce myself and ask for a clarification on an issue.

I teach courses in Scaffolding Safety to various contractors, industrial plants and other scaffolding users.

One of the issues that frequently comes up in class is that of putting a tag on the scaffold. My recommendations to our students is that a warning tag be placed on the scaffold while it is being erected because it is only partially complete and may not have all appropriate safety features installed.

After the scaffold is complete and final inspection is done, I recommend that a green "OK" tag or some other type of sign off sheet be placed on the scaffold to indicate that it is ready for use.

This is a widely accepted practice in industrial plants and tagging systems (including warning tags) are available from a variety of sources.

Many students ask me, "Is this required by OSHA?"

My answer is yes. My explanation is that 29 CFR 1926.200(h)(1) requires warning tags to be placed on any existing hazard such as a defective tool or equipment. A partially erected scaffold may not have guardrails, or ladders, and be properly tied or be fully planked. Therefore, it would be a hazard and fall under the defective equipment category until the scaffold is completed and all safety features installed. Consequently, a warning tag should be used.

Several compliance officers I have discussed this with have indicated that they are in agreement and might even use 1926.200(h)(1) for a citation if the situation warranted.

However, since my students like to see clarifications in black and white, I would appreciate a written response from your office.

Thank you for your time.

Sincerely,

John Palmer, Director
Scaffold Training Institute

FIGURE 4.3 Policy for warning tags on scaffolds *(continued on next page)*

John Palmer, Director
Scaffold Training Institute
1706 Center Street Deer
Park, Texas 77536-0067

Dear Mr. Palmer:

This is in response to your April 6 letter requesting a clarification of the Occupational Safety and Health Administration policy for warning tags on scaffolds. I apologize for the delay of this response.

Presently, there is no requirement that warning tags be placed on all scaffolds while they are being erected. Although there are instances where such tags are appropriate and a citation under 29 CFR 1926.200(h)(1) could be issued, there are also many instances where such would not be appropriate. An example of the latter is where a large scaffold is being built in phases or sections, employee access to the completed sections of the scaffold would not be prohibited. A tag which limits access to erection and disassembly crews on that portion of a scaffold being changed could be appropriate but a case by case determination is necessary.

I believe this is what the compliance officers meant when they indicated to you that they are in agreement and "might even use 200(h)(1) for a citation if the situation warranted".

If you have further questions, please contact [Mr. Noah Connell] of my staff in the office of Construction and Maritime Compliance Assistance at [(202) 219-7207].

Sincerely,

Patricia K. Clark, Director
Directorate of Compliance Programs

FIGURE 4.3 *(continued from previous page)* Policy for warning tags on scaffolds

differentiates the Competent Person from the qualified or authorized person. The "competent" person is the only designation allowed to stop work for this purpose.

Requirements for each designated "competent" person can vary from instance to instance. The "competent" person for scaffolding has duties and requirements which the "competent" person for excavation may not have, and vice-versa.

It is important to be aware, as a "competent" person, exactly what your duties entail. These duties are delineated by standard. The "competent" person may have duties as designated by the employer that exceed those duties outlined by standard.

In 1926.32(i), "Designated person" means "authorized person" as defined in paragraph (d) of this section. The "designated" person and "authorized" person are interchangeable terms. Again, an example of "designated" may be someone who has the authority and knowledge and as such is "authorized" to test equipment, repair equipment, etc.

1926.32(m), "Qualified" means one who, by possession of a recognized degree, certificate, or professional standing, or who by extensive knowledge, training, and experience, has successfully demonstrated his ability to solve or resolve problems relating to the subject matter, the work, or the project. "Qualified" can refer to a professional engineer, architect, or other person with extensive knowledge, training, and experience, who has been called in by the owner, employer, or others to perform a specific duty, such as design anchorages, design special-use scaffolds, etc. This person may not stop a job, but can recommend the job be stopped in order to correct a condition. Upon his recommendation, the owner, employer, or anyone else involved in the resolution of the problem may elect to stop the job:

For scaffolding the following are some examples of the use of the designations listed above.

- **1926.451(a)(6)** Scaffolds shall be designed by a qualified person and shall be constructed and loaded in accordance with that design. Non-mandatory Appendix A to this subpart contains examples of criteria that will enable an employer to comply with paragraph (a) of this section.
- **1926.451.(b)(10)** Scaffold components manufactured by different manufacturers shall not be intermixed unless the components fit together without force and the scaffold's structural integrity is maintained by the user. Scaffold components manufactured by different manufacturers shall not be modified in order to intermix them unless a Competent Person determines the resulting scaffold is structurally sound.
- **1926.451(b)(11)** Scaffold components made of dissimilar metals shall not be used together unless a Competent Person has determined

that galvanic action will not reduce the strength of any component to a level below that required by paragraph (a)(1) of this section.

- **1926.451(d)(10)** Ropes shall be inspected for defects by a Competent Person prior to each work shift and after every occurrence that could affect a rope's integrity. Ropes shall be replaced if any of the following conditions exist:
  - **1926.451(d)(10)(i)** Any physical damage that impairs the function and strength of the rope
  - **1926.451(d)(10)(ii)** Kinks that might impair the tracking or wrapping of rope around the drum(s) or sheave(s)
  - **1926.451(d)(10)(v)** Heat damage caused by a torch or any damage caused by contact with electrical wires.
  - **1926.451(d)(10)(vi)** Evidence that the secondary brake has been activated during an overspeed condition and has engaged the suspension rope
  - **1926.451(d)(11)** Swaged attachments or spliced eyes on wire suspension ropes shall not be used unless they are made by the wire rope manufacturer or a qualified person.
  - **1926.451(d)(18)** Two-point and multi-point suspension scaffolds shall be tied or otherwise secured to prevent them from swaying, as determined to be necessary based on an evaluation by a Competent Person. Window cleaners' anchors shall not be used for this purpose.

- **1926.451(e)(9)(i)** The employer shall provide safe means of access for each employee erecting or dismantling a scaffold where the provision of safe access is feasible and does not create a greater hazard. The employer shall have a Competent Person determine whether it is feasible or would pose a greater hazard to provide, and have employees use a safe means of access. This determination shall be based on site conditions and the type of scaffold being erected or dismantled.

Other responsibilities of the Competent Person include:

- **1926.451(f)(3)** Scaffolds and scaffold components shall be inspected for visible defects by a Competent Person before each work shift, and after any occurrence that could affect a scaffold's structural integrity.

- **1926.451(f)(7)** Scaffolds shall be erected, moved, dismantled, or altered only under the supervision and direction of a Competent Person qualified in scaffold erection, moving, dismantling or alteration. Such activities shall be performed only by experienced and trained employees selected for such work by the Competent Person.
- **1926.451(f)(12)** Work on or from scaffolds is prohibited during storms or high winds unless a Competent Person has determined that it is safe for employees to be on the scaffold and those employees are protected by a personal fall arrest system or windscreens. Windscreens shall not be used unless the scaffold is secured against the anticipated wind forces imposed.

This section supplements and clarifies the requirements of 1926.21(b)(2) as these relate to the hazards of work on scaffolds.

- **1926.454(a)** The employer shall have each employee who performs work while on a scaffold trained by a person qualified in the subject matter to recognize the hazards associated with the type of scaffold being used and to understand the procedures to control or minimize those hazards. The training shall include the following areas, as applicable:
  - **1926.454(a)(1)** The nature of any electrical hazards, fall hazards and falling object hazards in the work area
  - **1926.454(a)(2)** The correct procedures for dealing with electrical hazards and for erecting, maintaining, and disassembling the fall protection systems and falling object protection systems being used
  - **1926.454(a)(3)** The proper use of the scaffold, and the proper handling of materials on the scaffold
  - **1926.454(a)(4)** The maximum intended load and the load-carrying capacities of the scaffolds used
  - **1926.454(a)(5)** Any other pertinent requirements of this subpart
- **1926.454(b)(1)** The employer shall have each employee who is involved in erecting, disassembling, moving, operating, repairing,

maintaining, or inspecting a scaffold trained by a Competent Person to recognize any hazards associated with the work in question. The training shall include the following topics, as applicable:

- **1926.454(b)(1)** The nature of scaffold hazards
- **1926.454(b)(2)** The correct procedures for erecting, disassembling, moving, operating, repairing, inspecting, and maintaining the type of scaffold in question
- **1926.454(b)(3)** The design criteria, maximum intended load-carrying capacity and intended use of the scaffold
- **1926.454(b)(4)** Any other pertinent requirements of this subpart

- **1926.454(c)** When the employer has reason to believe that an employee lacks the skill or understanding needed for safe work involving the erection, use or dismantling of scaffolds, the employer shall retrain each such employee so that the requisite proficiency is regained. Retraining is required in at least the following situations:
  - **1926.454(c)(1)** Where changes at the worksite present a hazard about which an employee has not been previously trained
  - **1926.454(c)(2)** Where changes in the types of scaffolds, fall protection, falling object protection, or other equipment present a hazard about which an employee has not been previously trained
  - **1926.454(c)(3)** Where inadequacies in an affected employee's work involving scaffolds indicate that the employee has not retained the requisite proficiency

Your own company requirements may well exceed this definition as to when retraining is required.

OSHA believes that employees erecting or dismantling scaffolds should be trained in the following topics:

General Overview of Scaffolding

- Regulations and standards
- Erection/dismantling planning
- PPE and proper procedures

- Fall protection
- Materials handling
- Access
- Working platforms
- Foundations
- Guys, ties, and braces

Tubular Welded Frame Scaffolds

- Specific regulations and standards
- Components
- Parts inspection
- Erection/dismantling planning
- Guys, ties, and braces
- Fall protection
- General safety
- Access and platforms
- Erection/dismantling procedures
- Rolling scaffold assembly
- Putlogs

Tube and Clamp Scaffolds

- Specific regulations and standards
- Components
- Parts inspection
- Erection/dismantling planning
- Guys, ties and braces
- Fall protection

- General safety
- Access and platforms
- Erection/dismantling procedures
- Buttresses, cantilevers, & bridges

System Scaffolds

- Specific regulations and standards
- Components
- Parts inspection
- Erection/dismantling planning
- Guys, ties and braces
- Fall protection
- General safety
- Access and platforms
- Erection/dismantling procedures
- Buttresses, cantilevers, & bridges

Scaffold erectors and dismantlers should all receive the general overview, and, in addition, specific training for the type of supported scaffold being erected or dismantled.

Chapter 5

Supported Scaffolds — Materials and Methods

Scaffolds, as previously addressed, are of various types. They can however be first categorized into two general categories. The first category to be addressed is scaffolds that are ground supported. The other major scaffold category is suspended scaffolds. Both scaffold types present unique exposures. This chapter will focus on ground supported scaffolds and the various methods of constructing and using them safely.

Soil Bearing Capacities

Supported scaffolds are dependant upon the base support condition (where the scaffold meets the ground) for its support. The entire weight of the scaffold plus the weight of the men and any materials that are on the scaffold must be supported. For a simple four-leg supported scaffold, the weight can easily be divided equally between the legs. See Figure 5.1. Each leg supports the load imposed on the area adjacent to it. Leg A supports the load on area 1. For a welded frame, the legs of one frame, legs A and B, support the weight on areas 1 and 2. End frame legs C and D support the load on areas 3 and 4. For most conditions, one half of the weight being distributed to each end is a reasonable assumption.

For a continuous scaffold, i.e., one with more than four uprights divided into sections or bays, the entire weight of the men, materials, and equipment that is on the scaffold bay must be supported by the uprights. This is the cumu-

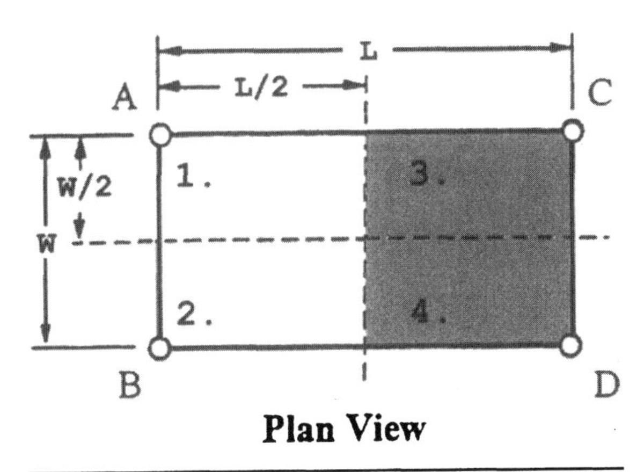

FIGURE 5.1 Plan view

lative load of one half of the bay on each side of the upright. See Figure 5.2. The drawing, Two-bay, illustrates a two-bay scaffold with six uprights, A, B, C, D, E and F. The two frames on each end, AB and EF, each support one half the load in the adjacent bay, as illustrated in the previous example. However, frame CD supports one half of the load of the bays on each side, loads from areas 3 and 4 as well as areas 5 and 6. This is twice the load that is imposed on the end frames.

This weight is transferred into the uprights and onto the ground or surface material. The weight of each scaffold component must be included with the men and materials on the scaffold to determine the total weight of the scaffold. The weight may be obtained from the manufacturer's catalog. This weight of the materials contained on the scaffold is determined by the intended use of the scaffold and the expected material load. In addition, the Occupational Safety and Health Regulations define criteria for light, medium, and heavy-duty scaffolds, and the expected number of personnel who may be in each bay, as a part of the analysis procedure. The non-mandatory Appendix A of the OSHA regulations provides a recommended intended load table and provides a method to determine the load.

Foundation/Support

Most scaffolding comes with a manufacturer-supplied bearing plate. Some of these plates are just individual bearing pads, while other plates come equipped

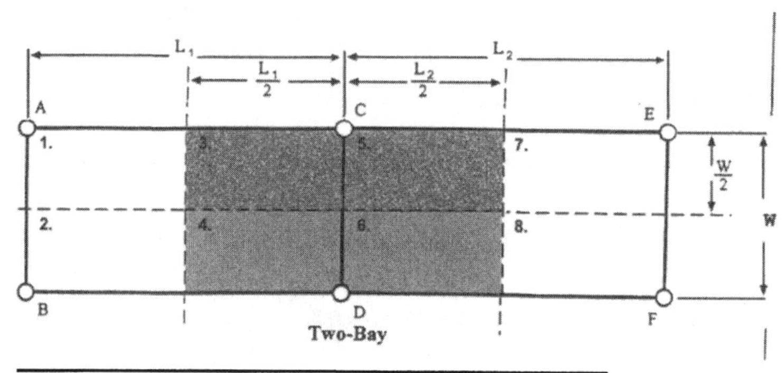

FIGURE 5.2 Two-bay

with a screw jack attached for leveling, while other manufacturers provide separate screw jacks. All scaffolds must have at least a bearing pad. These pads are typically furnished in two common sizes: 4½" square or 6" square. Because these pads support significant amounts of weight, they need to be provided with some form of cribbing, or means to distribute the loads on all but the most rigid of pavements, i.e., concrete.

There are numerous types of foundation materials that are encountered when placing supported scaffolds. On rare occasions, a nice, level concrete pad will be available to erect the scaffolding. This greatly simplifies the task of the scaffold designer/erector. Concrete can be depended upon (if it is a "slab on grade", i.e., concrete on ground) to provide a bearing capacity of several thousands of pounds/square inch depending upon the design strength of the concrete. This design strength may vary from 2000 PSI to in excess of 5000 PSI. 3000 – 4000 PSI are normal strengths.

A 4½" square base on the leg of the scaffold will provide 20.25 square inches of bearing area. A minimum strength supported concrete slab will provide sufficient capacity for all but the tallest, most heavily loaded scaffolds. This load, in many cases, will exceed the capacity of the scaffold frame or post. The slab on grade will provide adequate support for most reasonably sized scaffolding.

Asphalt and other bearing surfaces provide much less bearing capacity than does concrete. To provide adequate support for scaffolding on these materials, a means to spread the load over a larger area is necessary. This is most commonly done by using wood cribbing. Wood cribbing or sills are not always effective in spreading the load over the entire area of wood. For example, a piece of plywood will only increase the effective bearing area slightly. The full size of the plywood cannot be used to estimate the effective bearing area. The effectiveness of wood blocking can be estimated as the base plate dimension plus two times the thickness of the wood cribbing.

As shown in Figure 5.3, a 6″ × 6″ pad placed on a 1½″ thick piece of wood would effectively be 9″ × 9″, the 6″ plate plus two times 1 ½″ inches of the wood. The bearing area would be increased from 36 square inches to 81 square inches. 2″ × 8″ or 2′ × 10′ nominal sized lumber provides effective sills for most applications. This more than doubles the bearing area of the pad and correspondently reduces the bearing pressure on the soil. Sills larger than this "rule of thumb" do not contribute significantly to reducing the bearing pressure because of bending and shear forces induced in the wood. If larger areas are required, a qualified engineer should be consulted.

Another limiting factor is the bearing capacity of the wood perpendicular to the grain. The load imposed on the wood cribbing should never exceed this capacity times the area of the bearing plate. This is the maximum load that can be imposed by the leg of the scaffolding. A table of the capacity of various species of wood can be found in Table 5.1.

To provide stability, the wood cribbing or sills should be provided continuously between the two legs of the end frame or scaffold legs. Although they do not contribute significantly to increasing the bearing capacity, they do significantly increase the stability.

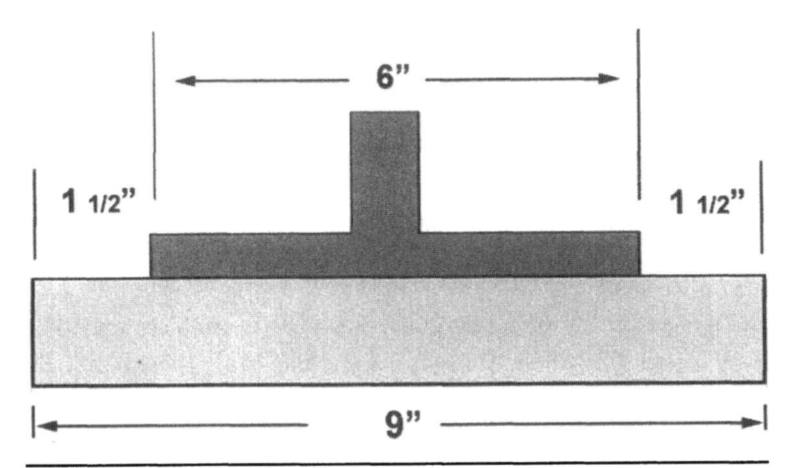

FIGURE 5.3 Bearing area

TABLE 5.1 Table of wood values

Species & Commercial Grade	Extreme Fiber Bending & Tension Parallel to Grain	Compression Parallel to Grain	Compression Perpendicular to Grain
Pine, Southern Long Leaf, Select Structural	2400 psi	1750 psi	455 psi
Pine, Southern Long Leaf, Prime Structural	2150 psi	1550 psi	455 psi
Pine, Southern Short Leaf, No. 1 Dimension	1450 psi	1075 psi	390 psi
Douglas Fir, Coast Region, 2150 f Dense Select Structural	2150 psi	1550 psi	455 psi
Douglas Fir, Coast Region, 1200 c No. 1		1200 psi	390 psi
Spruce, Eastern, 1450 f Structural Grade	1450 psi	1050 psi	300 psi
Spruce, Eastern 1200 f Structural Grade	1200 psi	900 psi	300 psi

Tall scaffolding, many stories high, or heavily loaded scaffolds need to be designed by a professional engineer who will analyze the base bearing conditions and may, in many cases, identify the need to supply additional reinforcement in the concrete pads being used. Supported scaffolds are limited to a maximum height of 125′ unless designed by a professional engineer.

Asphalt paving may be used for support. Asphalt is placed in various thicknesses and over many types of sub-bases. The base of the asphalt is usually compacted stone. Without a stone base, asphalt is unreliable as a support. The strength of the asphalt is primarily dependent upon the foundation material beneath the stone base course and its relative degree of compaction.

Asphalt is subject to problems if a weak or incomplete base is present and may be subjected to compression and/or actual softening or flowing of the material, particularly during the heat of summer when temperatures can reach in excess of 100 degrees. Scaffolds that will remain in place for a significant period of time need to be evaluated for both the effect of summer heat and winter freezing and thawing action. Both of these occurrences significantly impact the support provided to the scaffold and the ability of the scaffold to remain level.

Scaffolds placed on asphalt should always be placed on wood cribbing or stringers. The area on which the scaffold is to be placed should be carefully examined for holes, depressions, or patches, all of which may indicate weak sub-base material or recent excavation. These conditions must be fully investigated prior to erection of the scaffolding, as they may quickly become unstable.

Groundwater or ponding of surface water can have a very detrimental effect on the bearing capacity of soil. Table 5.2 illustrates that what appears to be a hard clay with the capacity of 6 tons/square foot, with the addition of water, becomes a soft clay, reducing the strength to one ton/square foot, one sixth of its previous value. Water should be kept drained away from scaffold bases and not allowed to pond or form puddles underneath or around the scaffold. Snow and ice must be removed from the area of the base of the scaffold during winter months.

A well-compacted granular, large stone or gravel base area is the next strongest support media for a scaffolding. If it is well compacted on a firm sub-base, the use of sills beneath the scaffold support legs will usually provide a firm stable support. Recent fill or poorly compacted material under the gravel base will allow significant settlements to occur and render the stone base ineffective.

TABLE 5.2 Typical bearing capacities *(subject to actual load test)*

Material	Weight
Massive Rock	100 Tons PSF
Laminated Rock	35 Tons PSF
Hardpan	10 Tons PSF
Compact Sands-Gravel (Mixtures)	5 Tons PSF
Loose Sands (Mixtures)	4 Tons PSF
Sand Coarse Loose	3 Tons PSF
Sand Fine Loose	1 Ton PSF
Hard Clay	6 Tons PSF
Medium Clay	4 Tons PSF
Soft Clay	1 Ton PSF

On a construction site, the recently occurring excavations and backfills should be investigated. The areas along foundation walls will have been recently back-filled and may not be adequately compacted when backfilled. Locating trenches for utilities and placement of scaffold legs on or near the excavation should be avoided. The plans for the project will indicate when and where utility and sewer lines were installed. Inadequate deep surface base support is not corrected by a few inches of stone or gravel. If there is a question of the quality of compaction of these or similar fill areas, a thorough investigation of the supporting capacity of this material needs to be confirmed by testing.

Most scaffolds are not supported on pavement, either concrete or asphalt, but on earth or soil. Soils are classified primarily by the size of the particles that make up the soil. The particle sizes range from the largest gravel or stone to the finest clays that are less than 0.005 millimeters in diameter. The most commonly utilized system is the triangular soil classification chart (Figure 5.4). This classification system groups soils by three sizes of soil: clay, silt, and sand — gravel and sand being coarser, silts finer, and clay being composed of the smallest of particles. The relative percentage of each size of particles determines the classification of the soil, i.e., clay, sand, or silty sand or sandy silt. See the chart for clarification.

The chart illustrates various simple materials. Most soils are a combination of the three sizes of particles, clay, silt, and sand and the bearing capacity varies accordingly. No natural soil is composed of entirely one-sized particles. To illustrate these properties, the triangular classification chart was developed. The points of the triangle represent a sample composed of 100% single grain size. Points within the center of the chart indicate relative proportion of the different size particles. The names listed in the various sub-divided areas of the

TABLE 5.3 Classification — grain size table

SOIL	DIAMETER OF PARTICLES (MM)
Fine Gravel (Grit)	2.0 to 1.0 mm
Coarse Sand	1.0 to 0.5 mm
Medium Sand	0.5 to 0.25 mm
Fine Sand	0.25 to 0.10 mm
Very Fine Sand	0.10 to 0.05 mm
Silt	0.05 to 0.005 mm
Clay	0.005 to 0.0001 mm

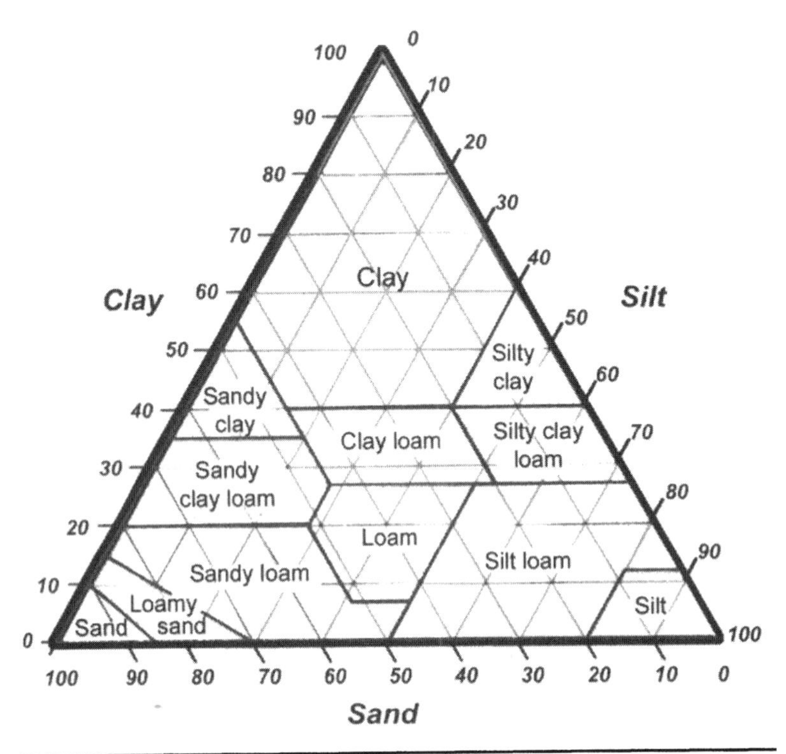

FIGURE 5.4 Soil classification chart

chart represent common names applied to various soil types. Each point within the chart totals 100% of the soil sample. For example, Figure 5.5 illustrates a soil sample that is 40% clay, 40% sand, and 20% silt.

The chart gives rule-of-thumb bearing capacities for various types of soils. These capacities are given in tons/square foot. One ton/square foot is equal to 13.89 pounds/square inch or 281 pounds, or a 4½" square base plate or 500 pound or a 6" base plate. It is easy to see why a scaffold supported on soil requires cribbing and stringers to distribute the load. These bearing capacities must be reduced by an appropriate factor of safety. The soil investigation, which is performed prior to construction, probably contains soil bearings that define the type of soil and also estimate the bearing capacity of the soil.

A well-compacted clay material is relatively dense when dry and provides excellent support for scaffolding. Clay is composed of the smallest particle sizes of soil and can very easily be displaced by running water. If the clay is

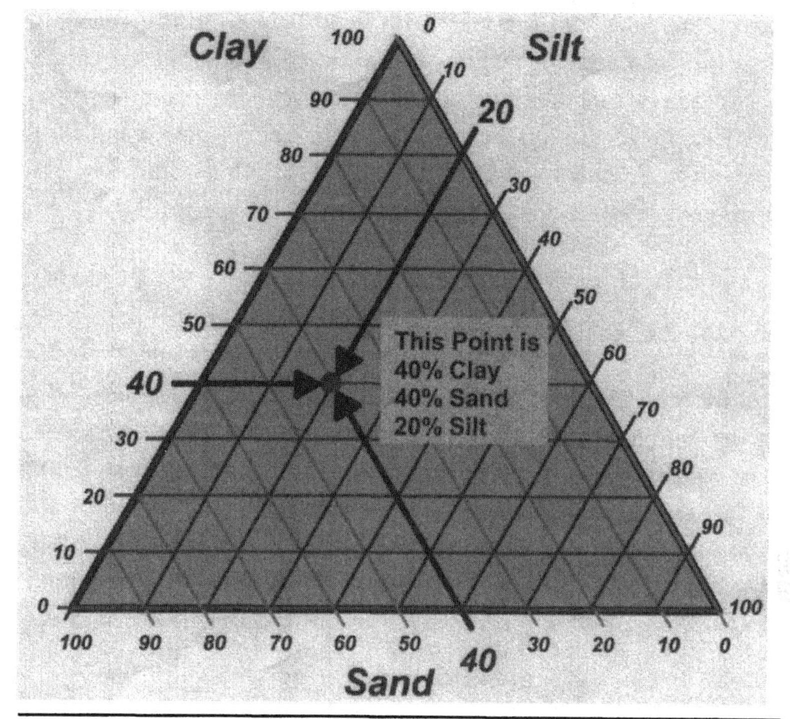

FIGURE 5.5 Soil sample

dry and well compacted, it has high compressive strength. If, however, it is subject to moisture in the form of rain or standing water on the job site, it can very easily and quickly lose its compressive strength and allow significant settlement of the scaffolding.

From the above comments, it is obvious that the bearing conditions for the scaffolding on most job sites can change from hour to hour, if not day to day, and these soil conditions require regular inspections by Competent Persons to determine:

- That the support is adequate
- If conditions have changed, to provide the necessary compensation in the form of adjustment of the scaffold legs and/or draining of water, etc

Once the scaffolding has been erected, it is very difficult to compensate for inadequate foundation materials. The only solution is to disassemble and repair the base conditions. An investigation of the soil conditions prior to erecting the scaffold is imperative and daily, if not more frequent, observations of the changes to each bearing point of the scaffold must be inspected and noted by a Competent Person. Inspections are required by the OSHA regulations before the start of each shift. Rain, snow, freezing, thawing, or other conditions may require more frequent inspections.

Scaffolds on Elevated Structures

Scaffolds are occasionally supported on an existing structure. The concrete slab of an existing building is perhaps the most common. These slabs are designed to support very limited loads. The slabs for most construction are designed to support loads that range from 50 pounds/square foot up to approximately 150 pounds/square foot maximum, depending upon the intended use of the building. Residential buildings are at the lower end of the scale with office buildings, industrial buildings, and parking garages on the higher end of the scale.

Most scaffolds and manually propelled mobile towers, if lightly loaded, can be utilized on supported slabs without fear of damage to the slab. Scissors lifts and boom lifts impose significantly greater floor load. The architect or engineer of the project should always be consulted before this type of equipment is operated on an elevated crane slab.

Most other structures, roofs, wood framed buildings, canopies, etc. will require an engineering analysis of the structure prior to the placement of a scaffold scissor lift or boom lift on the structure. Many older existing concrete structures have deteriorated conditions, such as spalled concrete, exposed rebar, or cracked slabs. Spalling or flaking concrete or exposed rebar cracks in floors or columns can result in a reduced capacity of the structure to support the imposed load. A careful examination of the existing structure is required before erecting a scaffold.

Planks

"Planking", boards spanning sections of scaffolding, allows workers to access areas and to perform construction activities at a height higher than ground level. According to OSHA and the federal labor statistics, more accidents occur as a result of planking giving way than any other cause. These same sources have documented that almost 25% of the work force receives no safety training for erecting scaffolding and for the installation of work platforms.

OSHA provides a Scaffolding E-Tool addressing the need for specific and accurate information about the lumber from which platforms are made. Topics covered include:

- Lumber grading
- Condition of the wood
- Allowable spans
- Allowable deflection

CalOSHA, in meeting the requirement for inspection of lumber, provided a Suggested Test for Scaffold Planks. Their suggested test, utilizing a simple impact loading method, suggests that:

- Select a span that is convenient for the length and size of plank to be tested. Obtain two (2) blocks about two feet long and of a thickness corresponding to the span and plank size. Set blocks and plank on a flat surface so that clearance between surface and plank is uniform.
- Have two men spring on the board, standing close together at midspan until the board deflects to the surface below several times. DO NOT apply a load so large that the plank is held against the floor. It should only touch the floor and only for an instant during the springing.
- Listen for cracking and look for splitting.
- If no cracking or splitting is noted, turn the plank over and repeat the test. If the plank cracks or splits, reject. In case of faint cracking sound, caused by stretching of fibers near allowable knots, repeat the test and accept the plank if no further cracking sounds are heard.

The main advantage of this system, in addition to simplicity, is that it will show up wood that is brash or contains compression failures, while not damaging good planks. OSHA has instituted a safe load-bearing capacity based on deflection or bending of the plank. This limits deflection under load to less than $^1/_{60}$ of the plank's span between supports. This is discussed later in this section.

Lumber Grading

Lumber grading is important. There are two kinds of lumber grading: construction-grade and scaffold-grade. Construction-grade lumber withstands only two thirds the capacity of scaffold-grade. Using construction-grade lumber on scaffolds violates OSHA standards and permits an unsafe practice inviting a deadly accident.

Scaffold-grade lumber is measured by rings per inch (six or more), slope of the grain (one inch to the side for every 16 inches along the length of the board for Douglas Fir, 1 inch to the side for every 14 inches along the length of the board for Southern Pine), and the number of defects, such as knots and notches.

Solid sawn wood used for scaffold planking should follow the grading rules of a recognized lumber grading association or an independent lumber inspection agency and be identified as "scaffold plank" by that agency or association's grade stamp. Organizations and their grading rules must be certified by the Board of Review of the American Lumber Standard Committee, per the U.S. Department of Commerce.

Figure 5.6 shows two stamps from different grading agencies. The one on the left specifically lists "Scaffold Plank". The stamp on the right indicates "Sel Str", Select Structural. Both are acceptable grades. NOTE: Construction grade lumber is NOT acceptable for scaffold planks.

OSHA does NOT inspect lumber and any scaffold planking stamped "OSHA Approved" is misleading and should not be assumed to meet the standard on the basis of that stamp alone. However, an agency CAN claim that their product meets OSHA requirements. See Figure 5.7. This stamp indicates the plank meets the OSHA standards. It also, above the OSHA stamp, has a grading stamp by a lumber-grading agency indicating "Scaffold Plank".

Because planks last longer than the markings, many contractors mark their scaffold planks by either painting the ends or side of the planking (NOT the surface) with a distinguishing color or by branding (yes, just like cattle) with a hot iron so that the plank can be recognized as:

SPIB® DNS IND 65
K019 S-DRY 7
SCAFFOLD PLANK

Grade stamp courtesy of Southern Pine Inspection Bureau

MILL 10
WC LB® SEL STR
SCAF PLK
D. FIR S. DRY

Grade stamp courtsey of West Coast Lumber Inspection Bureau

FIGURE 5.6 Two stamps from grading agencies. On the left is “Scaffold Plank”. The stamp on the right indicates “Select Structural”.

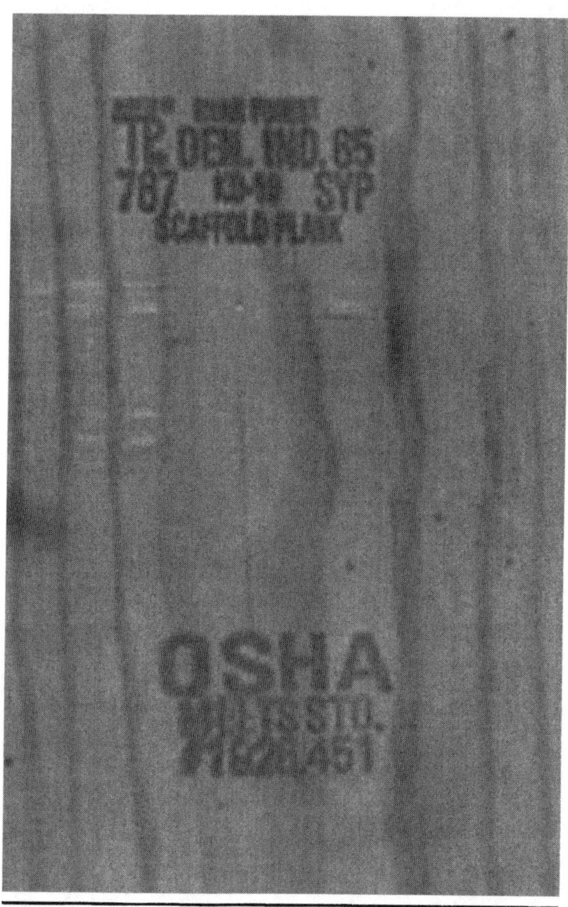

FIGURE 5.7 Stamp indicating that the plank meets OSHA standards

- Belonging to the contractor
- Being of scaffold grade when the stamping has worn off

These types of markings also help to avoid having a scaffold plank (a plank of better grade and more cost) being used for something not requiring scaffold grade planking and helps the scaffold contractor locate his materials at the end of the job. It is also easier for the Competent Person who is inspecting the scaffold because the designation for scaffold grade planking is easily discernable.

A scaffold plank may still be in use long after the grade stamp has faded and may not be marked, as above, and workers should pay attention to the quality and condition of the planking, whether it bears a stamp or not. My suggestion is that if you are not sure, really sure, that the plank is scaffold-grade, do not use it! Replace the plank with one you know is scaffold-grade and don't take any chances. The cost of the plank is negligible when compared to the cost of accident or injury caused by use of the wrong grade of plank.

Evaluating the Condition of Scaffold Planking

There are many visual clues indicating the condition of scaffold planking. As wood ages and reacts to usage, it will begin to show checks, splits, and notches. These can vary in degree depending on loads imposed, weather conditions, and length of service.

Splits (cracks going clear through the wood) more than a few inches in length should not remain in service. These splits may render the plank unable to maintain the necessary load-bearing capacity.

Also note, the following information is taken verbatim from OSHA INSTRUCTION STD 3-10.4 OCTOBER 30, 1978: c. 29 CFR 1926.451(a)(7) and (8) — "A scaffold plank is considered a component of the scaffolding. As such it must be capable of supporting four times the maximum intended load without failure (a safety factor of four). The fact that a plank might have a split in one end does not automatically mean that it must be removed from service, provided it does not otherwise create a hazard to the employees."

The Competent Person must examine all planks and determine that they are safe for use. Checks (cracks on the surface only and not clear through the wood) should be watched as checks may develop into splits over time. Notches (small checks at the ends of the plank) should also be watched as over time these notches can lengthen and deepen until they become splits.

NOTE: Scaffold planks that have accumulated layers of paint, plaster, etc., should not be permitted to remain in service because it is impossible to determine the condition of the wood. Dangerous splits may be hidden beneath the paint, plaster, etc. [See 1926.451(b)(9)]

If a scaffold plank has been used as a mudsill, it should NOT be returned to service on a platform. Moisture that the plank has absorbed from standing water, as well as point loading from the scaffold legs, may have weakened the plank, making it unable to bear the weight that will be placed on it.

Evaluating the Span of a Scaffold Plank

What is the span of a scaffold plank? The span is the distance it runs between supports. The longer the span, the more bend or deflection, it will have. Therefore, the longer the span, the less its load-bearing capacity will be.

Nominal thickness lumber is not cut to its exact dimensions. For instance, nominal 2 × 10 is really more like 1½ × 9¼. As a result, it does not have the same load-bearing capacity of full thickness lumber.

If fabricated planks and platforms are being used, maximum spans are to be as recommended by the manufacturer. To assure that scaffold planking remains within its safe load-bearing capacity, it shall not be allowed to deflect more than 1/60 of its span between supports [See 1926.451(f)(16)].

Workers should be aware of deflection because it can indicate when a platform is overloaded. Repeated use of a plank in excess of these allowed deflections overworks the plank and can cause failure much like repeated bending of a paper clip. Excessive deflection also makes the scaffolds difficult to walk on because of the springing action and makes storage of materials difficult. A Competent Person should not allow employees to occupy a platform exceeding the 1/60 ratio.

PLATFORMS

OSHA's Construction E-Tool for platforms covers how planks are utilized within the scaffold. They are specific in saying that except when used only as a walkway, the platform is the work area of the scaffold. Therefore an inspection of a scaffold platform requires safety checks of both the platform structure and how the platform is used by the workers. They also note that, except where indicated, these requirements also apply to manually propelled, pump jack, ladder jack, tube and coupler and pole scaffolds, as well as the specialty scaffolds described in the supported scaffolds portion. Items covered by this E-Tool are:

- Planking
- Working Distance
- Overlap
- Brackets
- Capacity
- Falling Object Protection

Platforms must be fully planked or decked between the front uprights and the guardrail supports. [1926.451(b)(1)] If the platform is used as a walkway, or is used during erection and dismantling, the planking used in these cases is only required to provide safe working conditions. [1926.451(b)(1)(ii)]

Gaps greater than 1 inch are not permitted between planks or deck units or between the platform and the uprights unless the employer can show that a larger space is necessary. If this condition should occur, the gap will be as small as possible and in no case may exceed 9½ inches. [1926.451(b)(1)(I) and (ii)]

Workers should be alerted to any change from the normal 1-inch gap. It might even be a good idea to flag or somehow mark the areas as being different, just as a reminder to the worker. If the scaffolding is "loaned" to anyone else on the site when it is in this configuration, care must be taken to advise the worker(s) who would not be aware of this exception to the 1-inch gap and that this condition is present. Each construction discipline may use the scaffold in a different way and should never, never assume that the planking and scaffold are set up in a way that would suit them.

Wooden planking covered with paint, plaster, opaque finishes, etc., (the platform edges, however, may be marked for identification) must not be used. Opaque finishes hide conditions such as splitting, checking, and notching, making identification of these conditions more difficult to ascertain. Platforms may, however, be coated periodically with wood preservatives, fire retardants, or slip-resistant finishes, provided they do not obscure the top or bottom wood surfaces. [1926.451(b)(9)]

Scaffold platforms and walkways must be at least 18 inches wide unless the employer can show that the areas they are using are so narrow that the platform must be less than 18 inches. Whatever width is used to access this narrow area must be as wide as feasible and fall protection must be provided. This looks like another case where marking the area for workers to be aware of this lack of general conformance might be a good idea. Also, identifying this area to workers through job meetings, tool box talks, etc., is important. Remember, this is an area that should be discussed and noted if loaning the scaffold to another employer on the jobsite.

When using platforms, it is important that anything that could cause a slip, trip, or fall (such as tools, scrap materials, chemicals, snow, ice, etc.) is NOT allowed to accumulate on the platform. [1926.451(f)(8) and 1926.451(f)(13)] This is why cleats or other means of connecting planks should be on the underside. Inadequately placing the planks with the cleats to the topside creates a tripping hazard and completely renders useless the reasons the cleats are there in the first place.

When moving platforms to the next level, the existing platform must be left undisturbed until the new end frames have been set in place and have been braced. [1926.452(c)(1)] There must be no more than a 14-inch gap between the scaffold platform and the structure being worked on. For lathing and plastering, a gap of 18 inches is permitted. [1926.451(b)(3) and 1926.451(b)(3)(ii)]

Platforms must be cleated or otherwise restrained at each end or overlap centerline support at least 6 inches. [1926.451(b)(4)]

Unless a platform is designed and installed to support employees and materials without tipping — or has handrails that block employee access — each end of the platform may not extend over its support more than 12 inches (for platforms 10 feet or shorter in length) or more than 18 inches (for platforms more than 10 feet long). [1926.451(b)(5)]

Where platforms are overlapped to create a long platform, the overlap:

- May ONLY occur over supports
- May not be less than 10 inches, unless the platforms are restrained (that is, nailed together, etc.) to prevent movement [1926.451(b)(7)]

Where platforms are abutted to create a long platform, each abutted end must rest on a separate support surface. (This does not preclude the use of shared support members such as "T" sections, hook-on platforms that rest on common supports, etc.) [1926.451(b)(6)]

When platforms must overlap because a scaffold changes direction (such as turning a corner), platforms resting on a bearer at an angle other than a right angle shall be laid first. Platforms that rest at right angles over the same bearer shall be laid second, on top of the first platform. [1926.451(b)(8)]

When brackets are used to support cantilevered platforms, they must:

- Be SEATED with side-brackets parallel to the frames and end-brackets at 90 degrees to the frames [1926.452(c)(5)(I)]
- Be used ONLY to support personnel, unless the scaffold has been designed for other loads by a qualified engineer and built to withstand the tipping forces caused by other loads [1926.452(c)(5)(iii)]

Scaffold platforms must be able to:

- Support their own weight, plus
- Four times the maximum intended load. [1926.451(a)(1)]

Scaffolds or their component parts should not be loaded beyond their maximum capacity (their own weight plus and 4 times their maximum intended load). A scaffold can be overloaded by:

- Too many people on the platform
- Too much material stored on the platform
- Point loading — concentrating too much of the load in one area [1926.451(f)(1)]

FALLING OBJECT PROTECTION

When considering falling object protection, two kinds of hazards associated with scaffolds exist. One is the employees on the scaffold and the other is the employees who may work in or enter the area below the scaffold. The employee on the scaffold must be protected from falling hand tools, debris, and other small objects by:

- Hardhats
- Toeboards, screens, or guardrail systems
- Debris nets or canopy structures to contain or deflect falling objects
- Placing potential falling objects away from the edge of the surface from which they may fall, when the objects are too large, heavy, or massive to be contained or deflected by any of the above listed measures [1926.451(h)(1)]

If there is danger of tools, materials, or equipment falling from the scaffold onto employees or others below, these people must be protected by:

- Barricading the area, so no one is able to enter the area below the scaffold
- Installing toeboards along the edge of the platform if it is more than 10 feet above lower levels
- Paneling or screening the area from the toeboard to the top of the guardrail to prevent tools, materials, or equipment piled higher than the toeboard from falling on people and objects below
- Installing a canopy structure, debris net, or catch platform over the employees who may be in the area underneath the scaffold of a strength to withstand the impact of potential falling objects

Toeboards used for falling object protection must be:

- Able to withstand a force of at least 50 pounds applied in any downward or horizontal direction — at any point along the toeboard
- At least 3½ " high from the top edge to the level of the walking or working surface.
- Securely fastened in place at the outermost edge of the platform and must not have a more than ¼-inch clearance above the walking or working surface
- Solid — with openings not over 1 inch [1926.451(h)(4)]

FYI: Appendix A of SubPart L lists some non-mandatory toeboard guidelines.

CAUSES OF SCAFFOLD ACCIDENTS

It has been my experience that one of the most frequent causes of scaffold accidents is related to planking and platforms. Following the regulations under SubPart L, as they relate to these matters particularly, would greatly increase the safety in working with scaffolds. Lack of training is apparent in looking at the accident figures and their causes. There is no excuse for not providing training. Even seasoned scaffold workers may be ready for a re-training or refreshing of their knowledge of scaffold rules and regulations.

Too often, workers assume that if they have done something that is considered unsafe by OSHA standards before and nothing horrible happened, that it is okay to continue those practices. Employers must provide training and provide proper supervision to end these unsafe practices before they become a very bad habit leading to an accident, injury, or death. Remember, OSHA requirements are the bare minimum acceptable safety requirements — the really good guys in construction strive for and attain safety requirements that are much, much better than OSHA!

WIRE SUSPENSION ROPES

Suspension rope is used as a means of hoisting and supporting suspended scaffolds. Wire rope is often used because of its strength. Adjustable suspension scaffolds include the following types:

- Multi-level
- Single-point adjustable
- Multi-point adjustable
- Interior hung
- Needle beam

- Catenary
- Float [ship] scaffolds

They are designed to be lowered and raised while occupied with materials and workers and must be capable of bearing their load, whether stationary or in motion. These scaffolds and their scaffold components must be capable of supporting (without failure) their own weight and at least four times their intended load. [1926.451(a)]

Intended load, what is that? Well, this certainly is no place to "guess". It is important to know the weight of the workers as well as the weight of any materials that may be stored or located on the scaffold. Remember, this number can change each day and should be recalculated in order to assure that the suspension rope can withstand the weight of the scaffold itself and its "intended load".

Each suspension rope, including connecting hardware, must be capable of supporting, without failure, at least 6 times the maximum intended load applied to that rope while the scaffold is operating at the greater of either the rated load of the hoist or 2 times the stall load of the hoist. [1926.451(a)(4)] Taking a scaffold apart before it is on ground level is NEVER a good idea.

All suspension scaffold support devices (outrigger beams, cornice hooks, and parapet clamps) must rest on surfaces capable of supporting at least 4 times the load imposed on them by the scaffold operating at the greater of either the rated load of the hoist or 1.5 times the stall capacity of the hoist. [1926.451(d)(1)] These support devices must be supported by bearing blocks [1926.451(d)] and be secured against movement with tiebacks. These tiebacks should be installed at right angles to the face of the building or structure or by opposing angle tiebacks installed and secured to a STRUCTURALLY SOUND POINT OF ANCHORAGE! Sometimes this means bringing in an engineer to help determine the structural stability. Believe me, this is a smart move and one that will prevent accidents, injuries, and fatalities. Don't take chances. Structurally sound points of anchorage include structural members, but NOT vents, electrical conduit, or standpipes and other piping systems! No more than TWO EMPLOYEES should occupy suspension scaffolds designed for a working load of 500 POUNDS. No more than THREE EMPLOYEES should occupy suspension scaffolds designed for a working load of 750 POUNDS. [1926 Subpart L, Appendix A (2)(p)(2)] This is non-mandatory, but makes good sense to me. Remember, this does not take into account anything more than the employees — any materials they may be taking with them must be added so that the weight of the employees and materials does not exceed the 500 and 750 pounds.

Another very important point — scaffolds should be altered ONLY under the supervision and direction of a Competent Person. [1926.451(f)(7)] Should workers need to have the scaffold altered, it is not to be done by them. They need to communicate their needs to the Competent Person.

Suspension ropes that support adjustable suspension scaffolds must be of a diameter large enough to permit proper functioning of brake and hoist mechanisms. The use of repaired wire rope as suspension rope is absolutely, positively prohibited. [1926.451(d)(7)]

Wire suspension ropes must not be joined together except through the use of eye splice thimbles connected with shackles or coverplates and bolts. [1926.451(d)(8)]

The load end of wire suspension ropes must be equipped with proper sized thimbles and secured by eyesplicing or equivalent means. [1926.451(d)(9)] See Figure 5.8. This wire rope has a thimble secured by an eyesplice, as required on the load end of suspension ropes.

The Competent Person must inspect the ropes for defects at the beginning of each workshift and after any and every occurrence that could affect a rope's ability to perform properly. [1926.451(d)(10)] NOTE: According to the OSHA E-Tool, analysis of Bureau of Labor Statistics data for suspended scaffold fatalities from 1992-99 found that over 20 percent of fall deaths were due to suspension ropes breaking. This fact underlines the importance of inspecting ropes before each and every workshift. Ropes are to be replaced when any of the following conditions exist:

- Physical damage impairing the function and strength of the rope [1926.451(d)(10)(I)]

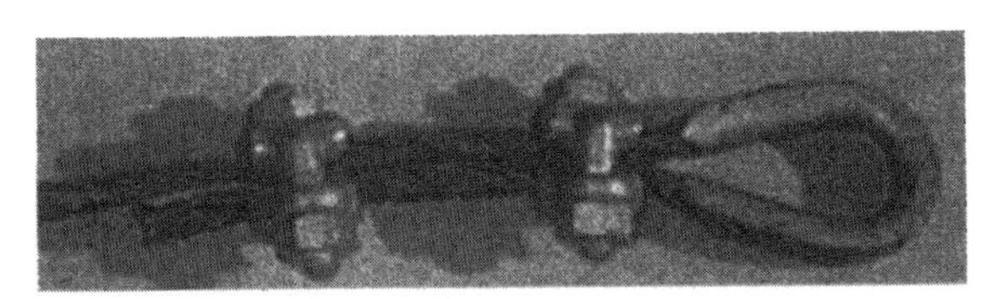

FIGURE 5.8 Wire rope with thimble secured by an eyesplice

- Kinks which might impair the tracking or wrapping of the rope around the drum or sheave of the hoist [1926.451(d)(10)(ii)]
- Six randomly distributed broken wires in one rope lay, or three broken wires in one strand in one rope lay [1926.451(d)(10)(iii)]
- Loss of more than one-third of the original diameter of the outside wires because of corrosion, abrasion, flattening, scrubbing, or peening [1926.451(d)(10)(iv)]
- Heat damage caused by a torch or any damage caused by contact with electrical wires [1926.451(d)(10)(v)
- Any evidence that the secondary brake has been activated during an overspeed condition and has engaged the suspension rope [1926.451(d)(10)(vi)]

Wire rope is constructed in many configurations for specific uses. Running ropes on cranes are constructed in one configuration while standing ropes such as those that hold the gantry or ropes for guying towers or derricks are constructed in other configurations. The properties of a given wire rope depend on many factors. The grade of steel is the primary factor determining the strength and there are many grades of steel used in wire rope.

A wire rope is essentially a group of strands laid helical around a suitable center that may be hemp or metallic. The strands themselves consist of a number of individual wires laid about a central wire, the number and arrangement of the wires in the strand, the number of strands in the rope, and the type of center.

The rope center serves as a support to the strands laid about it. It may consist of a hemp rope or an independent wire rope or strand. A hemp center may be used, but in cases where it would not give adequate support to the strands, an independent wire rope center (IWRC) is advisable. This IWRC would be used, for instance, where operation at high temperatures would deteriorate a hemp center or where winding in several layers on a drum would promote crushing, etc.

What is the lay of the rope (Figure 5.9)? The lay of a rope refers to the direction in which the strands pass around the axis or center line of the rope. In a right lay, the strands pass around the center line in a right-hand spiral. In a left lay rope, the strands pass around the center line in a left-hand spiral. You

FIGURE 5.9 The lay of a rope

may also hear such terms as regular lay and lang lay. Regular lay rope has the wires in the strands and the strands in the rope laid in opposite directions about their respective centers. In a lang lay rope, both the wires in the strands and the strands in the rope are laid in the same direction about their centers. Both regular and lang lay rope can be supplied in either a right or left lay.

The following are some of the more common types of wire ropes. A 6 × 7 rope consists of six strands of seven wires each wrapped around a core. It may have an IWRC or fiber core. A 6 × 19 rope consists of six strands of nineteen wires each wrapped around either an IWRC or fiber core. A 6 × 37 rope consists of six strands of thirty-seven wires each wrapped around an IWRC or fiber core. As you can see, the first number designates the number of strands — the second number designates the number of wires in each strand. (Figure 5.10)

Ropes with large wires around the outside, such as 6 × 7, are more resistant to abrasion, but are stiffer and have less reserve strength. One broken wire is $^1/_{42}$ of a 6 × 7 wire rope (2%) and $^1/_{222}$ of a 6 × 37 wire (0.4%), given equal wire size in all strands. In ropes of equal size, larger numbers of wires give increased flexibility and reserve strength but are less resistant to abrasion.

Because of the construction of wire rope, it is not perfectly round. The diameter of wire rope is defined as the diameter of a circle circumscribed around the rope. Measure this diameter that is outside to outside of the opposite strands – not across the "flat" of the strands. See Figure 5.11 and Figure 5.12.

FIBER ROPES

Fiber ropes are rarely used in connection with scaffolding. They may be used in connection with safety lines or boson's chairs, but generally speaking, fiber ropes are not used with scaffolding. However, for those applications that might be suitable for fiber ropes, we are including the following types:

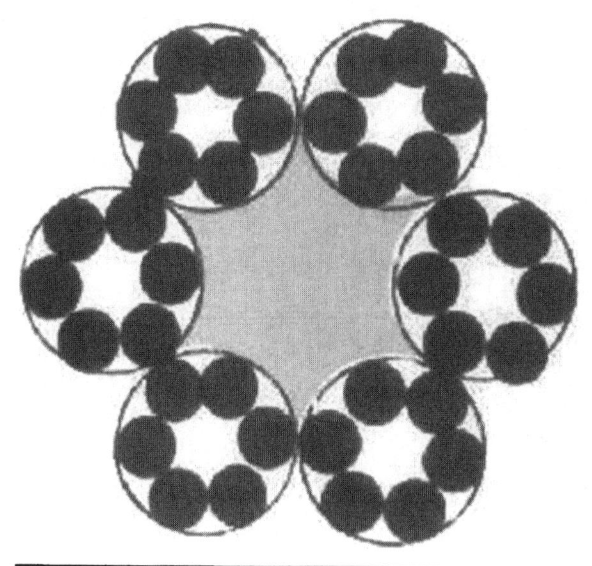

FIGURE 5.10 Common type of wire rope

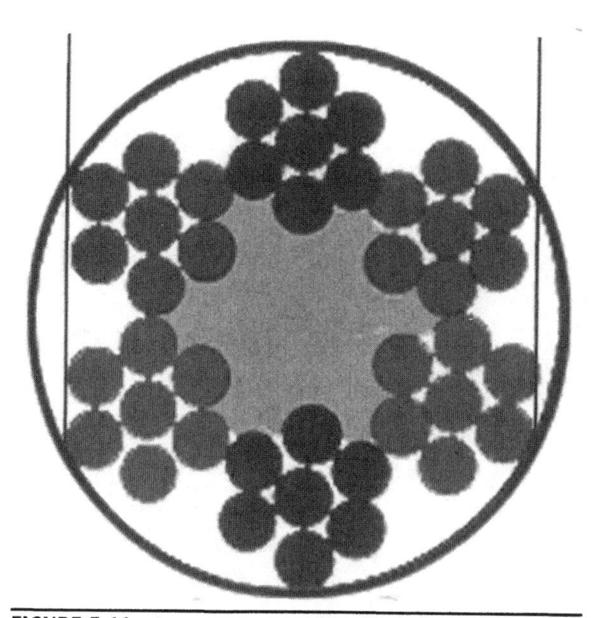

FIGURE 5.11 Incorrect way to measure wire rope

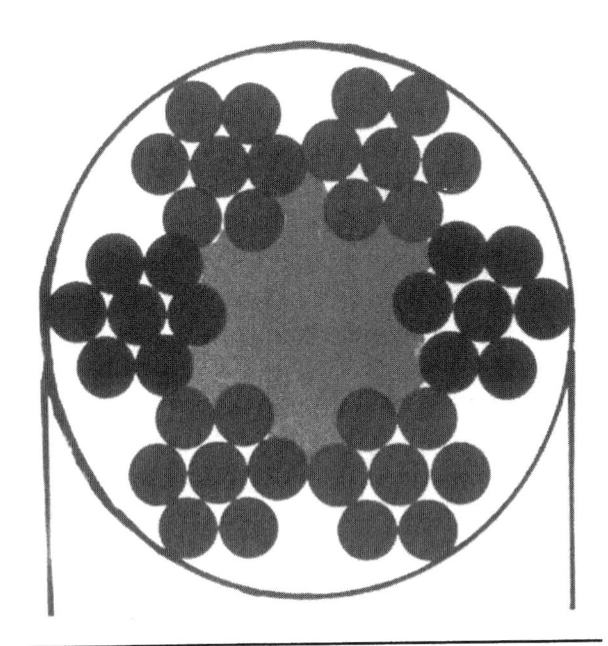

FIGURE 5.12 Correct way to measure wire rope

- Manila rope
- Nylon rope
- Polypropylene rope
- Polyester rope

Generally speaking, fiber ropes are less strong, lose strength easier, are less reliable under shock loading, less able to take sustained loads, and are more subject to damage from abrasion, chemicals, weather damage, and ultraviolet radiation.

Well then, why use fiber ropes at all? There are some applications where the user must handle the rope, and fiber rope is considerably softer and more pliable than wire rope. Also, fiber rope is less expensive than wire rope. So, where handling the rope is an issue, fiber ropes are more suited for this activity. Fiber ropes are available in several types of construction:

- Solid Braid — 9, 12, or 18 strands

- Hollow Braid — 8, 12, or 16 strands
- Double Braid — 16, 20, 24, or 32 strands
- Maypole/Diamond Braid — 8, 12, or 16 strands
- Twisted — 3 strands

Anytime a rope is being chosen where protection of life or property is involved, it is imperative that expert advice be sought. This is NO PLACE to "guess". I cannot stress this enough!

The following is a comparison of rope construction selection, showing characteristics of the types of ropes mentioned:

- Some of the types of construction for ropes may not be spliced. Solid Braid, Double Braid, and Twisted fiber ropes fall into this category.
- Strength in terms of weight is listed with the best first, with the poorest last: Hollow Braid, Double Braid, Twisted Braid, Maypole/Diamond Braid, and Solid Braid.
- Flexibility is listed with the best first and the poorest last: Hollow Braid, Double Braid, Solid Braid, Maypole/Diamond Braid, and Twisted Braid.
- Hollow Braid and Maypole/Diamond Braid have been shown to flatten under load.
- Twisted Braid has been shown to rotate under load.
- Mechanical elongation is shown with the highest first and the lowest last: Solid Braid, Twisted Braid, Maypole/Diamond Braid, Hollow Braid, and Double Braid.
- Cost per size is listed with the highest first and the lowest last: Double Braid, Solid Braid, Maypole/Diamond Braid, Hollow Braid, and Twisted Braid.
- All ropes as listed above show a working load of 5-20% tensile.
- Abrasion resistance is shown with the best first and the poorest last: Solid Braid, Double Braid, Maypole/Diamond Braid, with Hollow Braid and Twisted Braid showing the poorest abrasion resistance.

In getting some idea of fiber types (nylon, polyester, polypropylene, and manila) and how they compare to each other, the following provides some information:

- Strength of the above four fiber types is listed with the best first and the poorest last: nylon, polyester, polypropylene, and manila.
- Wet strength as compared to dry strength: nylon - 85%, polyester - 100%, polypropylene — 100%, manila - 115%.
- Shock load ability is listed with the best first and the poorest last: nylon, polypropylene, polyester, and manila.
- Nylon, polyester, and manila ropes all sink in water. Polypropylene ropes float.
- Enlongation at break (an approximation) are as follows: nylon 20-34%, polyester 15-20%, polypropylene 15-20%, and manila 10-15%.
- As far as water absorption, manila rope 100%, polypropylene 0%, polyester less than 1%, and nylon 6%.
- Melting points for the ropes are: nylon - 480 degrees F, polyester - 500 degrees F, polypropylene - 330 degrees F, and manila does not melt, but chars at 350 degrees F.
- Abrasion resistance is listed with the best first and the poorest last: polyester, nylon, manila, polypropylene.
- Degradation — resistance to sunlight: polyester - excellent, nylon and manila - good, polypropylene - poor
- Degradation — resistance to rot: nylon, polyester, polypropylene all show excellent resistance, with manila showing poor resistance.
- Degradation — resistance to acids: polyester and polypropylene both show good resistance, with nylon and manila showing poor resistance.
- Degradation — resistance to alkalis: nylon and polypropylene both show good resistance, with polyester and manila showing poor resistance.

- Degradation — resistance to oil and gasoline: nylon, polyester and polypropylene all show good resistance, with manila showing poor resistance.
- Electrical conductivity resistance — polyester and polypropylene show good resistance, with nylon and manila showing poor resistance.
- Flexing endurance is shown best to poorest: nylon, polyester, polypropylene, and manila.
- Specific gravity for the ropes is shown as: nylon 1.14, polyester 1.38, polypropylene .90 and manila 1.38.
- Storage requirements for the above ropes show that nylon, polyester, and polypropylene can be stored either wet or dry, but Manila must be stored in a dry condition only.

As with wire rope, any rope showing wear, abrasion, broken strands, change in dimensions of either in diameter or length (indicating overloaded conditions, etc.) or any rope which is been known to have been used in a situation which might cause permanent damage to the rope (shock loads, use of swivels under load, sustained loads, etc.) should be marked and REMOVED from service. Better than removal from service is to destroy the rope so there is no temptation to use it at all! Fiber ropes should be kept clean, as any dirt or debris collected within the fibers are a source of hidden abrasion. Always important: DO NOT EXCEED WORKING LOAD LIMIT OR 20% OF THE PRODUCT'S BREAKING STRENGTH! Doing so only asks for trouble!

Chapter 6

Fall Protection During Erection and Use of Scaffolds

Introduction

Fall protection, as all aspects of life, needs to be preplanned to be successful. The earlier you plan, the more options will be available and the better and more cost effective your fall protection plan will be. Preplanning for the erection and use of the equipment covered in this book needs to start with choosing your means and methods to accomplish your work and then the selection and use of the appropriate systems. Once you have determined which work platforms, scaffolds, aerial lifts, stairways, or ladders will be appropriate to accomplish your tasks, then you need to plan on where and how they will be used and make sure that the design of your systems takes into account the function and design loads of the intended work. After you have planned your work sequence and where the equipment is to be used, then you need to plan the scheduling of the appropriate equipment to arrive on time and to have a site-specific plan for the safe erection, use, and dismantling of the equipment in a timely manner.

Identifying Hazards – The Job Hazard Analysis Form

First it is important to understand that fall protection is not just equipment! Actually, the best fall protection is not the fall protection equipment itself but the elimination of the hazard by either designing the hazard out or perhaps by changing the erection plan or the means and methods. Identifying potential fall hazards is the first priority in developing your approach to problem solving.

One of the best methods of identifying, and capturing specific hazards is by examining the plans and memorializing each identified hazard on a Job Hazard Analysis Form, Figure 6.1. This simple one-page form allows you to list each Activity/ Operation and how it is to be accomplished. In other words, here is your opportunity to determine how you are going to accomplish the task, how cost effective is it, how many people are involved, and the potential problems and opportunities. Each task needs to be described in brief simple language. Remember, from this description, you will need to be able to communicate your expectations to the foreman and workers, train them, and be able to determine your time and costs as well as potential hazards.

The next area on the form is to list the Unsafe Condition, Action or Hazard. Here you concisely define what is unsafe and perhaps what laws, regulations, or standards these practices or conditions may violate. Also you need to take into account any specific requirements of the contract documents. In this section we are merely stating the problem, not solving it. Actually at this point you may decide that you can eliminate this hazard as we discussed above.

Next is the Preventative or Corrective Action section. Now is the time to determine specifically how the condition, action, or hazard will be handled. Should it be eliminated or managed? This is a good place to define and discuss the various methods of fall protection solutions.

The next section of the Job Hazard Analysis is the Discussion. Here an in-depth discussion outlines how the task will be completed and what means and methods will be employed to manage the potential fall hazards. The contractor will appoint the responsible supervisor as the Competent Person for the setting up, moving, and removal of the system. Workers' training and under whose direction they will work will also be decided.

The Job Hazard Analysis Form is then signed by the Responsible Supervisor and the Competent Person(s). There very well may be more than one Competent Person. For example, you may have a Competent Person who designed the system, another to do the training, and another to do the erection and dismantling. Further inspections may be required and, if so, what frequency needs to

JOB HAZARD ANALYSIS

Page ________ of ________

Date ________________

Contract No. ________________

Phase No. ________________

Contractor ________________

Location ________________

ACTIVITY/OPERATION	*UNSAFE CONDITION, ACTION OR HAZARD*	*PREVENTATIVE OR CORRECTIVE ACTION*

FIGURE 6.1 Job Hazard Analysis Form

be listed. Finally the Job Hazard Analysis must be approved by the approving authority and signed and dated. Following this format will ensure that you have a plan and record of how you solved your work-related fall hazards.

Fall Protection Solutions

Fall protection solutions need to be site- and task- specific! After eliminating as many fall situations as you can, then you need to manage the ones you cannot eliminate. Fall protection solutions are divided into two general categories: passive solutions and active solutions. Many solutions are a combination of using both methods.

Now let us turn our attention to what goes into designing a passive or active fall protection system. These systems must be designed by a Qualified Person or engineer and must be installed, inspected, and repaired under the supervision of a Competent Person. Training must be site- and task-specific, developed from the manufacturer's instructions by a Competent Person, and given under his supervision.

All active or passive systems are either pre-engineered or engineered. Pre-engineered systems are usually designed and manufactured by a third party and are rented or sold to the end user for applications permitted by the manufacturer. Engineered systems are usually designed and made for a specific task or application. Let us now look at some of the components or subsystems that are evaluated when designing passive or active fall protection systems.

Passive Solutions

Passive systems are designed by a qualified person and erected and put in place under the supervision of a Competent Person. Careful consideration needs to be given to the adequacy of the type of system and the ability to protect the user in the intended application. For instance, in the case of handrails, guard rails, and platforms, design and installation should take into consideration the intended use and the maximum intended loads. Personnel and debris nets need to be installed as close as possible to the work area and need to minimize the fall distance as well as consider the maximum intended loads. Where it is appropriate, methods of rescue need to be determined and personnel need to be trained.

Passive solutions, as the name implies, do not require the exposed persons to be proactive in protecting themselves from a fall situation. The fall protection equipment or plan is put in place before the potentially exposed workers are allowed to perform their tasks. The passive equipment is usually both designed and engineered by a Qualified Person to meet the job specific requirements or is purchased from a manufacturer as a pre-designed system that is installed under the direction of a trained Competent Person. During the erection and dismantling of the passive systems it is important that, where hazards exists, a Job Hazard Analysis is completed and that those workers have fall protection where needed.

Some examples of passive systems that can be used are guardrail systems, personal and debris net systems, and guard railed mobile platforms. Most applications of the equipment described in this book, such as work platforms, scaffolds, aerial lifts, stairways, and ladders are examples of passive systems. Refer to Figure 6.2a and Figure 6.2b.

Passive solutions, where they can be used, are preferable for eliminating hazards and preventing a potential severe fall. Furthermore, they take out most of the human elements, such as having "a bad day" and not properly using your active fall protection solutions.

Active Solutions and Equipment

Active solutions are systems, again as the name implies, that require the user to do something to prevent a potential fall. These systems are usually divided into three categories: fall restraint, positioning, and fall arrest. All three systems have similar component parts. Each needs an anchorage, a connector, and a full body harness or in a few cases a body belt (Figure 6.3). It is important to note that, because of the potential misuse of the equipment, almost all manufacturers and users require the full body harness in all applications. Furthermore, the anchorage and connectors will vary in strength and type depending on the application and must be designed and chosen under the supervision of a qualified or Competent Person. Where the equipment is purchased from a manufacturer it must also be installed, inspected, and used in accordance with the manufacturers' instructions. It is also important to note that training of employees must be done under the supervision of a Competent Person and in accordance with the manufacturers' instructions and the task-specific application.

What will you be able to do with this knowledge? Since a picture is worth a thousand words, I am including below some pictures of scaffolding jobs that have been erected by the authors following the same safety procedures.

Twin Towers 250' with Staircase

Commercial Construction (Hospital)

The tank shown above had a cone shaped bottom and was roughly 40' diameter by 100' high. The unusual part was that the scaffold had to be designed and sonstructed before the tank was built, because the scaffold was used to weld the tank walls together.

FIGURE 6.2A Examples of passive systems

The frames shown above are heavy duty shoring frames and are used to support the concrete slab when poured.

This is an industrial storage tank. The insulation was torn off during a hurricane and is being replaced.

The airport control tower had a cantilevered top. (Frame scaffolding)

This tube and clamp scaffold is suspended from the top center.

FIGURE 6.2B Examples of passive systems

Active fall protection systems need to address the compatibility of the component parts as well as the ability of the system to adequately protect the user in the intended application.

Anchorages are either fixed or portable. Fixed anchors embedded in the structure or attached to the structure or some other suitable fixed point. Portable anchors are removable and can be moved from point to point. Careful consideration needs to be taken to ensure that the user is not exposed to a fall sit-

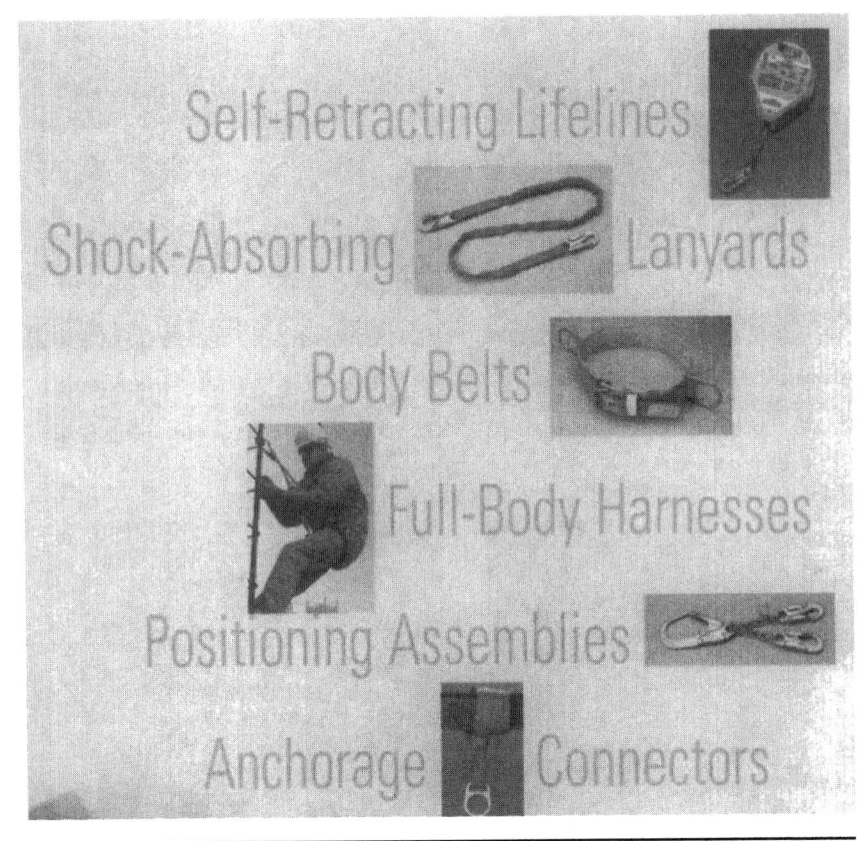

FIGURE 6.3 Examples of active solutions and equipment

uation during erection or movement of the anchorage. Anchorages are an important element in the use of active fall protection solutions. All parts of the engineered systems need to conform to OSHA and ANSI Standards and manufacturer's instructions. Anchorages should be preplanned and their locations and strength should be known to foreman and workers as part of the site-specific fall protection plan and the training program.

The connector, as the name implies, is the means of attaching the anchorage to the harness. Some of the more common connectors are rope or web lines between the anchorages and the harness. Some connectors contain shock absorbers; others are adjustable in length depending on the application.

Retractable lifeline connectors are sometimes used and keep the line taut to reduce the fall distance of the user. Finally many systems incorporate horizontal or vertical lifelines that allow continuous movement of the user while being fully attached. Connectors are the devices that connect the anchorage to the full body harness. Connectors come in many different configurations and the type and function is part of the design and use of the different active fall protection systems: fall restraint, positioning, and fall arrest.

Full body harnesses come in many different varieties for different applications and uses. A suitable harness should be selected by a Competent Person for the work to be performed and the types of active fall protection systems that the worker will be utilizing. Individual harnesses should be sized to the specific individual and worn in accordance with the manufacturer's instructions and the training performed by the Competent Person. Full body harnesses are used in almost all active fall protection applications and are required by OSHA in all fall arrest applications after January 1, 1998. There are many different harnesses available from many manufacturers. You need to take these three areas into consideration when making your decision on an application: fall restraint, positioning, or fall arrest. You may want to choose a harness that will allow you to utilize it in all applications or choose for the task at hand.

Now that we have discussed the different components of active fall protection, we shall explore some of the different configurations of each type of system.

Configurations for Active Fall Protection

Fall restraint systems, as the name implies, do not allow the restrained worker from reaching the potential fall area and thereby prevents the hazard of a fall. This is accomplished by ensuring that the anchorage and the length of the connector will not allow the worker to reach the fall hazard. This system prevents the user from falling any distance. Other components typically include a lanyard and may also include a lifeline and other devices. Careful planning, installation, and training are necessary to insure that a fall restraint system works to properly restrain a fall in all conditions that it is used. When used by workers have been trained in a location where the system has been properly designed and installed, this is an excellent means of protection.

Positioning systems, sometimes referred to as positioning device systems, are used to allow an employee to be supported on an elevated, vertical surface, such as a wall or column, and work with both hands free while leaning. These systems are used at the workstation allowing the worker free movement to do his work and still be supported in place with little or no risk of fall distance. Typical examples of these devices would be rebar chain assemblies and form hooks attached to concrete or other formwork. See Figure 6.4. These applications certainly need to be designed by a Competent Person and are usually a part of an overall fall protection plan.

Fall arrest systems are potentially the most dangerous of all fall protection systems because they allow a planned freefall of up to six feet and a force of up to 1800 pounds applied to the body. Personal fall arrest systems are used to arrest an employee in a fall from a working level. A personal fall arrest system consists of an anchorage, connectors, a full body harness and may include a lanyard, deceleration device, lifeline, or a suitable combination of these (Figure 6.5). The use of a body belt is prohibited. A Competent Person needs to consider carefully alternate choices before using this method which can be considered the choice of last resort. Further consideration needs to be given to designing systems that reduce the fall distance and the force on the body as much as possible. Also one needs to consider the total fall distance from the working surface and to insure that the worker will not impact any structure or swing into any obstruction during the fall.

If you examine the following illustrations (Figure 6.6 and Figure 6.7), you will notice that, depending on the components used in the fall arrest system, substantial total fall distances exist. These distances can be critical if not properly addressed. For example the distance between floors in a multi-tiered steel structure is typically 15 feet. In one case, the worker almost impacts the

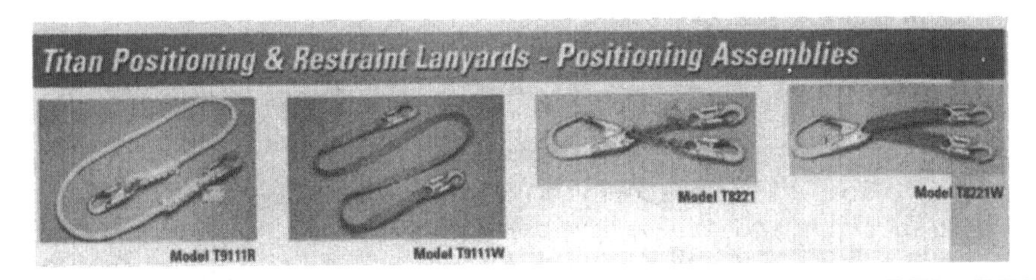

FIGURE 6.4 Positioning device systems

FIGURE 6.5 Personal fall arrest system

floor below and in the other case, the system does not arrest his fall before impact. Furthermore these two illustrations are only examples of fall distances. You must consider all of the components that make up your system: the anchorage, the connector, and the harness. Consider that if the anchorage is a horizontal lifeline, not fixed, deflection of the line adds fall distance that can be critical. The total extended length, including the extended shock absorber, needs to be considered in total fall distance. Finally the length of the body from the Dee Ring to the bottom of the feet, including any stretch of the harness, completes the considerations needed to determine total distance.

Rescue Plans

A plan for immediate rescue is imperative because the worker can be safely suspended in a full body harness for only a very short time. Help in designing these systems can be found in the ANSI Standards and from a multitude of manufacturers of fall protection equipment (Figure 6.8).

Finally you will find that each activity/operation described will lend itself, based on your means, methods, and culture, to a combination of different solutions. After each job you will want to capture your solutions, critique them, and make any changes or modifications that will improve your future solutions.

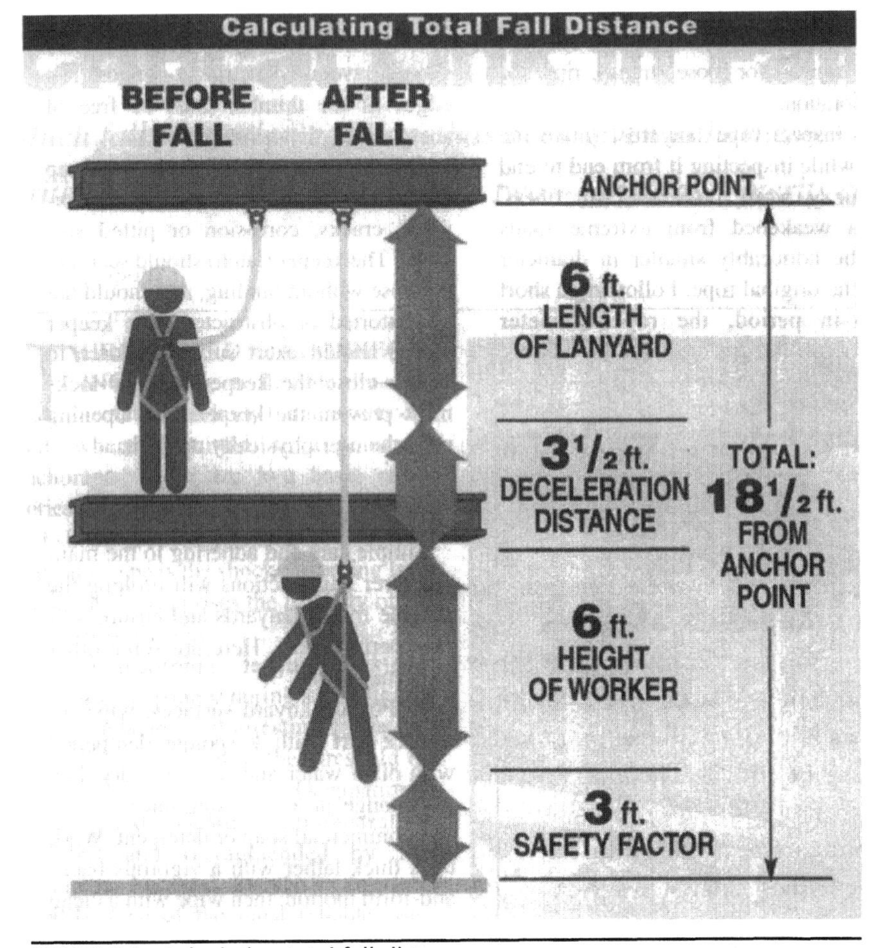

FIGURE 6.6 Calculating total fall distance

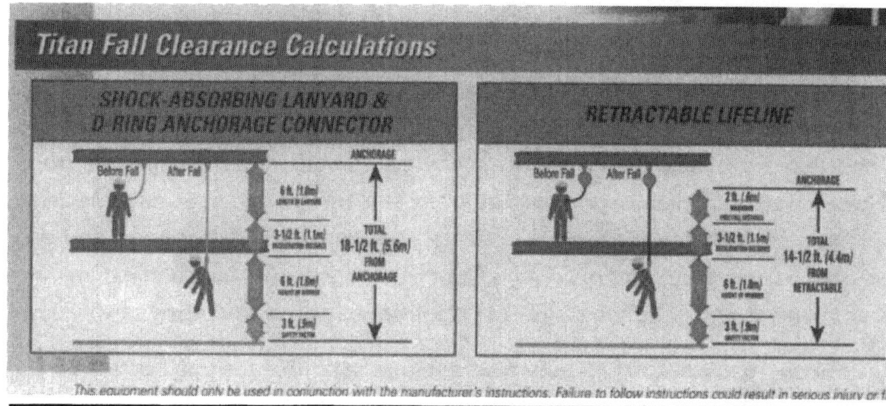

FIGURE 6.7 Fall clearance calculations

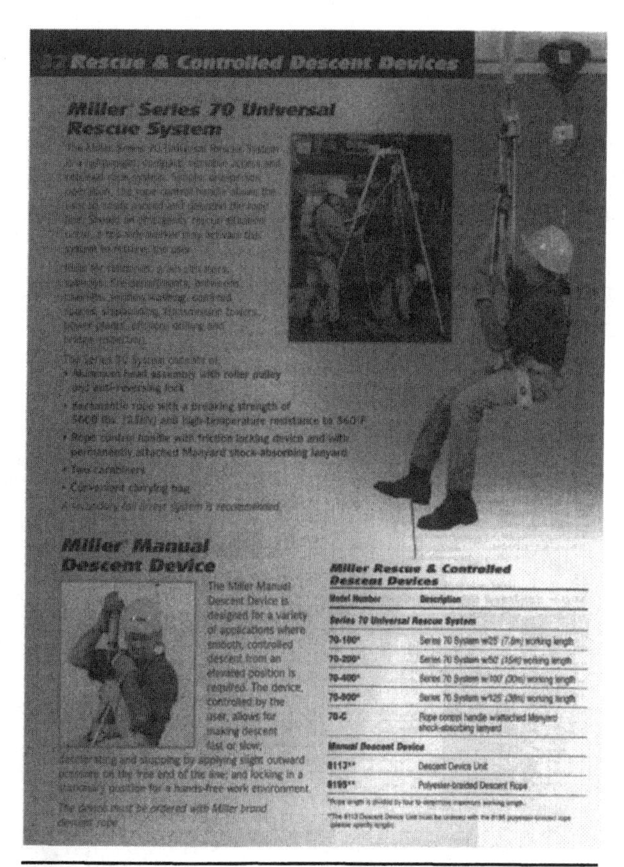

FIGURE 6.8 Fall protection system

SUMMARY

Having outlined how to solve fall protection problems, explore the different considerations as you make your selection of appropriate fall protection systems that are available for use with the equipment described in this book. These systems, as already pointed out, must take into consideration federal, state and local laws and regulations, the contract documents, and the owner and general contractor requirements. Obviously, your own company safety plan, site-specific plan, and culture also affect your final selection or choice of solutions.

Further you need to carefully read and understand the fall protection equipment manufacturer's manuals for design, installation, use, and maintenance. You also need to pay particular attention to all cautions, warnings, and danger labels and ensure that proper labels are displayed, not damaged or covered.

Wherever possible, workers who are going to be exposed to a potential fall need to be evaluated to determine that they are in good physical condition and do not have medical limitations that will put them in jeopardy.

Remember, designing the proper fall protection system for the proper application is very important; but equally important is proper training of the users. Only qualified people should use the systems and only qualified and trained people under the direction of a Competent Person should install, maintain, and take down the equipment or systems. This will ensure that you have a safe, profitable, and well-run application.

Chapter 7

Electrocution on Scaffolds

INTRODUCTION

Supported scaffolds, because they are metal and are built and used in proximity to overhead power lines, often present a dangerous risk of electrocution. Often the worker who is shocked by the faulty equipment survives the electrical shock, but loses his/her balance and is killed in a fall from the scaffold. This is one more really good reason for always utilizing fall protection. Using equipment you know is safe is half the job — the other half is using this equipment in a safe manner!

NIOSH (National Institute for Occupational Safety and Health) in May of 1988 produced the publication "Worker Deaths by Electrocution". This publication consisted of a summary of surveillance findings and investigative case reports. Their findings show that the industries with the highest number of electrocutions between the years 1980 and 1992 were:

- Construction (40%)
- Transportation/communication/public utilities (16%)
- Manufacturing (12%)
- Agriculture/forestry/fishing (11%)

The construction industry had a rate of 2.4 per 100,000 workers, followed closely by mining, which had a rate of 2.2. (Figure 7.1)

SAFETY FACTORS

The industries with the highest number of electrocutions during the years 1982 to 1994 were: Construction (121), Manufacturing (40), Transportation-Communications-Public Utilities (30), and Public Administration (19). See Figure 7.2. In 35% of the incidents investigated, no safety program or established written safe work procedures existed! Worse yet, when the fatality investigations were conducted as a result of the NIOSH Fatality Assessment and Control Evaluation (FACE) program, investigators noted that more than once during these investigations involving sign technicians, tree trimmers, utility line workers, and telecommunication workers, co-workers interviewed did not know the power lines posed a hazard, i.e., they thought the power lines were insulated.

In this study, it was noted that at least one of the following five factors was present in the incidents evaluated by the FACE program:

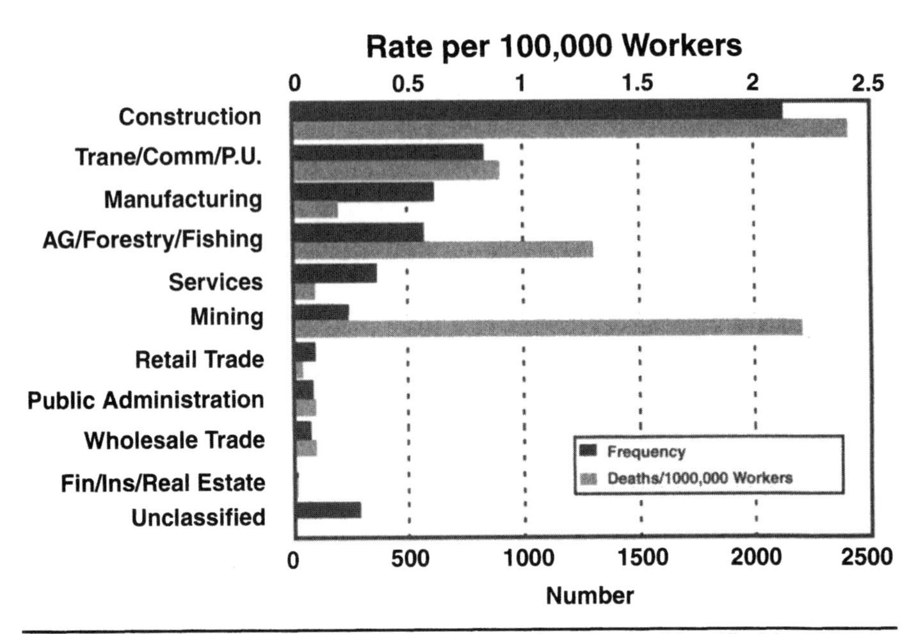

FIGURE 7.1 Frequencies and rates of electrocution deaths identified by NTOF by industry, 1980-1992

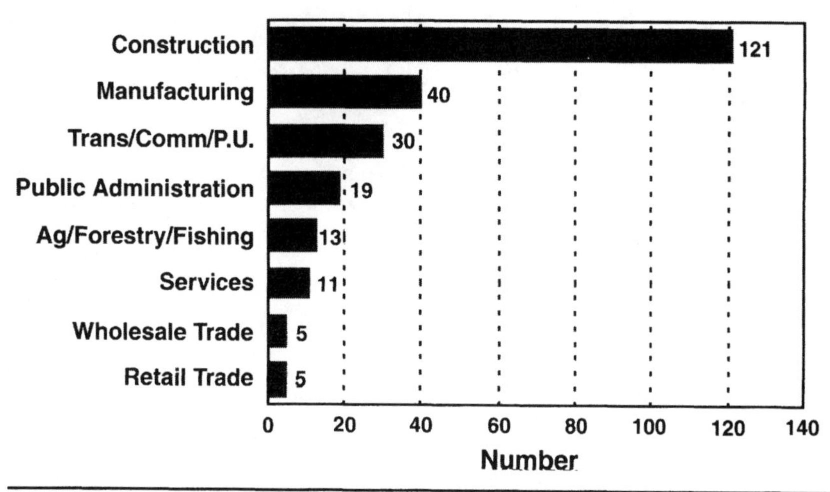

FIGURE 7.2 Frequencies of electrocution deaths identified by FACE by industry, 1982-1994

- Established safe work procedures were either not implemented or not followed.
- Adequate or required PPE was not provided or not worn.
- Lockout/tagout procedures were either not implemented or not followed.
- Compliance with existing OSHA, NEC (National Electrical Code), and NESC (National Electrical Safety Code)regulations were not implemented.
- Worker and supervisor training in electrical safety were not adequate.

Determining Location of Power Lines

Care must be taken that scaffolds are not located close enough to overhead power lines for the scaffolds themselves and any conductive materials, such as building materials, paint roller extensions, scaffold components, to come closer than 10 feet to the power line. This additional distance of 10 feet increases 4 inches for every 10 kv when voltage is higher than 50 kv.

I find it difficult to determine where that 10-foot line is, considering that power lines sway and move with the breeze, plus the fact that other power-conducting pieces, such as transformers, etc., can affect that 10-foot distance according to the size of the unit. I also have great concern for "arcing" which can occur in some instances. While the rules and standards say 10 feet, I feel there are better ways to control electrical contact besides just staying ten feet plus the length of any equipment extending beyond the scaffold away from power lines.

Preventing Contact

Electrical lines can be seen right from the first time a contractor goes on site and puts his figures together for the initial bid. If the contractor sees that the lines are close enough to create problems, there are several questions he should ask before preparing his bid:

- First, can the situation be "engineered out"? In other words, is there a way of totally avoiding working in that area? If the job is in the planning stages, it may be possible to re-arrange the site or place the building or structure in a location far enough away to completely avoid this problem. (Scaffolds are not the only equipment on the site that needs consideration about power lines.)
- Second, if that doesn't work, can those lines be de-energized and normal electrical traffic going through them be re-routed? During the construction process, the lines would be rendered useless, thereby assuring employees that they are safe working in that area.
- Third, if this is impractical, can the lines be shielded or in some way be protected from contact by scaffolds, equipment, etc.?

Scaffolds are not the only equipment in the area that need to be concerned about contact with electrical lines, and all are subject to the same requirement of 10 feet, plus the length of beam, equipment, etc., which may extend beyond the scaffold or other equipment, for safe working conditions. Since electrocution is a real concern for many on the jobsite, more and more owners are looking into the above concerns and are choosing to use one of the three methods of avoiding contact mentioned above.

There is an exception to the rule of 10 feet plus equipment which might extend past the area when utilizing the equipment. Insulated power lines of less than 300 volts have a safe distance of only 3 feet. It is not easy to determine if a power line is insulated, or what the exact voltage of that line is, and a good practice is to use the 10-foot rule plus extending equipment or items in all cases. This is one instance where it is truly better to be safe than sorry!

The OSHA E-Tools say that scaffolds may be closer than specified above IF such proximity is necessary for the type of work being done and IF the power company or electrical system operator has been notified and has either:

- De-energized the lines
- Relocated the lines
- Installed protective coverings to prevent accidental contact with the lines [1926.451(f)(6)]

Use of Electrical Power Tools on Scaffolds

Use of power tools, cords, etc., that could suffer insulation failure can electrify the entire scaffold — posing a risk of electrocution not only to the worker holding the tool, but also to all using the scaffold. Workers should always check electrical tools for indication of possible internal damage, deformed or missing pins, cord damage, and insulation damage.

All portable electric equipment used while on scaffolds MUST be protected by:

- GFCIs (Ground-Fault Circuit Interrupters), the preferred method, or
- An AEGCP (Assured Equipment Grounding Conductor Program) in accordance with 1926.404(b)(1)(I)]

What is the difference between the Ground Fault Current Interrupter (GFCI) and an Assured Equipment Grounding Program (AEGCP). A GFCI is equipment that serves as a circuit breaker if continuous ground continuity in an electrical tool is not present. The AEGCP is a scheduled program for testing construction site electrical tools and extension cords to assure their proper grounding, polarity, and resistance. A company's AEGCP is a written program and should include a log of the items inspected and the date of the testing which is kept by the authorized person who is competent to recognize electrical hazards. Color coding of the electrical system can serve as the documentation of the completed testing procedure.

GFCI

What is a GFCI and how does it work? The following is taken directly from an OSHA E-Tool covering the topic:

"A ground-fault occurs when there is a break in the low-resistance grounding path from a tool or electrical system. The electrical current may then take an alternative path to the ground through the user, resulting in serious injuries or death. The ground-fault circuit interrupter, or GFCI, is a fast-acting circuit breaker designed to shut off electric power in the event of a ground-fault within as little as 1/40 of a second. It works by comparing the amount of current going to and returning from equipment along the circuit conductors. When the amount going differs from the amount returning by approximately 5 milliamperes, the GFCI interrupts the current." (www.osha.gov E-Tools)

The GFCI is rated to trip quickly enough to prevent an electrical incident. If it is properly installed and maintained, this will happen as soon as the faulty tool is plugged in. If the grounding conductor is not intact or of low-impedance, the GFCI may not trip until a person provides a path. In this case, the person will receive a shock, but the GFCI should trip so quickly that the shock will not be harmful. The GFCI will not protect you from line contact hazards (i.e., a person holding two "hot" wires, a hot and a neutral wire in each hand, or contacting an overhead power line). However, it protects against the most common form of electrical shock hazard, the ground-fault. It also protects against fires, overheating, and destruction of wire insulation.

Construction Applications of GFCI

For construction applications, there are several types of GFCIs available, with some variations:

- The Receptacle Type incorporates a GFCI device within one or more receptacle outlets. Such devices are becoming popular because of their low cost.

- Portable Type GFCIs come in several styles, all designed for easy transport. Some are designed to plug into existing non-GFCI outlets, or connect with a cord and plug arrangement. The portable type also incorporates a no-voltage release device that will disconnect power to the outlets if any supply conductor is open. Units approved for outdoor use will be in enclosures suitable for the environment. If exposed to rain, they must be listed as waterproof.
- The Cord-Connected Type of GFCI is an attachment plug incorporating the GFCI module. It protects the cord and any equipment attached to the cord. The attachment plug has a non-standard appearance with test and reset buttons. Like the portable type, it incorporates a no-voltage release device that will disconnect power to the load if any supply conductor is open.

Because GFCIs are so complex, they require testing on a regular basis. Test permanently wired devices monthly, and portable-type GFCIs before each use. All GFCIs have a built-in test circuit, with test and reset buttons that trigger an artificial ground-fault to verify protection. Ground-fault protection, such as GFCIs provide, is required by OSHA in addition to (not as a substitute for) general grounding requirements.

AEGCP

The following is taken directly from the OSHA E-Tool. E-Tools are available on www.osha.gov.

If an Assured Equipment Grounding Conductor Program (AEGCP) is used in place of GFCIs for ground-fault protection, the following minimum requirements apply, though additional tests or procedures are encouraged:

- Keep a written description of the program at the jobsite. Outline specific procedures for the required equipment inspections, tests, and test schedule, and make them available to OSHA and to affected persons upon demand.
- Designate one or more Competent Persons to implement the program. OSHA defines a Competent Person as someone who is a) qualified to identify hazards, and b) authorized to take prompt corrective measures.

- Visually inspect all cord sets, attachment caps, plugs and receptacles, and any equipment connected by cord and plug, before use each day. If you see any external damage, such as deformed or missing pins, damaged insulation, etc., or discover internal damage, take the equipment out of use until it is repaired.
- Perform two OSHA-required tests on all electrical equipment, a continuity test (see below), and a terminal connection test as described later in this chapter. Tests are required:
  - Before first use
  - After any repairs, and before placing back in service
  - After suspected damage, and before returning to use
  - Every 3 months [see 1926.404(b)(1)(iii)(E)(4) for exceptions]
- Maintain a written record of the required tests, identifying all equipment that passed the test and the last date it was tested (or the testing interval). Like the program description, make it available to OSHA inspectors and affected persons upon demand.

The continuity test ensures that the equipment-grounding conductor is electrically continuous. Perform this test on all cord sets, receptacles that are not part of a building or structure's permanent wiring, and cord- and plug-connected equipment required to be grounded. Use a simple continuity tester, such as a lamp and battery, bell and battery, an ohmmeter, or a receptacle tester.

The terminal connection test ensures that the equipment-grounding conductor is connected to its proper terminal. Perform this test with the same equipment used in the first test.

Suspended Scaffolds and Welding

Welding is more likely to be performed on suspended scaffolds than on supported scaffolds. Suspended scaffolds follow the same rules and regulations as supported scaffolds. When welding is performed on scaffolds, the user must be aware of certain exposures and take the following precautions, as they apply, to reduce the possibility of welding current arcing through the suspension wire ropes: [1926.451(f)(17)]

- An insulated thimble must be used to attach each suspension wire rope to its hanging support (such as cornice hook or outrigger). Excess suspension wire rope and any additional independent lines from grounding must also be insulated. [1926.451(f)(17)(i)]
- The suspension wire rope must be covered with insulating material at least 4 feet (1.2 m) above the hoist. [1926.451)(f)(17)(ii)]
- If there is a tail line below the hoist, it must be insulated to prevent contact with the platform. The portion of the tail line that hangs free below the scaffold must be guided or retained, or both, so that it does not become grounded. [1926.451(f)(17)(ii)]
- Each hoist must be covered with insulated protective covers. [1926.451(f)(17)(iii)]
- In addition to a work lead attachment required by the welding process, a grounding conductor must connect from the scaffold to the structure. See the second accident report that follows. The size of this conductor must be at least the size of the welding process work lead, and this conductor must not be in series with the welding process or the work piece. [1926.451(f)(17)(iv)]
- If the scaffold grounding lead is disconnected at any time, the welding machine must be shut off. [1926.451(f)(17)(v)]
- An active welding rod or uninsulated welding lead must not be allowed to contact the scaffold or its suspension system. [1926.451(f)(17)(vi)]

Case Studies

OSHA provides several case studies regarding electrocution and both suspended and supported scaffolding.

Death Due to Lack of Ground-Fault Protection

A journeyman HVAC worker was installing metal duct work using a double-insulated drill connected to a drop light cord. Power was supplied through two extension cords from a nearby residence. The individual's perspiration-soaked clothing/body contacted bare exposed conductors on one of the cords, causing an electrocution. No GFCI's were used. Additionally, the ground prongs were missing from the two cords.

Frayed Insulation Causes Arc, Damages Suspension Rope

An employee was arc welding from a suspended scaffold. The work lead on the welder had frayed insulation in one place, and the bare conductor was exposed. The frayed section of welding cable was tied around the metal guardrail of the scaffold. Since the scaffold was not grounded, it became energized. This caused arcing between the scaffold and the building. The welding current passed through the wire rope supporting the scaffold, and the wire rope separated. The employee and the scaffold fell 50 feet to the ground. The employee was hospitalized for his injuries.

Employee Dies After Contacting Power Line

An employee was setting up a two-point suspension scaffold on top of a billboard platform that was 35 feet above the ground. As he was installing a 21-foot-long metal guardrail, it contacted a 34-kilovolt overhead power line located 8 feet from the billboard. The employee received an electric shock and fell to the ground. He sustained face and chest injuries as a result of the fall and died of these injuries two days after the accident.

The above instances are clear violations of regulations. They are easy to see after the fact. Knowing what the regulations are and thinking the job through before starting work would have prevented these unfortunate instances.

Summary — Working in the Vicinity of Power Lines

What can be done when working in the vicinity of power lines? Be sure to take into account all the points below:

- Expect all power lines on the site to be energized with the highest voltage possible.
- Survey the site for power lines when making your initial site visit.

- Realize you may have to build-in safety for your people by asking if lines can be moved, de-energized, or shielded. If these items cannot be done, your employees must be sure they understand the rules, regulations, and hazards in working around electricity and that they are observing the 10-foot plus rule.
- Don't guess where 10 feet is when setting up your scaffolding. And remember to add additional distance when you are using any materials, tools, equipment which will extend past your scaffold and into that 10-foot area. Add a good margin for safety.
- Have an observer working with you if you are in a position where you must work near power lines. That person needs to be aware of any materials, tools, and equipment extending past your scaffold so that he/she can alert you if you are inadvertently crossing that line.
- If working on scaffolds where power tools are being used, be sure you have either GFCI connections or have an Assured Equipment Grounding Conductor Program in place.
- If working on scaffolding where welding is taking place, be sure the procedure is in conformance with 1926.451(f)(17). Welding on metal or aluminum scaffolds with metal suspension cables can be tricky and not following the standards and regulations can result in dire circumstances.
- Workers on scaffolds need to be extremely aware of any movement of the scaffolding, use of heavy equipment in relation to power lines, or ground slope of floor conditions that could change the elevation and change the clearance between the power lines and the scaffold.
- Take a good look at your safety program and assure that electrical safety is addressed.
- Provide workers with the information they need, through training, to recognize the hazards of electrical contact as well provide first aid training, CPR, and rescue training as part of the program for your scaffold workers.
- Conduct inspections at the jobsite — sometimes, scheduled inspections, sometimes, unscheduled inspections. Assure that your workers are knowledgeable. Working safe becomes a habit, when workers know you expect it from them.

- Conduct job hazard analyses and implement procedures that will protect, insulate, and isolate workers from electrical energy.

Management has to take a strong hand in implementing safety programs and assuring that those programs are completely implemented. A strong commitment to safety by both management and workers is essential in the prevention of severe occupational injuries and death due to contact with electrical energ

Chapter 8

Emergency Response and Rescue

INTRODUCTION

Firefighters Rescue Window Washers (AP) Atlanta - A firefighter repelled 400 feet down the side of a downtown high-rise hotel Thursday to rescue ...

Scaffold Collapses, Worker Holds On (AP) Atlanta - Firefighters rescued a maintenance worker at a high-rise ...

Three Construction Workers Injured After Fall (Click10.com) - Three construction workers suffered injuries after a fall from scaffolding Friday ...

Bridge Workers Fall Into River (The Detroit News) - Two survive, one missing from scaffold on Ambassador Bridge ...

Two Hurt in Scaffold Accident (TheBostonChannel.com) - Two people were injured Monday afternoon when a scaffold on which they were working collapsed on ...

Scaffolding Collapses at Construction Site (Click2.Houston.com) - A massive scaffolding collapsed around 1:15 p.m. Tuesday at a construction site in southwest Houston ...

Worker on North Side Condo Killed After Scaffold Collapses (Chicago Sun-Times) - A man fell to his death after a scaffold gave way at a Gold Coast condominium Wednesday, but firefighters were able to rescue his co-worker as he...

How many times have you seen headlines like this? These kinds of rescues are difficult, often unsuccessful, and spectacular in their efforts. Take a look at the small amount of information given with each headline — man falls from scaffolding off bridge. Do you think that the fact that something or someone might possibly fall from scaffolding was an issue when planning the job? Apparently not, or proper precautions would have been used to prevent such occurrences. It would appear that the thought that someone might have to be rescued from a fall or collapse had never occurred to anyone. And yet, such occurrences do happen. Who is responsible for preparing for rescue operations? The contractor, the renter of the scaffold, the owner, the general?

THE CONTRACT

The answer to the above lies in the language of the contract. The owner has to provide a safe workplace for workers, according to the OSHA general duty clause. How the owner does that is determined contractually. He may assign that duty to the general or prime contractor. Does that let the owner off the hook? No way. The owner is still liable for providing a safe workplace in that he/she/it ("owner" can be many different entities) has to protect himself/herself/itself by requiring reports, checklists, or some kind of proof that the general or prime contractor is in fact doing the job according to the owner's instructions by contract.

Responsibility for emergency response and rescue is also determined contractually. Suppose the owner has assigned that responsibility to the general or prime contractor. The general or prime contractor puts an emergency response and rescue plan into effect. The general or prime contractor must then make this information known to the subcontractors, like the scaffold contractors. The scaffold contractors must then come back and let the prime or general contractor know if they require more effort toward rescue than the general is providing.

For instance, the scaffold contractor is working over water, or on a high-rise, or in an area that is difficult to bring a fire truck into for rescue from a hook and ladder. The area is highly excavated, or there are trenches and pipes all along the trenches waiting to be installed. This type of activity would greatly impair rescue efforts for scaffolding accidents or problems. The scaffold contractor has to make these conditions known to the general. The general contractor may say, "yes", we see you have a problem. Then it is up to the general to either designate an area for rescue equipment to access adjacent to the scaffold site (making sure to accommodate the fact that scaffolding may well move around a building on the inside as well as the outside) and propose to keep that area in a condition for easy access. This can be done by providing cribbing or some kind of matting for rescue equipment so that it will not sink into mud or be hampered by pipes, materials, etc. Or, the general can say, "Fine, Mr. Scaffold Contractor, this is your problem and you must provide the cribbing and/or matting." The scaffold contractor may come back with a back charge for this because the contract did not specify his performing that duty and therefore he could not add the charge for this when submitting his bid.

Now we can see how job costs rise. Having thought all this out, or even leaving it to the individual contractor to think out ("...Scaffold contractor is responsible for protection of his area, to be coordinated with the prime contractor, so that rescue equipment, etc., have a means of easy egress and exit") by including such possibility in the bid specifications. By doing that, all scaffold contractors bidding the job are forced to make the same concessions in computing their bids.

Possible Scenarios

What kinds of situations can arise on a jobsite for the scaffolding contractor which might necessitate rescue operations?

- A person on the scaffold could suffer from heat stroke, heart attack, stroke, diabetes-related problem, etc. One of the workers could have the flu and become so ill, he just can't walk or even move around. How does that person advise someone of his condition? What does that person do if he is many feet high in the air working on a high-rise? Is there a means of communication with someone on the ground in case of emergency?

- How can working over water or alongside an excavation affect rescue? Is there a means of getting quickly to someone on a scaffold from the ground? Should rescue people have to use a manlift or job type elevator and are those facilities available and adequate? What if someone falls from the scaffold into an adjacent excavation? How would that person be rescued? Is the excavation safely accessed in a quick manner?

- What if something happens and materials fall to the ground? How can workers below be protected? Does the location of the scaffolding any impair rescue efforts?
- Scaffolding is used for all kinds of activities, including window washing. Should that person have to be rescued by means of a window, is that window able to be opened so the person can be accessed? Is the window a fixed window which must be broken? How can that window be broken quickly and not injure the worker on the outside as well as the persons inside the building?
- What if the windows are somewhat vertical, like skylights? How would someone stepping or falling through a skylight be rescued? Are people working under the skylights and in their vicinity in any danger if the worker falls into the area below? What if the person falls through, but does not fall to the ground, and is caught by structural members, is bleeding from the cut glass and has broken bones, etc and can't do anything to help with his rescue? How would such a rescue be effected?
- What if there is a chemical leak, some kind of exposure to gases and vapors, rendering the person unconscious? How would you rescue that person and what type of PPE would you use if you were the one rescuing the worker? How would you know that such an exposure may exist? Would you set the job up differently if you were aware of possible exposure? Would you call on the same rescue system for this type of exposure as you would for a different type of emergency?
- If the scaffolding fails in some way and starts to collapse, how would you notify persons below you? Do you provide any kind of protection against falls for this type of occurrence? If you do, how would that person get to the ground safely? Are they trained in this procedure?
- A fire breaks out on an upper floor of a construction site. Your people are working above that floor on scaffolding. How would you alert them of the hazards below? What would you do if the smoke pouring from the building makes it impossible for the workers on the scaffolding to exit the scaffolding? What means of egress would you recommend for them — utilize the upper floors of the building and hope

they can make their way down avoiding the area which is ablaze? Stay right where they are and await a fire rescue unit?

- Do you know which individuals are on the scaffold when a rescue effort is necessary so that rescue efforts can take that into account?
- If a sudden windstorm comes up, do your people know when to leave the scaffold? If they don't and they get stuck up there, what do you do?

You can look at all of the above scenarios and say, "Sure, what are the chances that 75% of that stuff is going to happen?" I guess I would have to tell you that if any of those scenarios occur and you are not prepared, the chances of a successful rescue are very poor.

Part of jobsite planning is to be aware of jobsite conditions and see how they can affect you, the contractor. Multi-employer jobsites further compound the problems of rescues because each company is looking at his section of the work. The general or prime contractor needs to be aware of any problems you see for your workers, so he can alert others. You don't want to endanger your workers. You don't want your workers endangering any other workers. You don't want other workers endangering your workers. Communication would seem to be the key here.

Workplace Emergencies

Let us now shift to workplace emergencies. What is a workplace emergency?

- Floods
- Hurricanes
- Tornadoes
- Fires
- Toxic gas releases
- Chemical spills
- Radiological accidents
- Explosions

- Civil disturbances
- Terrorism
- Workplace violence resulting in bodily harm and trauma

In determining when rescue activities may be necessary, it is important to brainstorm the worst possible scenarios. What would you do if the worst happens? Now you must develop an emergency action plan. This plan is the best way to protect yourself, your employees, and your business during an emergency. Ideally, the owner or owner's representative (the general or prime contractor) would do this and make your input as a scaffold contractor a part of the overall plan. It is important for each and every job your employees are on to include:

- Names, titles, departments, and telephone numbers of individuals both inside and outside your company to contact for additional information or explanation of your duties under the emergency plan
- Procedures for employees who remain to perform or shut down critical operations, operate fire extinguishers, or perform other essential services that cannot be shut down for every emergency alarm before evacuating
- Rescue and medical duties for any workers designated to perform them

While it is not required by OSHA, you may find it helpful to include in your plan the following:

- The site of an alternative communications center to be used in the event of a fire or explosion
- A secure onsite or offsite location to store originals or duplicate copies of all records kept on site

Alarms

A way to alert employees on the jobsite to evacuate or take other actions must be determined and communicated to all workers. Remember, the hearing impaired and other disabled workers must be considered in setting this signal up. As a scaffold contractor, you must consider that your employees may be in areas which are not easily or quickly evacuated. How does the worker in that situation get the word and has he/she been trained to evacuate in every situation involving scaffolding?

Alarms must be distinctive and recognized by all employees. The hearing impaired must be considered and an alternative method visually alerting employees should also be installed.

There should be an emergency communication system, such as a public address system, portable radio unit, etc., available for use by all on the site. Be sure any site visitors, vendors, etc., are alerted, too. This would probably be security's job, as they are responsible for who enters and who leaves the site. Who is to call local law enforcement, the fire department, and others in case of an emergency? Have back-up people been alerted in case of the designated person's absence?

Having a system in place is good, but if electrical power is out, how will these communication systems work? Is there a back-up system for alternative power? Is it checked regularly?

What about off-duty hours? Consider any people who might wander into the site, invited or not, and how are they and the neighboring area to be alerted? What would occur if something on your site ignited overnight and caused an explosion releasing vapors, gases and etc., in the area?

On a construction jobsite, the general or prime contractor would probably have the responsibility of sounding the evacuation alarm. Are there back-up people available in case that person cannot be found? Is that person easily visible on the site (marking on hardhat or special ID badge) so that any employee can bring a problem to the person designated to determine if an evacuation alert needs to be sounded? Do you have evacuation wardens who will be responsible to assist in evacuation and account for personnel? As a scaffold contractor, one of your people should be assigned this responsibility, with a back-up person also designated. If you are working in shifts, these designations need to be made for all shifts.

Evacuation

Are specific evacuation procedures communicated to all subcontractors and all people on site? Are routes and exits clearly marked? Are each contractor's employees fully aware of the procedures for evacuation? Have considerations been made for those who do not read or speak English, so they may recognize emergency procedures, emergency exits, etc.?

The coordinator should be responsible for assessing the situation to determine whether an emergency exists requiring activation of your emergency procedures. As a scaffold contractor, you need to know what criteria they use in deciding and you need to communicate to them that your employees may need time to descend safely from scaffolds and that certain disasters (wind, tornado, etc.) are more hazardous for your people on scaffolds. They should also be aware of your designated evacuation warden so that your people are among the first to be notified.

In coordination with the prime or general contractor, your scaffold employees should know, once they have left the site, where to meet for a head count. This information will need to be relayed to the prime or general contractor so they know if anyone has inadvertently been overlooked.

Resolution

What do you do if this emergency is not easily resolved? Someone, most likely the prime or the general contractor, will have to make the decision regarding the next step, which may be as simple as asking all employees to return to their homes and check with their respective employers in the morning to see if work will commence that day.

I would think that as part of the orientation process, the general/prime contractor would make the rescue, emergency response plan information available, along with plans to inform any new or replacement personnel with the same information. This general information should include: roles and responsibilities; threats and hazards as well as protective actions; notification, warning and communications procedures; means for locating family members in an emergency; emergency response procedures; evacuation, shelter and accountability procedures; location and use of common emergency equipment; and emergency shutdown procedures. This information may very well be communicated to the scaffold contractor with his bid, and training for this procedure may fall to the scaffold contractor.

It would be prudent to assure all first aid training, including protection against bloodborne pathogens, respirator protection (including the use of an escape-only respirator), and methods of preventing unauthorized access to the site have been covered and can be verified as to date and extent of training.

Scaffolding may, in differing circumstances, require training specific to the equipment for evacuation, and those needs should be identified and training provided.

Part of an evacuation plan needs to be an awareness of various chemicals and hazardous substances on the worksite. As a scaffold contractor, it might be wise to alert your people through MSDSs what effects may be expected, what to do if they come in contact with any hazardous substances, as well as, what PPE would be necessary in case of emergency related to a substance on the site. Certainly, if work involved on the scaffolding included such things as cleaning of brick or block with chemicals, etc., this information should be transmitted to employees before the start of the job. As part of jobsite safety, acknowledgement of these chemicals on site, along with a copy of the MSDS and appropriate measures taken for storage and disposal, should be presented to the general/prime contractor. I am sure he will ask for the information, but even if he doesn't, you should make this information available. If you as a scaffold contractor are utilizing toxic substances or solvents, appropriate PPE should be provided to all of your employees along with a training course as to proper fit and use. Making an Emergency Plan Checklist is an important issue.

Part II

Supported Scaffolds

General Definition and Requirements

Supported scaffolds are work platforms that are supported by legs, outrigger beams, brackets, poles, uprights, posts, frames, or similar rigid support. They depend upon the support from the ground or the existing structure. Whatever load is imposed on the scaffold and its various work elevations, the entire load is transmitted by the structure to the ground. There may be several planked working levels as well as additional unplanked areas that are unused. The weight of the equipment, materials, and personnel that is anticipated on the scaffold, as well as the weight of the scaffolding itself, must be determined and these total loads analyzed in accordance with the allowable loads on the supporting structures.

The general requirements for supported scaffolds are set forth in the Occupational Safety and Health Regulations in paragraph 1926.451(c). The non-mandatory specifications published as Appendix A of this text provide guidance for standardized maximum loads for various capacity scaffolds. These loadings, plus the weight of the scaffold components, may be utilized to determine the load transmitted to the supporting structure. It is possible to have multiple levels of work platforms. The load imposed by each planked level must be included. The weight of the planks on any planked, but unused, areas must also be included in the total weight calculation. The total load of all the planked work areas must be added to the weight of the scaffold components to determine the total load on the bottom support elements. The load-carrying capacity of these bottom components must not be exceeded. Manufacturers' catalogs must be consulted to determine the capacity of the various components.

MAINTAINING STABILITY

One of the major hazards with supported scaffolding is that of maintaining stability to prevent overturning. Scaffolds that are free standing must never exceed a base width to height ratio of 4-1. For example, a base width of 8' allows a total height of the scaffold of 32'. A base width of 4' will allow a working height of 16'.

For free standing scaffolds, the easiest and sometimes simplest solution is to extend the width of the base of the scaffold through the use of additional scaffold sections, outrigger attachments, or other means to provide a sufficient width to support the required height of scaffold and to assure that the 4-1 dimension required for stability is maintained.

Guy Wires

In lieu of using an extended base width, guy wires may be attached from the scaffold to a ground anchor or other structure securing the scaffold. Guy wires can be very effective but must be installed in pairs on opposite sides of the scaffolding that fail to meet the 4-1 dimension. When utilizing guy wires, a means of equally tensioning the wires on both sides of the scaffold is necessary. Guy wires, if not horizontal, impose a load on the scaffold. The weight of wires plus any tension applied to the wires must be included in the scaffold vertical load when computing the total load imposed upon the scaffold. For example, a guy wire installed at a 45-degree angle imposes a load equal to 0.707 times the weight of the wire plus the tension in the cable

Guy wires must be attached to the scaffold where there is a horizontal member connecting to the vertical member so that both sides of the scaffolding are tied together. The regulations, as well as good practice, indicate that guy wires should be installed starting with the horizontal member closest to the 4-1 ratio without exceeding it. Additional sets of guy wires must be attached for every additional 20' in height for scaffolds less than three feet wide and every 26' for scaffolds greater than three feet. Guy wires need to be installed as a minimum one for each 30' length of the scaffold.

Tie Wires

Scaffolds that are erected adjacent to the building or a structure can be secured to a building or a structure with ties. Common practice has been to use number nine tie wire, tying the scaffold to the structure. Number nine tie wire does not provide a rigid connection between the scaffolding and the building and will restrain the scaffold only from movement away from the building, not movement towards the building. If a tie wire type connection is used, a rigid brace or block must be placed between the scaffolding and the building to prevent movement of the scaffolding towards the building. See Figure II.1. Tie wires used to secure the scaffolding to the building must be carefully twisted to avoid kinking, stretching, or damaging of the wire. The twisting mechanism, such as a paddle, stick of wood, or piece of metal, should be without sharp edges and must be left in place and secured to keep the wire from untwisting under the load as the scaffold might attempt to shift from one direction to the other. In addition, the blocking between the scaffolding of the wall must be secured in place so that movement of the scaffold, temperature changes, weather conditions, etc. do not cause the block to fall out. Although the twisted tie wire and block provide an acceptable means of tying the scaffold to the structure, it must be properly installed and maintained. Difficulties associated with achieving proper installation and the requirement for detailed, constant inspections of this type of securement make a rigid connection between the scaffold and the building a much more practical means of providing securement to the building.

Rigid Braces

Scaffold manufacturers provide many types of rigid braces for securing their scaffold to structures. These braces are compatible with their scaffold and easily attach. Braces from other manufacturers should not be used. Depending upon the structure, rigid braces may be attached through window openings or to both sides of the window opening to prevent movement in either direction. They may be attached to fittings or attachments placed in masonry walls or may be directly bolted to the steel structure of steel buildings. Refer to Figure II.2. Masonry wall ties may be broken off flush when the scaffold is disassembled. Rigid pre-fabricated braces may be reused. Once installed at the correct spacing and distances, these ties must be left in place while the scaffolding is in use. If there is the need to remove a tie for some reason, other ties in suitable

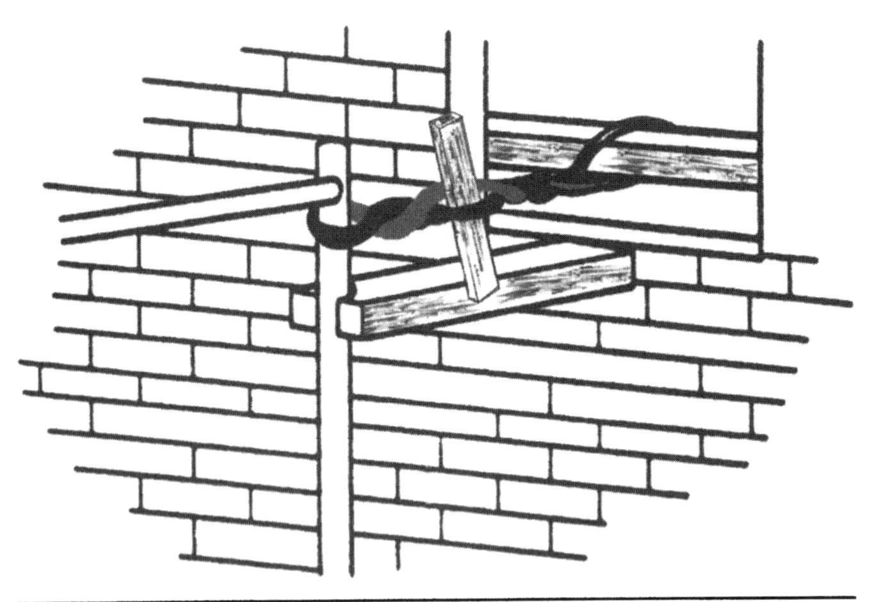

FIGURE II.1 Tie wire connection

locations must be installed before the tie needing to be removed is disconnected. The security and condition of these ties must be inspected as part of a required daily inspection of these scaffolds.

Preventing Contact with Electrical Lines

Many scaffold accidents occur because of contact by scaffolding components with electrical lines during movement or erection or by persons handling materials on scaffolding coming in contact with electrical wire. Prior to erection of a supported scaffold, the erectors must look upward and determine the location of any power lines. They should locate and mark their location on the ground so that when they begin laying out the scaffold they have a reference point to assure that the required clearances for use of the scaffold as specified in the Occupational Safety and Health regulations are maintained.

Most power lines which are encountered, except for house drops, are uninsulated lines and although from a distance they may look like they are

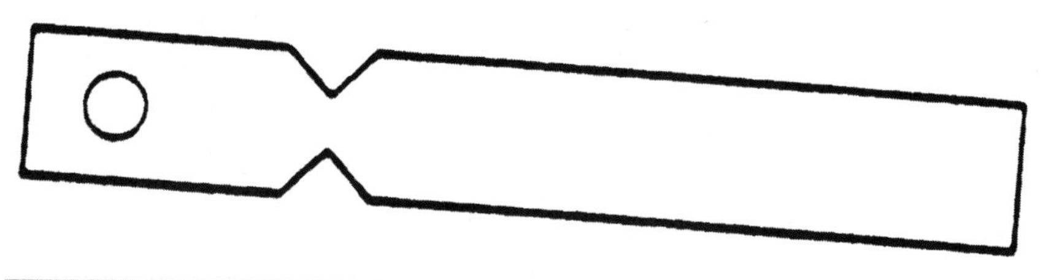

FIGURE II.2 Wall tie

covered with a black insulation, that is merely a corrosion or patina on the power line and they are uninsulated. Unless evidence to the contrary is determined, all lines should be assumed to be uninsulated and scaffolds should not be any closer than 10 feet from those existing power lines. The regulations require a minimum of a 10-foot distance between conductive materials which will be handled on the scaffolding and the power lines, not just a 10-foot distance to the power line. This insures that, if workers are erecting a scaffold and handling 6- or 8-foot long components of a scaffolding, they remain well clear of the line. It is prudent that the clearance between the workers and the scaffold be 10 feet in addition to the length of the piece in which they are handling, i.e. a 10-foot piece of cross bracing would require that the scaffolding be 10 plus 10 feet away from the power lines or a total of 20 feet. This clearance is also a major consideration, not only during erection, but also when workers are handling conductive materials on a scaffolding, such as highly conductive pieces of aluminum gutter or siding, steel components of a building, or other conductive materials. Adequate clearances must also be provided for any overhead pipe obstructions or building obstructions such as parapets, etc., that set out.

Scaffolds may only be erected, moved, dismantled, or altered under the supervision and direction of a Competent Person, qualified in scaffold erection, moving, dismantling, or alteration. In addition, there are specific training requirements for each employee who is working erecting a scaffold.

Access

All supported scaffolds must provide for a means of safe access. The Occupational Safety and Health regulations, as promulgated under 1926.451, paragraph (e), require this access and differentiates between those who are erecting or dismantling the scaffold and those employees who are working on a scaffold. For employees working on a scaffold, access must be provided when a platform is more than two feet above ground surface, which means that almost any scaffolding must be provided a means of access. There are many means of providing access to a scaffold.

Portable ladders, straight ladders, or extension ladders may be utilized. The requirements for the construction and use of the portable ladders are described in Subpart X of the Code of Federal Regulations entitled Stairways and Ladders. Of particular concern when utilizing a portable ladder to access the scaffold is that the slope of the ladder produces horizontal forces where it leans or is attached to the scaffold. These forces can be sufficient to overturn single or multiple level scaffolds. The position and securement of this ladder must be carefully considered when placing it against a scaffold. In addition, the requirements for the ladder extending above the work platform must be met and a means of access from the ladder to the work platforms from the guardrails must be provided. Climbing over, under, or through the guardrails from a ladder to the work platform is not an acceptable access way. In addition, when using portable ladders, there must be provided a secure foundation for the ladder as it rests on the ground and the conditions of use, proper slope and access to and from the work platform from the ladder, must be provided.

Most scaffold manufacturers provide premanufactured hook-on or attachable ladders that may be attached directly to the scaffold to provide access. Utilization of a manufacturer-provided ladder on the scaffold is usually the simplest and easiest way to provide access to medium height and short scaffolds. The ladder must be from the same manufacturer as the scaffold. It must be designed to attach to the scaffold securely, much like the use of the portable ladder. When these hook-on or attachable ladders are positioned outside the perimeter of the scaffolding, the weight of persons climbing the ladder will create an overturning moment on the scaffolding. Particularly small, light scaffolding is subject to significant overturning forces when workmen of normal weight are climbing outside the perimeter. It is important that the method of securement to the scaffold is sufficient to withstand this force and that the scaffold itself is completely assembled and of sufficient size to provide the necessary counterbalancing effect. Access must also be provided from the ladder through the guardrail to the work platform.

Many scaffold manufacturers provide for attachment of the ladder on the interior of the scaffold and access ways from the interior of the scaffold through to the work platform (Figure II.3). This arrangement requires that the work platform have an access way or trapdoor through to the work platform. In either arrangement, it is important that a safe means of access be provided from the ladder through to the work platform and that there be a means to secure or close the access area once the workers are on the platform. Most hook-on and attachable ladders supplied by manufacturers meet the requirements for rung spacing and rung length. However, if one is constructing a ladder for special conditions, certainly that section of the OSHA regulations should be consulted to be sure the ladder is constructed in accordance with these regulations. For larger scaffolds or those with significant employee traffic, separate stair towers can be constructed from components supplied by the manufacturer or as an individual construction. These stair towers can be separate structures with stairways and adequate handrails internal to the structure providing access to the scaffold. It is also possible to provide direct access from an adjacent structure, personnel hoist, or other scaffold.

FIGURE II.3 Access for a supported scaffold

SQUARE AND LEVEL

Once the scaffold erector is satisfied that he has accommodated the subsurface conditions, the next major concern is the topography. As mentioned before, it is rare when scaffolds can be erected on a level surface. Scaffold components need to be both plumb and level, as well as square. If the site is sloping, accommodations must be made through the use of the screw jacks or adjustable legs at the base of the scaffold. The lowest point of the scaffold system must be located and adjustable jacks or adjustable legs utilized to provide level work platforms and connection points for the bracing tube.

Tube and coupler scaffolds are somewhat more accommodating to a change in elevation. Mason's frame scaffolds and other types of pre-manufactured scaffold sections are not so forgiving, as the leveling capabilities of the screw jacks are limited — usually in the range of 12 inches. Provisions must be made to level these scaffolds around the building or across the building so that the scaffold frames can sit level and can interconnect at some level. In some cases this requires a significant amount of preparation in leveling the areas and determining the elevations of the sills. In extreme cases, it will require the utilization of various sized sections in order to achieve an even height for the working scaffold platforms.

CRITERIA FOR FALL PROTECTION ON SUPPORTED SCAFFOLDS

- Supported scaffolds are platforms supported by legs, outrigger beams, brackets, poles, uprights, posts, frames, or similar rigid support. 29CFR 1926.451 (b) The structural members, poles, legs, posts, frames, and uprights, must be plumb and braced to prevent swaying and displacement. 29CFR 1926.451 (c) (3)

- All employees must be trained by a qualified person to recognize the hazards associated with the type of scaffold being used and how to control or minimize those hazards. The training must include fall hazards, falling object hazards, electrical hazards, proper use of the scaffold, and handling of materials. 29CFR 1926.454(a) Employers must provide fall protection for each employee on a scaffold more than 10 feet (3.1m) above a lower level. 29CFR 1926.451 (g) (1)

- After September 2, 1997, a Competent Person must determine the feasibility and safety of providing fall protection for employees erecting or dismantling supported scaffolds. 29CFR 1926.451 (g) (2) Fall Protection includes guardrail systems and personal fall arrest systems. Guardrail systems are explained in the OSHA Regulations. Personal fall arrest systems include body belts (not acceptable after January 1, 1998), harnesses and components of the harness/belt such as Dee-rings, snap hooks, lifelines, and anchorage points. 1926.451(g)(3) Vertical or horizontal lifelines may be used. 1926.451(g)(3)(ii)-(iv)

- Lifelines must be independent of support lines and suspension ropes and not attached to the same anchorage point as the support or suspension ropes. 1926.451(g)(3)(iii)&(iv)

- When working from an aerial lift, attach the fall arrest system to the boom or basket. 1926.451(b)(2)(v)

- In addition to meeting the requirements of OSHA CFR 1926.502 (d), personal fall-arrest systems used on scaffolds are to be attached by lanyard to a vertical lifeline, horizontal lifeline, or scaffold structural member. Note: Vertical lifelines may not be used on two point adjustable suspension scaffolds that have overhead components such as overhead protection or additional platform levels. 29CFR 1926.451 (g) (3).

- When vertical lifelines are used, they must be fastened to a fixed safe point of anchorage, independent of the scaffold, and be protected from sharp edges and abrasion. Safe points of anchorage include structural members of buildings, but not standpipes, vents, electrical conduit, etc., which may give way under the force of a fall. 29 CFR 1926.451 (g) (3) (i)

- There are many types of rope and cable grabs that can be used with vertical lifelines. One must use them in accordance with the manufacturer's instructions and recommendations. Some general advice to keep in mind is to keep the connector length as short as possible and to allow no more than a maximum free fall of 6 feet. Remember, if your rope grab is the follow-along type, then the length of the lanyard can be only 3 feet to insure a maximum free fall of 6 feet. Carefully examine the rope grab before attaching it to the line to ensure that it is put on the line with the top up. If it is put on the line upside down it will not arrest your fall. Make sure the rope grab that you use is the correct type and size of recommended line. Generally rope grabs should be positioned as far above the user as possible when in the fixed or working position.

- It is dangerous and therefore impermissible for two or more vertical lifelines to be attached to each other, or to the same anchorage point. 29CFR 1926.451 (g) (3) (iv)
- When horizontal lifelines are used, they are to be secured to two or more structural members of the scaffold. 29CFR 1926.451(g)(3)(ii)
- When lanyards are connected to horizontal lifelines or structural members, the scaffold must have additional independent support lines and automatic locking devices capable of stopping the fall of the scaffold in case one or both of the suspension ropes fail. These independent support lines must be equal in number and strength to the suspension ropes. 29CFR 1926.451(g)(3)(iii)
- On suspended scaffolds with horizontal lifelines that may become vertical lifelines, the devices used to connect to the horizontal lifeline must be capable of locking in both directions. 29CFR 1926.502(d)(7)

Guardrail Systems

- Guardrail systems must be installed along all open sides and ends of platforms, and must be in place before the scaffold is released for use by employees other than erection/dismantling crews. 29CFR 1926.451(G)(4)(i)
- Each toprail or equivalent member of a guardrail system must be able to withstand a force of at least 200 pounds applied in any downward or horizontal direction, at any point along its top edge. 29CFR 1926.451(g)(4)(vii)

- The top edge height of toprails on support scaffolds must be between 36 inches and 45 inches. When conditions warrant, the height of the top edge may exceed the 45-inch height, provided the guardrail system meets all other criteria. (Note: The minimum top edge height on scaffolds manufactured or placed in service after January 1, 2000 is 38 inches). 29CFR 1926451.(g)(4)(ii)
- Midrails, screens, mesh, intermediate vertical members, solid panels, etc., must be able to withstand a force of at least 150 pounds applied in any downward or horizontal direction, at any point along the midrails or other member. 29CFR 1926.451(g)(4)(ix)
- When midrails are used, they must be installed at a height approximately midway between the top edge of the guardrail system and the platform surface. 29CFR 1926.451(g)(4)(iv)
- When screens and mesh are used, they must extend from the top edge of the guardrail system to the scaffold platform, and along the entire opening between the supports. 29CFR 1926.451(g)(4)(v)
- When intermediate members (such as blausters or additional rails) are used, they must be no more than 19 inches apart. 29CFR 1926.451(g)(4)(vi)
- Guardrails must be surfaced to prevent punctures or lacerations to employees and to prevent snagging of clothing, which may cause employees to lose their balance. 29CFR 1926.451(g)(4)(xi)
- Ends of rails may not extend beyond their terminal posts, unless they do not constitute a projection hazard to employees. 29CFR 1926.451(g)(4)(xii)
- In lieu of guardrails, crossbracing may serve as a toprail or midrail, providing the crossing point is:
  - Between 20 and 30 inches above the work platform for a midrail
  - Between 38 and 48 inches above the work platform for a toprail 29CFR 1926.451(g)(4)(xv)

Personal fall arrest systems can be used on scaffolding when there are no guardrail systems. 1926.451(g)(1)(vi

TABLE II.1 This table illustrates the type of fall protection required for specific scaffolds.

Type of Scaffold	*Fall Protection Required*
Aerial lifts	Personal fall arrest system
Boatswains' chair	Personal fall arrest system
Catenary scaffold	Personal fall arrest system
Crawling board (chicken ladder)	Personal fall arrest system, **or** a guardrail system, **or** by a ¾ inch (1.9cm) diameter grabline or equivalent handhold securely fastened beside each crawling board
Float scaffold	Personal fall arrest system
Ladder jack scaffold	Personal fall arrest system
Needle beam scaffold	Personal fall arrest system
Self-contained scaffold	**Both** a personal adjustable scaffold arrest system **and** a guardrail system
Single-point and two-point suspension scaffolds	Both a personal fall arrest system **and** a guardrail system
Supported scaffold	Personal fall arrest system **or** guardrail system
All other scaffolds not specified above	Personal fall arrest system **or** guardrail systems that meet the required criteria

Chapter 9

Tubular Welded-Frame Scaffolds - Mason's Frames

Introductory Description

Welded-frame scaffolds consist of a series of components that can be assembled in various configurations to form the scaffold as necessary. A Mason's frame scaffold consists of two end frames, two cross braces, and pads or supports for the bottom of each of the four legs, as well as necessary planking and guardrails for the top of the scaffold. See Figure 9.1. The end frames for a welded-frame scaffold come in numerous sizes and shapes for various purposes. Each manufacturer has their own standard designs, but in general they conform to a standard series of even heights. The heights start at two feet and increase in one-foot intervals up to eight-foot-high scaffold frames (Figures 9.2 and 9.3).

The width of the frames also varies. Widths begin at two and a half feet or 30 inches. Four-, five-, and six-foot width frames are common sizes. There are numerous configurations of end frames. The most common is the ladder type. See Figures 9.4 and 9.5. The horizontal members are spaced approximately 12 to 18 inches apart and are suitable for use as a ladder for access. These frames can be assembled and planked at any of the horizontal members, providing flexibility in establishing a work level. However if these frames are assembled as part of a continuous scaffold, they prohibit walk-through at elevations other than the top.

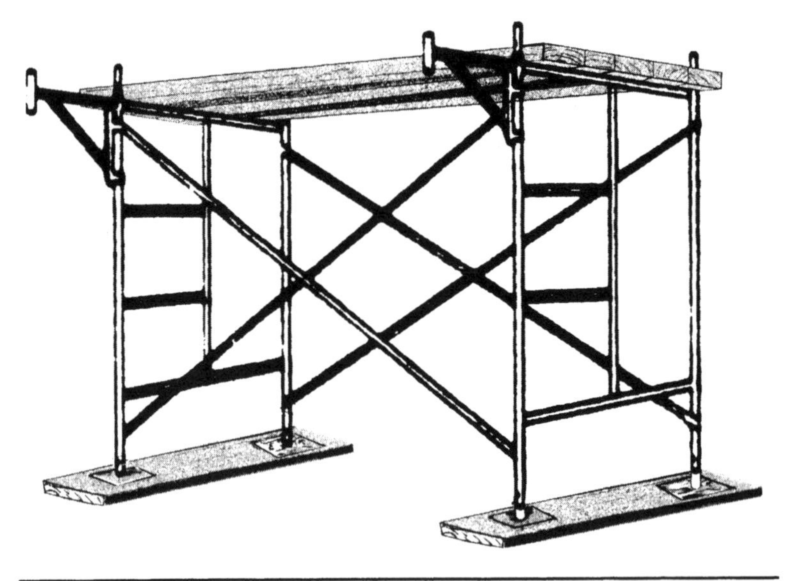

FIGURE 9.1 Example of Mason's frame scaffold

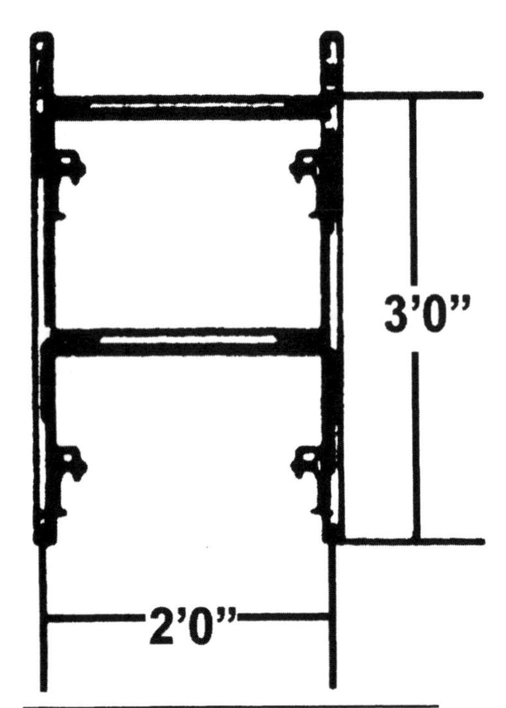

FIGURE 9.2 Example of mason's frame scaffold

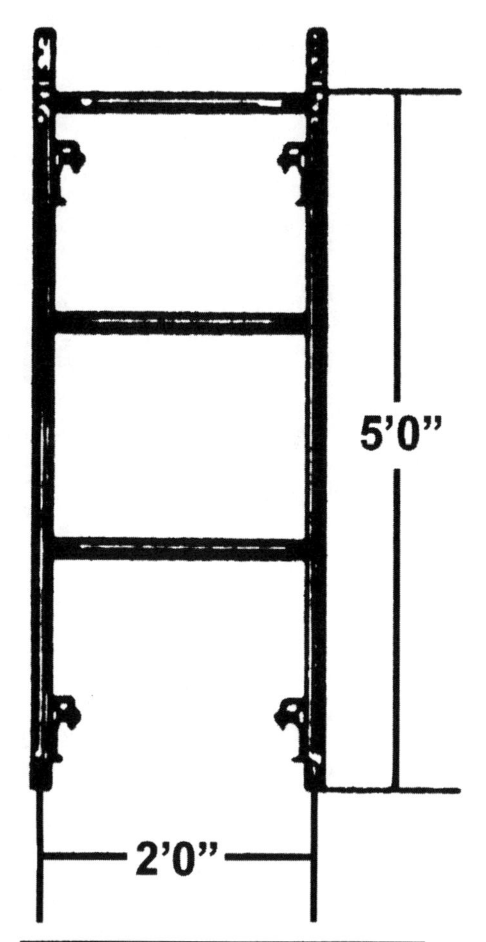

FIGURE 9.3 Example of mason's frame scaffold

The Half-Ladder

Another type of frame (Figure 9.6) is the half-ladder where the horizontal members run across one half of the scaffold and the remainder of the sections open except for the top and bottom members. This allows planking on half of the scaffold at the intermediate stages, but does allow for a pass-through or a walk access on the opposite side. On smaller frames of four feet and less in height, it is very difficult for a person to pass through the opening. Materials may be placed and passed through that area. The larger frames, six feet in height or more, provide somewhat easier access through the frames.

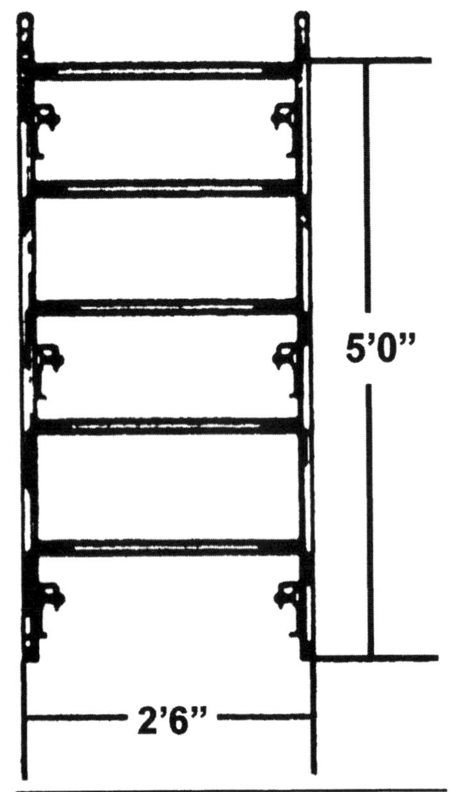

FIGURE 9.4 Ladder-type scaffold

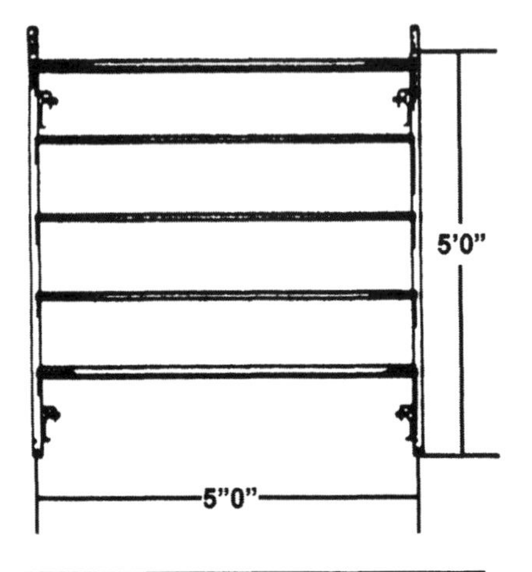

FIGURE 9.5 Ladder-type scaffold

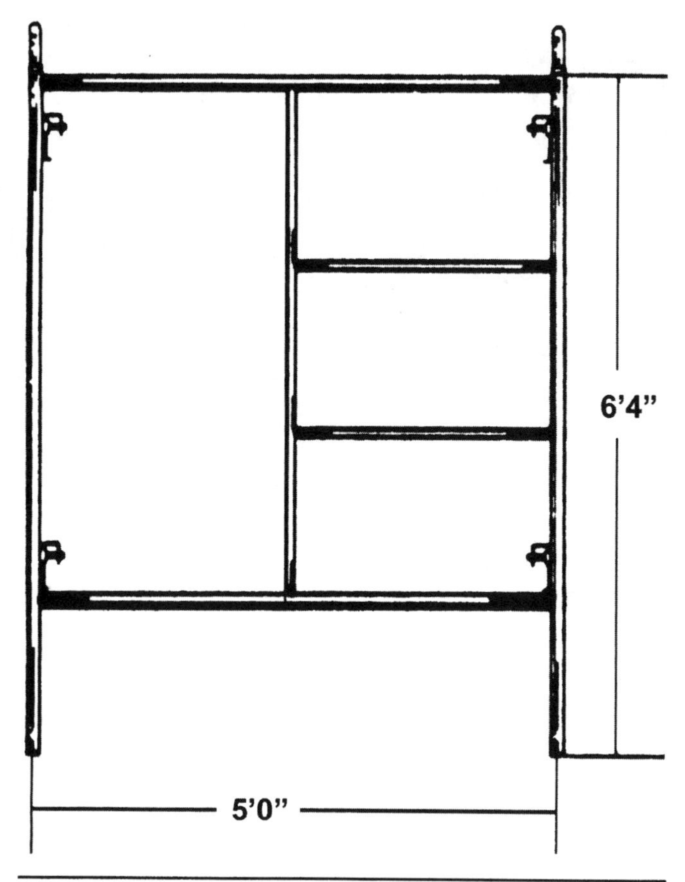

FIGURE 9.6 Half-ladder frame

The Walk-Through Frame

Another type of frame is the walk-through frame. The side legs are reinforced with bracing, however there is an area in the center (Figure 9.7) immediately between the legs that provides sufficient height for men and materials to move through. These are commonly used as mason scaffolds where the planked area is used for material storage and a walkway. They may also be used as a sidewalk or protective shed, where the top area is planked over and pedestrian traffic is allowed to pass through the walk-through frame. The height of walk-through frames generally starts at six feet and are available up to eight feet. Some walk-through frames have a small ladder on one side. Refer to Figure 9.8.

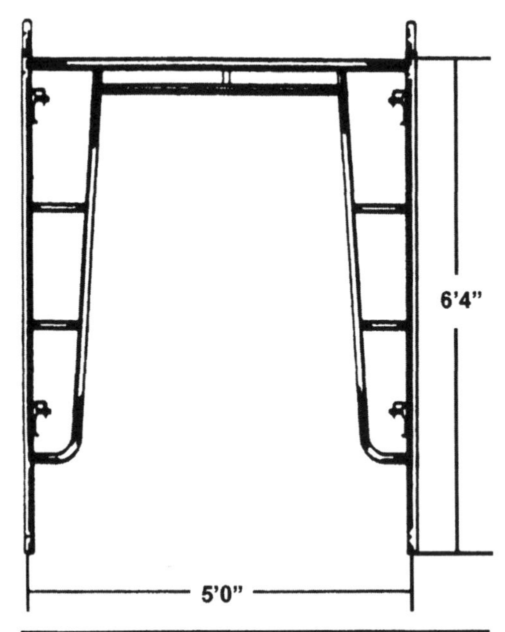

FIGURE 9.7 Walk-though frame

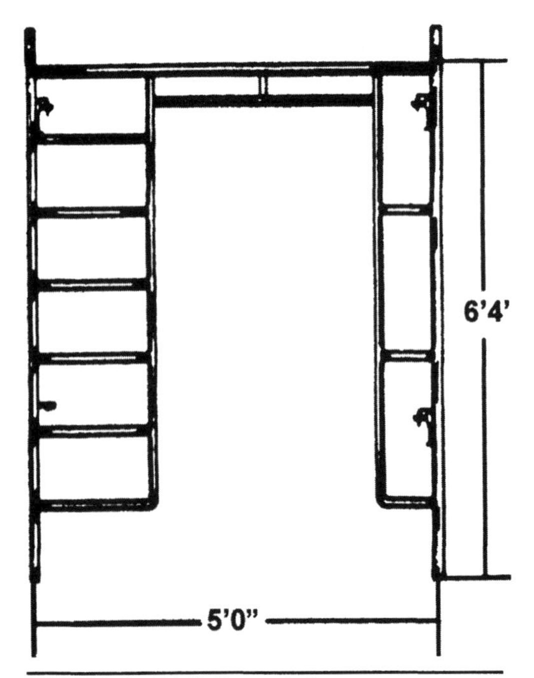

FIGURE 9.8 Walk-through frame

End Frames

All manufacturers of welded-frame components are able to supply various types of end frames in various sizes. Many of the manufacturers produce end frames of various load capacities. End frames from different manufacturers or of different capacity should not be interchanged in the same scaffold. It is never recommended to interchange any of the various manufacturers' scaffolding components, particularly end frames, cross bracing, and outrigger brackets within a single scaffold. There may be differences in the dimensions of the frames and in the rated capacities. Also the locking lugs which interconnect the frames and the cross braces may be different.

The two end frames are connected by "x" or cross bracing or diagonal bracing which attaches to two lugs on the inside of each end frame forming a rectangular structure. By changing the lengths of the cross bracing, the spacing of the end frames can be varied. The longer the cross braces, the further apart the frames. Cross bracing is provided in lengths that will provide spacings of four to eight feet of the end frames. The spacing of the lugs on the end frame also determines the center-to-center spacing of the end frames. Spacing of the lugs is consistent and therefore most frames at a given height of x-bracing provide the same spacing, regardless of the height of the frame. Manufacturers try to provide X-braces, which provide the same spacing on different height end frames. The larger end frames maintain a consistent vertical spacing between the lugs. However, shorter end frames are not capable of maintaining that and usually require a different length cross brace to maintain the same spacing as taller frames.

Frames may be stacked on top of each other vertically through the use of a steel connecting pin. This is usually a formed or cast connecting pin that fits into to the top of each leg of the bottom frames and into the bottom of each of the upper frames. These pins are provided with a locking pin, which goes through the leg of each section and secures the two frames together, prohibiting uplift. This connection is particularly important to make the scaffolding function as a continuous unit. It is also very critical, when outrigger brackets are attached as a work platform for masonry construction, that the outrigger bracket is secured to the scaffold frame.

Some manufacturers provide a ladder which connects to the outside of the scaffold for climbing and also provide a well wheel attachment for hoisting materials or other operations. These attachments induce eccentric or off-center loads, which may cause the scaffold to lift up. If the pins are in place, the entire scaffold resists the uplift. These connecting pins should always be installed.

The Pad

A pad is needed, at a minimum, at the level each frame is supported. Pads are supplied in two sizes, a 6 x 6- inch or 4 ½ x 4 ½- inch pad area metal foot that fits into the bottom of each scaffold frame. Pads are required on all installations. The legs of the end frame require a means to distribute the load. This pad is also provided with locking pins to secure the pad to the leg.

Jacks

For purposes of leveling the scaffold, manufacturers provide jacks to be inserted into the base of the scaffold frame set on the pad. The jacks may be adjusted to level the scaffolding in both directions, across the end frame first and then in the direction of the scaffold.

Prior to beginning the erection of a scaffold, it is necessary to determine the difference in elevation of the support structure between ground, building, or pavement from one end of the scaffold to the other. Although the jacks will provide leveling capabilities, most jacks are limited in a total extension to between 12 and 14 inches. It is always advisable to keep the jacks extended as little as possible. If there are significant differences in elevation, the erector may utilize a shorter frame to start, for example a 4-foot frame. When a 12-inch extension of the jacks is required to maintain a level platform, a 5-foot high frame can be used and the jacks can be retracted to their minimum extension again. By using different sized frames in the bottom section of frames, differences in ground elevation can be accommodated. By using this erection procedure, significant differences in elevation can be accommodated and unsafe blocking or excessive cribbing under the support legs of the scaffold can be eliminated

Planning

During the planning stage, the total anticipated load must be determined. Knowing the loads, the required bearing capacity of the support structure may be determined. The total load must include the weight of the scaffolding, the scaffolding materials, the equipment, and men. Sufficient width and depth of sills shall be provided. Each end frame should be placed on a continuous sill across the length of the end frame with sufficient length to adequately support the pad.

It is always necessary to ensure there is sufficient clearance from the inside of the end frame to protruding parts of the structure including roof eaves. It is also necessary to provide for clearances around overhead pipes and utilities, and as always, overhead power lines should be avoided whenever possible. Most components of welded-frame scaffolds are designed to be placed in a straight line and offsets and turns are difficult to form with standard components. Special provisions need to be made to deal with these obstructions.

Some manufacturers make outrigger brackets and at least one manufacturer supplies a put-log extension to accommodate offsets in the building structure such as pilasters or alcoves. The spacing and clearances for these devices need to be accommodated in the planning stage before the placement of sills and or scaffolding has begun.

Pre-Erection Inspection

Each component of the scaffold structure needs to be inspected prior to the beginning of erection. Damaged or defective components, once installed in the scaffold, can be very time consuming and expensive to remove and replace. All end frames should be individually inspected for damage. Areas of excessive rust which indicate a thinning of the members, bends in the tubular sections, dents or distortions of the members which would preclude the legs from fitting together properly, and cracks in the welds where the frame members are welded are all conditions which result in weakened components. In particular the condition of the cross bracing lugs and locks must be inspected. They must be in proper operating condition and functioning properly to secure the x-brace to the frame. The ends of the tubular sections must not be deformed so that the pins connecting the vertical members are easily placed. Incomplete insertion of the pins will result in the upper section not being aligned with longer sections. In addition, when weight is placed on the scaffold, the joist may suddenly close, causing movement in the upper portion.

Additional horizontal sections or bays may be added to the welded-frame scaffold by attaching additional cross braces to the end section end frame and attaching another end frame to the opposite end of the cross braces. Each bay on the end frame will accept cross braces. Welded-frame scaffolds can be extended indefinitely by this technique. The end frames must be at the same elevation for the cross bracing lugs to interconnect. For the cross bracing to fit properly, the frames must be at the same elevation and square.

OSHA Regulations

The OSHA regulations require that scaffolds shall only be erected under the supervision of a competent person. The definition of a competent person in OSHA 1926.451(a) is one who is capable of identifying existing and predictable hazards in the surroundings or working conditions which are unsanitary, hazardous, or dangerous to employees, and who has authorization to take prompt corrective measures to eliminate them. The competent person, in addition to other duties, has a specific duty assigned during erection operations. All scaffolds require safe access. Safe access must be provided for all employees and workers using the scaffold on a regular basis as defined previously. There are specific provisions for access for employees erecting and dismantling the scaffolding. They are also required to be provided with safe access in accordance with the access requirements for erecting and dismantling a scaffold. However there is an exception to this requirement if the access creates a greater hazard. This needs to be determined by the competent person and his determination is based on the site conditions and the type of scaffold. This determination states that a defined means of access presents a greater hazard and regulations require that the competent person direct employees to a safe means of access. Prior to allowing people to use an alternative access, the reasons and basis for this determination should be recorded. The access provided in this manner should be clearly defined and explained to the employees.

Climbing of Tubular Welded-Frame Scaffolds

There is also a specific exception for tubular welded-frame scaffolds. If the end frames have horizontal members that are level and not more than 22" apart, they may be used as climbing devices provided the method of erection creates a usable ladder and provides a good handhold. This is a more lenient standard than that required for the general usage of the scaffold. These frames are to be used only during the erection of the scaffold and great care must be taken on the part of the employer to emphasize to the employee that this use is limited only to erection of the scaffold, and once completed, normal access use is required. In no case may the cross or x-braces be utilized as a means of access, either during the erection process or the use of the scaffold. Many falls have occurred to employees attempting to climb these sloping braces. Many of the braces are not adequate to support the weight and will bend or distort. Not all securing locks on the legs are strong enough to withstand the forces imposed by a person climbing and the lock may break and allow the cross brace to pull from the lug. This can result in not only an employee falling, but also a collapse of the scaffold.

Outrigger Brackets

Tubular welded-frame scaffolds are typically equipped with outrigger brackets (Figure 9.9). These brackets hook on to the side of the frames and provide a work platform of two or three planks in width. These outrigger platforms are to be utilized for persons only. Storage or stacking of materials is not to be permitted without the manufacturer's recommendation. The bracket itself is adjustable in height, hooking over a horizontal bar on the end frame, and the

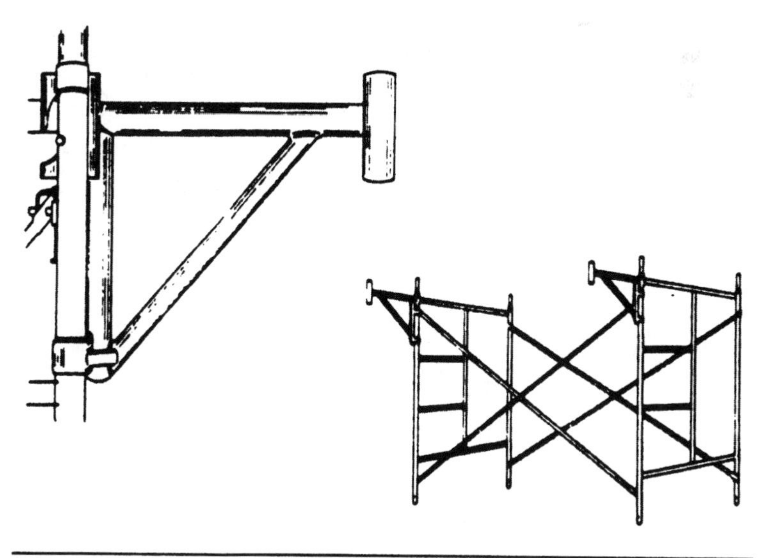

FIGURE 9.9 Outrigger brackets

bottom of the bracket rests on the side of the end frame post. Because of the protruding hooks and brackets, these brackets can be easily damaged by overloading. They must be carefully inspected prior to erection and on a daily basis. Many scaffold-related injuries and deaths occur when the bracket becomes dislodged, causing the planks to fall.

When long planks that span three supports are utilized, the support points are not always in firm contact with the planks because of the need to overlap the planks. This causes constant loading and unloading of the bracket as persons walk along the plank and can cause the bracket to dislodge. In addition, raising the planks and moving the brackets can cause dislodgement. Manufacturers are recently adding a pin and drilled hole to secure the outrigger bracket to the end frame. Outrigger brackets that are not provided with a means of securing should not be used.

Chapter 10

Tube and Coupler Scaffolding

Definition

Tube and coupler scaffolds are erected from individual lengths of steel or aluminum tubing and several types of connectors or coupling devices. There are generally no prefabricated components in a tube and coupler scaffold. Because of the nature of their construction, tube and coupler scaffolds are very versatile in that they can be assembled into almost any configuration. This allows easy assembly of irregular shaped structures into alcoves of buildings, underneath overhangs, around corners, and around circular structures.

Important Features of Tube and Coupler Scaffolding

The spacing of the posts or uprights is infinitely variable. Posts may be connected to runners or bearers at any position along their length. This allows the post spacing to be arranged to accommodate the supporting structure or ground conditions, even though it may not be regularly spaced in pattern.

Another important feature of the tube and coupler scaffolding is that the various work platforms can be located at precise elevations to accommodate the needs of the work. The bearers for the work platforms can easily be located at any elevation along the support posts.

This flexibility of the tube and coupler scaffold also requires that the erectors be highly skilled and well trained. Tube and coupler scaffolds can be very complex with many turns and elevation changes. The location of the bracing must be carefully evaluated prior to the start of erection and the plan for erection must be detailed and implemented completely. It allows for innovation and creativity in solving problems presented by the structure.

Figure 10.1 shows a typical tubing and coupler scaffold illustrating the primary components. There are posts that are the vertical members, runners that are horizontal along the length of the scaffold, the bearers, which run in a perpendicular direction and are used primarily to support the platformed work areas, and diagonal members that run both longitudinally and across the scaffold to provide the stability. The posts have available a base plate or pad, however no jacks or screw adjustments are utilized in a tube and coupler scaffold post because the bearers and the runners can be clamped to the vertical member at any elevation.

The primary members, the vertical posts and the horizontal runners and bearers, are connected by a rigid clamp. See Figure 10.2. This is a clamp that is fixed at a 90-degree angle and the various components, vertical and horizontal, are attached through the clamp forming a rigid, square connection. Although these connections are rigid and upon connection provide some rigidity in the plane of the coupler, they may not in any circumstances be relied upon to provide the necessary bracing or longitudinal support that is required on a tube and coupler scaffold. They are only to be relied upon to resist forces in the directions of the member being clamped, i.e. vertical or horizontal.

The required diagonal and longitudinal bracing are secured by means of a swivel clamp in which the joint between the two clamping mechanisms is free to pivot. See Figure 10.3. These swivel clamps connect the posts, bearer and runner, to the bracing at a 45-degree angle. As with any scaffold, it is necessary to erect the tube and coupler scaffold plumb, i.e. the vertical members vertical in both planes and each bay of the scaffold-erected square. The squareness of the individual bays is easy to control by installing horizontal, diagonal bracing between the vertical uprights at opposite corners throughout the scaffolding.

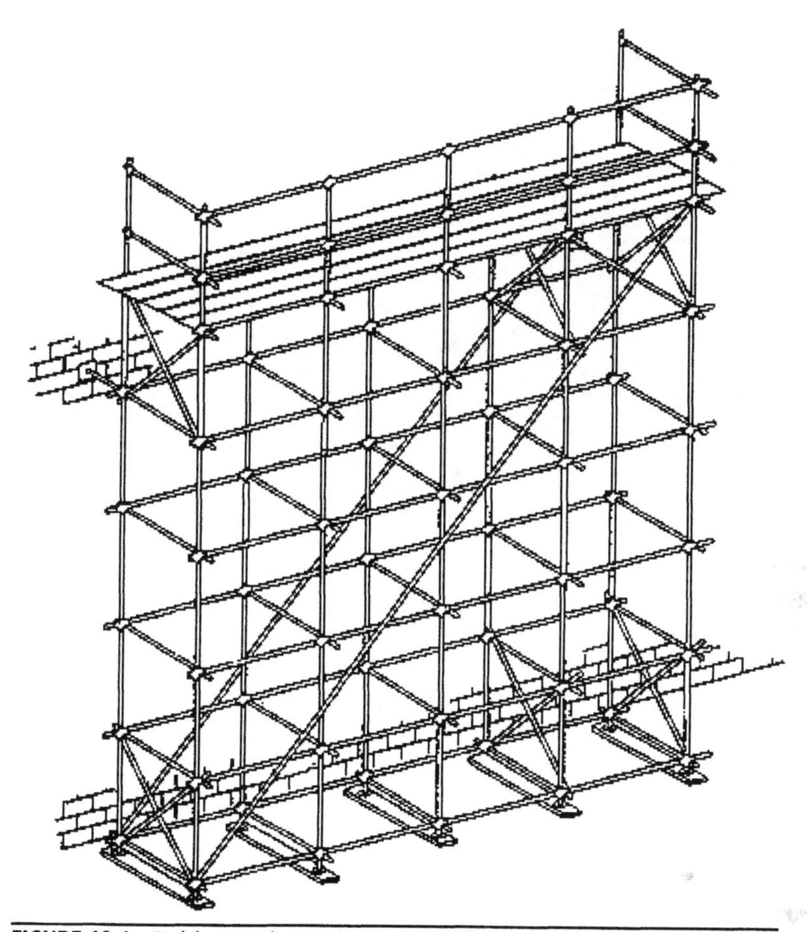

FIGURE 10.1 Tubing and coupler scaffold

The vertical support members or posts are generally of two types. Although all posts consist of hollow, steel or aluminum tubing, the original tube and coupler scaffolding posts were without end fittings and were connected either through the use of a coupling pin, inserted at the end of each post and tightened by means of an expansion bolt, or were coupled adjacent to each other through the use of a special connecting clamp. While these open tube systems are still in use, the more popular systems, particularly in North America, consist of steel or aluminum tubes with connecting fittings on each end. These fittings are typically a bayonet type with a male/female coupling with a "twist to lock" feature.

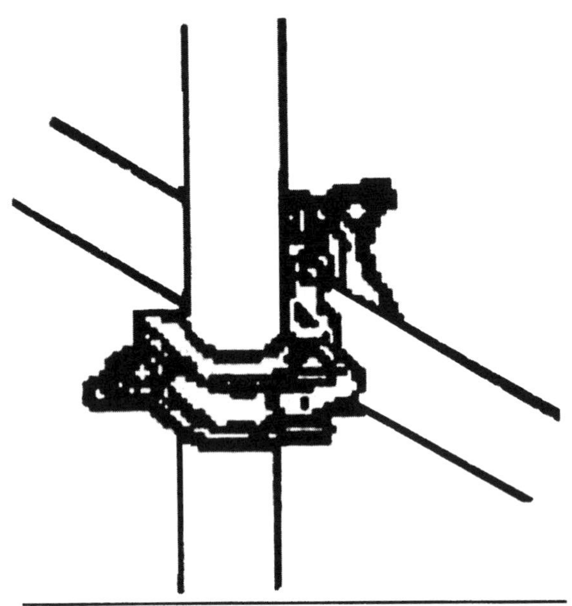

FIGURE 10.2 Rigid clamp

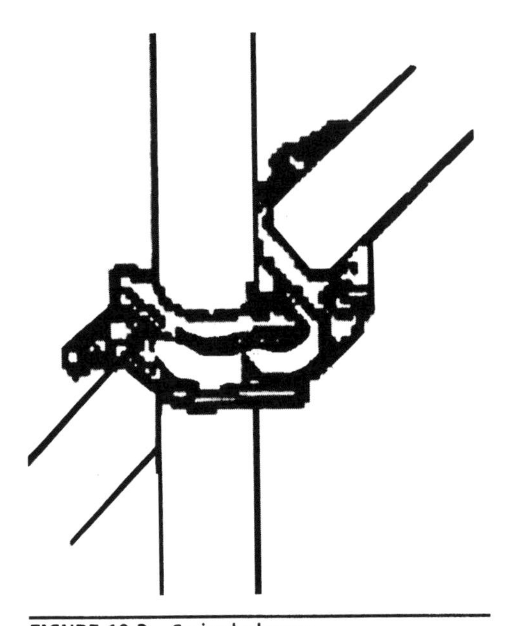

FIGURE 10.3 Swivel clamp

STANDARD TABLES FOR MINIMUM CONSTRUCTIONS

The standard tables for the minimum constructions allowed in tube and coupler scaffolds, as published in the Occupational Safety and Health act, non-mandatory Appendix A, minimum size of members, provide for three duty ratings of scaffolds: light, medium, and heavy, with prospective loads of 25 pounds per square foot, 50 pounds per square foot and 75 pounds per square foot. These tables all specify posts, runners, and braces of a nominal two-inch diameter steel tube or pipe. The light and medium duty scaffolds also use nominal two-inch diameter steel tube or pipe and vary the spacing to provide for the distance in rating of the two scaffolds. In the light and medium duty, the width requirement or length of the runner spacing remains at four feet, however the length of the platform, i.e. the longitudinal direction of the scaffold, is reduced from ten feet at a light duty scaffold, to seven feet for a medium duty. Heavy-duty scaffolds require bearers to be two and a half-inch nominal in diameter with a maximum post spacing of six feet by six feet. If bearers of two and a half inch diameter are utilized on a medium-duty scaffolding, the post spacing may be increased to six feet by eight feet. The spacing of posts and bearers may be decreased, but may not exceed the tubular values. Scaffolds, which are in excess of these maximum spacings and exceed the rated capacities and standard configurations, must be designed by a professional engineer. Although this table is entitled "Minimum Size of Members," it provides the maximum spacing of members both horizontally and vertically. Any scaffold erected in excess of 125 feet in height must also be designed by a professional engineer.

There are also provisions in the Occupational Safety and Health act, non-mandatory Appendix A, maximum number of planked levels, which limit the number of working levels and additional plank levels for each of these grades of scaffolding. This should be carefully consulted prior to the design and erection of a tube and coupler scaffold. Although couplers come in two basis configurations, rigid and swivel, there are many types of couplers and materials of which they are constructed. The OSHA regulations recognize drop-forged steel or malleable iron. They specifically preclude the use of couplers from gray cast iron. All couplers depend upon the compression of the fitting against the pipe to prevent slippage of the connection. Over-tightening of couplers can cause crushing or deformation of the pipe with subsequent loss of strength and stability. The manufacturer's recommendations must be followed when tightening the clamps. It is important that couplers which use screw type connections, T-bolts or eyebolts, have threads of the bolts that are clean and freely threadable in order that the correct manufacturer specifications can be followed. Manufacturers of clamps and accessories have specifications for tightening the bolt and most supply ratchets or wrenches for this purpose and prescribe a procedure. Employees should be trained in the manufacturer's recommended procedure. Manufacturer's specifications for tightening are based on clean undamaged threads. Dirty, corroded, or damaged threads increase the force required to property rights and will make the manufacturer's recommendations unusable.

Forged couplers consist of a center casting with two hinged flaps secured by either eyebolts or T-bolts. These come in either swivel or rigid and most manufacturers provide right- and left-handed flap connections. They should always be installed so that direction of the load imposed by the tube on the coupler is against the solid portion of the casting or the flap and not against the bolt portion. These couplers should always be arranged so that the bolt is in tension.

Forged couplers come in various sizes to allow clamping of various sized materials, variations from diameters of slightly over an inch to up to three inches. Couplers connecting various sizes of tubing such as two-inch nominal OD tubing post to a nominal two and a half inch diameter bearer will require couplers with different sizes on each side of the coupling. Care must be taken to ensure that not only is the proper sized coupler provided, but that they are installed in the correct position.

The bracing of tube and coupler scaffolds is critical to their stability. OSHA requirements, which set for the minimum accepted, require x-bracing at each end of the scaffold and at every three intermediate posts along the length. Only two posts may be unbraced. It also required x-bracing at every fourth runner vertically. This bracing must extend diagonally form one post to the adjacent post. Attached directly to the post is advisable, however it may be attached to the runner at the post. It should not be attached to the bearer.

Longitudinal bracing is required on both sides extending from the base upwards at a 45 degree angle to the opposite end post. Scaffolds, which are taller than they are long, require that the longitudinal bracing then return on the opposite angle upward to the end. The bracing should alternate until reaching the top of the scaffold.

The runner should be installed on the inside of the scaffold posts and the diagonal bracing installed on the outside of the posts. This allows both the runner and the longitudinal brace to be clamped to the post. On large area scaffolds, multiple posts with the same pattern should be followed. The bearers should be placed on top of the runner and clamped to the post. The planks or platform must always rest on the bearer. The bearer is the member that runs the shortest distance between the post spacing as indicated in the table. For example, a light duty scaffold with 4′ x 10′ post spacing would have the bearers in the 4′ direction not the 10′ direction. In medium and heavy-duty scaffolds the bearer's length would be 6′ if the 2 and a half-inch bearer is utilized.

It should be noted that the maximum span for scaffold grade nominal sized lumber is 8′, therefore nominal sized lumber cannot be used on the maximum span of 10′ for light duty scaffolds. Nominal sized lumber may not be utilized on heavy-duty scaffolds at any span length.

Chapter 11

Wood Pole Scaffolds

Wood pole scaffolds, as the name implies, are constructed entirely of wood and come in two general types. Refer to Figure 11.1. The double pole, which is supported entirely by its own structure, is commonly referred to as an independent pole scaffold. It is constructed much like a tube and coupler with poles, bearers, and cross bracing, except that the members are wood rather than tubular aluminum or tubular steel. The second major type is a single pole wood scaffold which relies upon a structure to support one side of the scaffolding and consists of a row of poles with bearers, runners, and cross bracing connecting to the building.

Use of Wood Pole Scaffolds

Wood pole scaffolds have been almost entirely replaced in the construction industry by metal welded-frame scaffolds and tube and coupler scaffolds. Except for minimal special uses where the properties of wood are necessary or where the shape of the area to be scaffolded is extremely unusual, they are generally not used. Wood pole scaffolds are limited in height to 60 feet unless they are designed by a licensed professional engineer.

Wood Pole Scaffold Requirements

The guying or bracing requirements for a wood pole scaffold require a guy wire with a tie to a structure every 25 feet in height and at intervals no greater than

FIGURE 11.1 Wood pole scaffold

25 feet vertically. The independent wood pole scaffold needs to be provided with diagonal bracing at each support or bearing level. This diagonal bracing shall be provided between the inner and outer sets of poles. The longitudinal bracing shall be provided parallel to the wall on all faces. On wood pole scaffolds, this diagonal bracing must be in both directions forming an "X" across the face and should span at least two poles.

In Appendix A, tables are given for single pole wood scaffolds and independent wood scaffolds, which define the sizes of the material and the span lengths in both directions.

Note these tables require that all the wood bearers must be reinforced with a 3/16-inch thick by two-inch steel strip or equivalent on the lower edge of the bearer for its entire length. Wood pole scaffolds can be very efficiently and effectively erected to provide scaffolding in irregular or unusually shaped areas. If the selection of the materials is carefully monitored and the bracing properly attached to the upright members, wood pole scaffolding of either the independent or single pole type provides a very effective scaffolding method for construction sites.

Please Note: Appendix A, 1. General Guidelines, Section a., ... assumes that all load-carrying timber members (except planks) of the scaffold are a minimum of 1,500 lb-f/in2 (stress grade) construction grade lumber. All dimensions are nominal sizes as provided in the American Softwood Lumber Standards, dated January 1970, except that, where rough sizes are noted, only rough or undressed rough lumber of the size specified will satisfy minimum requirements.

Chapter 12

Pump Jack Scaffolds and Ladder Jack Scaffolds

Definition

A pump jack scaffold is a type of scaffolding utilizing a single pole with moveable brackets attached that move vertically (Figure 12.1). These brackets support a work platform and, in some cases, a workbench on the opposite side of the pole. This type of scaffold is very versatile for certain kinds of work. Specifically, it is very popular in the residential construction industry and remodeling industry for working on the sides of residences, installing siding, installing windows, and other purposes. Pump jack scaffolds are erected with a minimum of two poles and two pump jacks. However, they may be expanded to include multiple supports. One of the advantages to this type of scaffolding is that the worker on the scaffold can elevate the scaffold to the desired work elevation merely by pumping on a lever or foot control. This allows the scaffold to be positioned vertically very precisely. Pump jack scaffolds are limited to a maximum load of 500 pounds, which limits their use to not more than two persons in any one span.

The Importance of the Structure and Poles

The pump jack scaffold depends entirely on the structure to which it is erected to provide for the stability. It is secured to the building by metal brackets. The

FIGURE 12.1 Pump jack scaffold

poles may be as long as 30 feet and made from either a metal, such as aluminum, or, more predominantly, wood. Commonly two 2″ x 4″s, nominal size, are spliced together with their flat sides to form a nominal sized 4″ x 4″. The seam between the two 2″x 4″s is placed perpendicular to the face of the building so that the full strength of the 4″ dimension of the 2″ x 4″s is developed in bending. On some occasions a nominally sized solid 4″ pole may be used. Aluminum poles are generally fabricated in nominal sized 4″ x 4″ extruded aluminum and may be substituted for the wood post. Aluminum poles have many advantages, including the ability to achieve longer lengths and an increased useful life. Aluminum poles are not subject to damage due to the assembly and disassembly process as readily as wood poles.

The bottom of the pump jack has an arm or platform extension that is designed to support wood planks or an aluminum or prefabricated pick or platform. Many ladder jacks have a similar arm at about 42 inches in height extending on the other side of the pole that may be planked to form a workbench or platform.

Structure of Pump Jack Scaffolds

Over the years, the predominant failure mode for pump jack scaffolds has been the failure of the pole in bending. All of the scaffold's platform extends toward the building and the load is on one side of the post. This creates a severe bending movement in the wood members. Constant assembly, nailing and re-nailing weakens the poles. To resist this bending and minimize the pull-out forces on the nails, it is critical that the joint between the 2" x 4"s face the building. In order to minimize these bending forces, a support between the pole and the building structure must be provided at a maximum of once every ten feet in height. The pole needs to be attached to the building structure at the base of the pole, at the top of the pole, and at every 10-foot interval in between. If wood is used to construct the poles of the pump jack, it is required to be kiln dried, straight-grained fir or the equivalent. Wood that meets this definition is a high quality wood. Normal construction grade 2"x 4"s are not of sufficient quality to meet this requirement. Wood utilized to construct the poles for a pump jack scaffold should be a structural grade wood, straight-grained, and without knots or defects.

Mending Plates

The Occupational Safety and Health regulations require a "mending plate" at all splices, which develops the full strength of the members. If mending plates are utilized, they must be inset into the wood in order to ensure clearances for the jack to pass by. It should be noted that the industry consensus standard, ANSI A10.8, specifically precludes the splicing of the poles.

Poles

Industry recommendations are that the poles be continuous pieces. The splicing of the poles is a critical issue with the use of pump jack scaffolds. Poles that are assembled and disassembled by the nailing operation very quickly lose their strength, thereby becoming extremely dangerous. The industry- recommended practice of using continuous lumber for the pump jack scaffolds is by far the safer means. The bracing of the poles at 10-foot intervals requires that some form of metal braces or fixtures be attached between the pump jack pole and the building. These braces restrict the elevation of the pump jack and must be removed to allow the jack to pass. For safety and to comply with the regulations, a supplementary brace must be installed within four feet of the original brace while the original brace is being removed. The pump jack then can be elevated past the brace location. The temporary brace must remain in place until the original brace or bracket is reinstalled. Most of the uses of pump jack scaffolds are to install finished wall surface materials. The installation of the bracing, unless it is well planned, will cause damage or marring to the wall and therefore the effective practical elevation at which pump jack scaffolds may be used is 10 feet. Pump jacks in excess of 10 feet impose special requirements for obtaining the correct length of lumber and also for providing the necessary bracing. For longer heights of poles, up to the maximum of 30 feet, aluminum poles with adequate bracing installed at the same ten-foot intervals are the more practical means of utilizing pump jack scaffolds.

Guardrails

As with any scaffold, guardrails must be provided for pump jack scaffolds. This is particularly necessary on the ends of the scaffold planks. Guardrails may not be attached to the planks themselves, but must be attached to the poles or pump jack structure. If a pump jack is equipped for a workbench, the workbench may be considered the top rail of the back scaffold.

Ladder Jack Scaffolds

Ladder jack scaffolds (Figure 12.2) utilize a pair of standard extension ladders and, by attaching a ladder jack, allow the plank or platform to span between the two ladders. The ladder jacks may be attached to the face of the ladder that is away from the building or structure that is supporting the ladder, or on the backside of the ladder, between the ladder and the structure itself. The ladders that are utilized for this type of scaffolding must be type I ladders, rated at 250 pounds, or preferably type one A ladders which are rated at 300 pounds.

The working height of the ladder jack scaffold is limited to 20 feet above the base of the ladder. It is important that both of the ladders be set at precisely the same angle in order that the plank or platform bears evenly on both of the ladder jacks. If the angle between the ladder jack varies, the platform will rock. The use of the ladders and their securement shall be as outlined in the chapter on ladders and in accordance with the applicable ladder standards as published in ANSI A14. One major difference when using a ladder with ladder jacks is that both the top and the bottom of the ladder must be secured.

Bridging from one ladder jack scaffold platform to an adjacent ladder jack is not permitted. The ladder jack scaffolds are restricted to one span, i.e. one plank or platform between two ladders.

Ladder jack scaffolds are limited to not more than two persons occupying the platform at any one time. The 250 or 300 pound rating for the maximum rated load for the ladders indicates that if two persons are on the scaffolding platform, they should remain at opposite ends of the scaffold, as their combined rate most likely exceeds the rated capacity of the ladder

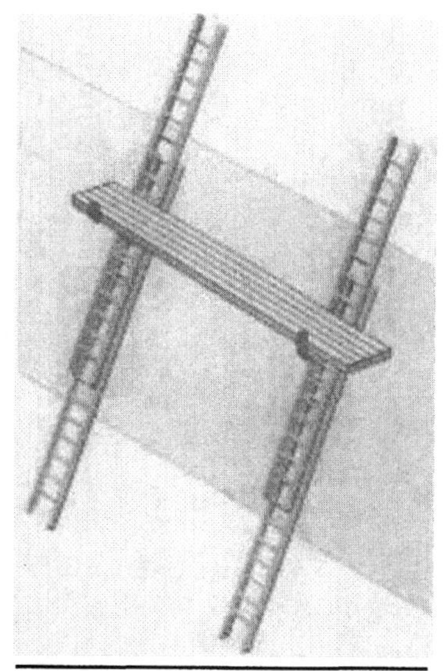

FIGURE 12.2 Ladder jack scaffold

Chapter 13

Job-Manufactured Scaffolds and Form Scaffolds

The Need for Forming Systems

One frequently performed construction operation is the building of formwork and placing of concrete for reinforced concrete walls. In order to perform this operation, it is necessary for the carpenters, laborers, and other workers to be on the wall during the process of construction and during the placement of concrete. In the past many of these workers climbed and stood on the 2″ by 4″ or 2″ by 6″ whalers (horizontal members running along the length of the wall) and were subject to significant fall hazards.

The advent of "forming systems" has allowed manufacturers of pre-fabricated formed panels to include in their inventory a metal bracket that can attach to the formwork at specified locations. These brackets are securely attached by a key or pin. This bracket allows the placing of planks from one bracket to the next. The bracket also includes a post for a guardrail system to provide the workers a safe platform. These are increasing in popularity and provide safe work platforms for the workers, both during the construction and the placing of the concrete.

A major concern with this system is that the brackets and formwork depend on the stability of the formwork itself against overturning. During the time period that the formwork is under construction, the formwork must be adequately braced to a supporting structure or to the ground in order to accommodate the eccentric load placed on it by the weight of the workers and equipment on the scaffold.

These brackets qualify as a scaffold and require the services of a Competent Person to erect and inspect the scaffolds. They also require specific training of the workers in the hazards associated with scaffold. Because of the fast moving nature of this type of construction due to continual construction of the formwork, it is necessary to inspect these types of scaffolds on a continual basis as they are erected and also continuously during the course of the day.

These scaffolds are rated for light duty work with a maximum load rating of 25 pounds per square foot. During the placement of the concrete, all of the activities are occurring at one location and it is not unusual to see several workers congregated around that area performing their tasks. The Competent Person must assure that these scaffolds are not overloaded by this concentration of workers. It is also quite common to have equipment such as vibrators on top of the formwork. This adds significantly to the weight imposed on the platform. An electric or gas concrete vibrator and two or three workers can easily overload a platform between supports. Care must be taken to keep these materials evenly distributed throughout this operation.

For non-system forms, there are other metal brackets that are designed to be attached to the 2" by 4" whalers or the 2" by 6" whalers on the formwork. These brackets can be placed on the formwork, planked, and provided with guardrails to provide a safe working platform. It is important that the whalers to which these brackets are attached are supported vertically through nailing or other means of securement. The whalers are normally used to provide strength and straightness to the work in a horizontal dimension and, unless they are specifically nailed or supported in the vertical direction, they are not adequate for the attachment of the form brackets. The wall ties used to hold the formwork together are sufficient to hold the bracket against the form but must be additionally supported to resist the vertical load. Remember, the wall ties are called "snap ties" and are designed to break off after the concrete is placed. It is also imperative that the whalers be secured with the form ties and dogs or clips on both sides of the formwork prior to the attachment and utilization of the form brackets.

This is a critical area for the Competent Person. As indicated above, formwork erection and the placement of concrete is a quickly progressing operation and the scaffolding must be inspected prior to allowing persons to use it. It is very easy for workers to hook these metal brackets on whalers that have not been properly nailed or supported and which do not have the snap ties secured on both sides of the wall. This can result in the scaffold collapsing. Again, these are designed as relatively light duty scaffolds and the allowed spacing of the men in control of the equipment on these platforms must be carefully controlled. Both the system form bracket and the bracket for conventional formwork contain provisions for inclusion of guardrails along the length of the scaffold. At the end of the scaffold or where corners are turned, special attention must be devoted to properly supporting the guardrail and closing off the ends of the scaffolding.

On occasion, wood is used to construct these form brackets. Although these wooden brackets are normally constructed of 2″ by 4″ material on 4′ centers, the possibility of the 2″x 4″ failing by bending and shearing in the middle is greatly increased. The outrigger bracket or ledgers should be nailed to the form stud above a whaler that is properly secured. An upright perpendicular to this is attached to the bracket and a knee brace is utilized to support both the outrigger bracket and the upright for the guardrail. The guardrail is extended sufficiently above the outrigger to provide attachment of two guardrail rails.

An easier method of construction which provides a more substantial work platform is to extend two 1″ by 6″ boards for the bearer, nailed one on each side of a form stud and on top of the whaler. The 2″ by 4″ upright is placed between these two 1″ by 6″ boards and two 1″ by 6″ knee braces are nailed from the formwork up to the bearer attaching to the upright. These knee braces should be at approximately a 45-degree angle and should be cut to fit flush under the bearer and flush to the formwork. To provide stability for the vertical guardrail post, a 2″ by 4″ can be cut at a reverse 45-degree angle down to the bearer of the upright. This provides a significant stability to the guardrail posts. Constructed in this manner, center to center spacing of 8′ is adequate. The planking requirements for this are the same for any other scaffold platform. The considerations are the same in that the formwork must be adequately supported from overturning and that the concentration of workers and equipment in any one area is carefully monitored and limited to the rated load of 25 pounds per square foot.

A similar configuration of brackets is used in what's referred to as a metal bracket through the wall use. In this configuration a pre-fabricated metal bracket or wood structure is provided with a through bolt. A hole is drilled through the wall, the bolt is extended, and a nut and bearing plate is secured on the inside of the wall. For a wood stud wall, a member must be provided between the studs in order to provide adequate support. This is normally two 2″ by 4″s.

The four-inch ledge is placed in the direction of the wall, with the bolt extending between them and secured. It is also necessary to securely fasten the 2" by 4" to the studs in a vertical direction so that they may help in assisting the support. The sheeting on the wall must be of substantial nature and structural quality. A minimum of half-inch plywood should be utilized for wall sheeting in order to provide the vertical support. The bracket is required to be nailed to the plywood. Depending upon the bolt for vertical support is not adequate. The bolt only supplies a horizontal tension force to hold the bracket against the wall, and nails or other securement must be made in the vertical direction.

These brackets are also equipped with provisions for installing guardrails of 2" by 4" posts and adequate back rails. As with formwork scaffolds, the stability of this unit depends on the stability of the wall. Tall walls must be adequately braced to resist the overturning moment.

Roof Brackets

Roof brackets are metal supports that are placed on a pitched roof to provide the ability to support planking that is level. These provide a work platform for workers on the roof who are working around or adjacent to a vertical protrusion through the roof such as a chimney or dormer. These are secure through the roof to a structural member such as a rafter or metal piece. The fall height from the eave of the roof on which they are working exceeds the allowable requirements for fall protection. Either guardrails or a fall arrest system must be employed when roofing brackets are being used.

Most of the roofing brackets are designed so that they can be placed under the shingles and nailed into a rafter and then be disconnected from the nail through a keyhole-type slot and the nail driven flush and left in place. This, thereby, provides a seal for the roof and does not affect the watertightness of the roof. Adequate nailing for these roof brackets must be provided. Because the nailing is not visible underneath the shingles without additional effort to inspect and raise the shingles, the installer must carefully assure himself that the nail is located within the narrow end of the keyhole type slot. The Competent Person should inspect these nails on a regular basis.

The Occupational Safety and Health Administration has produced an OSHA instructional standard entitled, "Interim Fall Protection Compliance

Guidelines for Residential Construction." It is standard STD 3-0.1a and as a version of the instructional standard 3.1. This standard provides the fall protection requirements when working on residential structures and is significant in that it permits the use of a slide guard on pitched roofs. Slide guards are similar to the roof brackets. However, instead of providing a working platform that is level, a slide guard is attached in much the similar manner as a roofing bracket and holds a wooden member, in most cases, a 2" by 6" perpendicular to the roof. This is intended to prohibit the sliding of materials and workers off the roof. This instructional standard, that is different than the OSHA regulations, provides guidance in determining what elevations these slide guards are used and in what configurations they may be used. When reading the instructional standard, remember these apply only to what is defined as residential construction and they do have restrictions prohibiting use in certain configurations and above certain eave heights.

Part III

Suspended Scaffolds — General Information

Introduction

As the name implies, suspended scaffolds are suspended from overhead supports by wire rope or, in a very few cases, manila rope may be used. There are numerous types and configurations of suspended scaffolds. They share a commonality in that they are suspended and depend upon the strength of the rope to support the loads imposed upon them. The reader who is working with suspended scaffolds should study the materials section to understand the properties of wire rope and the fittings associated with wire rope, as well as the types of connections.

Support

Suspended scaffolds all depend upon the overhead structure for support. The weight of the scaffold and components can be more accurately determined than the load. The regulations 1926.451(a) require: "Each scaffold and scaffold component shall be capable of supporting without failure its own weight plus 4 times the maximum intended load applied or transmitted to it." This regulation separates the load of the scaffold itself from the intended load imposed on it. This permits the utilization of materials for the scaffold without the application of a safety factor. This presents no concern for most scaffolds, however, if the weight of the scaffold becomes a significant portion of the total load, the overall safety factor is reduced. For example, if the weight of the scaffold and its components, hoists, cables, platform, etc., equals ½ of the intended load, the general safety factor is reduced from 4 to 3. Designers and erectors should be aware of this and carefully monitor the condition. The weight of the scaffold and components must be carefully determined. In some cases it may be beams or columns in a building, girders in a bridge, or cornices or parapets on the edge of a building. Each of these structures must be carefully analyzed in order to determine that proper and adequate connections are made between the wire rope and supporting structure.

Structure Attachment Points

There are numerous ways to attach the support ropes to an overhead structure. Most commonly, beam clamps are used on overhead beams and girders. Clamps with fittings for attaching to the rope are attached to the beam and provide a secure attachment point.

Cable or wire rope may be wrapped directly around the beam. In these cases, padding or softeners are required around the beam to prevent the cable from becoming kinked or cut as it passes around the beam.

It is also possible to drill holes in the structural members and attach special fittings into the beams or other important parts of the structure. It is important that if there is any question about the ability of the above structure or the attachment method to support the load, qualified professional engineers should be consulted to determine an adequate means of attachment to the existing structure and to review the existing structure for adequacy of the attachment.

Loads

The load imposed on the support cables of a 2-point suspended scaffold can easily be determined. The uniform load will distribute equally to each cable. Concentrated loads will be distributed to the cables on the basis of their distance from the cables.

Scaffolds supported by three or more wire cables in a straight line require complex calculations to determine the load distribution. If the work platform is rigid across all three supports, slight movement of one support cable can transfer significant loads to the other cables unless a means of equalizing this load is provided. This can be accomplished by the use of a hinge in the support structure that provides support, but does not allow significant weight transfer. For scaffolds supported by four or more cables without a means to equalize the load, it must be considered that the entire load is transferred to just two cables. Changes in the length of the rope can effectively leave the scaffold balanced on only two ropes, thus doubling the expected load on each rope.

WEIGHT AND STABILITY

Because all of these scaffolds are supported from wire rope, which is a flexible material, the stability of the scaffold in a horizontal direction is of major concern, hence the reference used to "swing stage". On all suspended scaffolds, a means of providing stability in both horizontal directions must be provided and utilized by the workers. Each type of scaffold requires unique precautions and will be discussed as part of specific analyses in subsequent chapters.

Each suspension rope and its connecting hardware for directly supported scaffolds, those without a hoisting mechanism, must be capable of supporting without failure six times the maximum intended load. This requirement does not exclude the weight of the scaffold from the factor of safety of 6. The total weight of the scaffold, platform, and the intended load must be included to determine the load on the suspension ropes.

Hoisting Devices

Scaffolds that are used with an adjustable mechanism, hoists, etc., are governed by different requirements. In these scaffolds, the required suspension rope capacity is determined by the rated capacity of the hoist. The weight of the scaffolds and its components do not enter into the calculation. An additional factor to be considered is the stall load of the hoist.

Many suspended scaffolds have some kind of powered device to adjust the elevation of the scaffold in a vertical direction. These motors are rated and controlled by the Underwriters Laboratory Standard 1323. This standard can be purchased from Underwriters Laboratory, www.ul.com. There are several important numbers that must be considered when selecting the hoist and specific standards for the rating of these hoists. One of the important numbers to consider is the rated capacity of the hoisting device. This must be at least equal to the load imposed by the scaffold components, the workers, and the materials.

Stall Force

There is another very important number, which is the stall force. This is the maximum amount of force the winding motor can deliver at stall. This force must be considered because it is the maximum amount of force that the motor can put on the anchorage if the scaffold becomes jammed or stuck and the motor continues to run. The anchorage must always be able to support a minimum of two times the stall speed of the device used for hoisting. Hoisting devices are limited to stall load of three times its rated load. They are governed by Underwriters Laboratory Standard 1323. Installing a large over-capacity winding hoist, while it may appear to be a safety factor, can actually become a major hazard if the stall force of the motors is greater than the anchorage system can accommodate. The scaffold may stall and cause the motor to overload the anchorage resulting in failure.

Fall Protection on Suspended Scaffolds

According to OSHA, the number one scaffold hazard is worker falls. Fall protection consists of either personal fall arrest systems or guardrail systems, and must be provided on any scaffold 10 feet or more above a lower level. Two-point scaffolds require both a Personnel Fall Arrest System (PFAS) and guardrail systems. This is especially critical with suspended scaffolds, because they are often operated at extreme elevations. Note: Except where indicated, these requirements also apply to multi-level, single-point adjustable, multi-point adjustable, interior hung, needle beam, catenary, and float (ship) scaffolds.

In addition to meeting the requirements of OSHA CFR 1926.502 (d), personal fall-arrest systems used on scaffolds are to be attached by lanyard to a vertical lifeline, horizontal lifeline, or scaffold structural member. Note: Vertical lifelines may not be used on two-point adjustable suspension scaffolds that have overhead components such as overhead protection or additional platform levels.29CFR 1926.451 (g) (3)

- When vertical lifelines are used, they must be fastened to a fixed safe point of anchorage, independent of the scaffold, and be protected from sharp edges and abrasion. Safe points of anchorage include structural members of buildings, but not standpipes, vents, electrical conduit, etc., which may give way under the force of a fall. 29 CFR 1926.451 (g) (3) (i)
- There are many types of rope and cable grabs that can be used with vertical lifelines. One must use them in accordance with the manufacturer's instructions and recommendations. Some general advice to keep in mind is to keep the connector length as short as possible and to allow no more than a maximum free fall of 6 feet. Remember if your rope grab is the follow-along type, then the length of the lanyard can be only 3 feet to insure a maximum free fall of 6 feet. Carefully examine the rope grab before attaching it to the line to ensure that it is put on the line with the top up. If it is put on the line upside down it will not arrest your fall. Make sure the rope grab that you use is of the correct type and size of recommended line. Generally rope grabs should be positioned as far above the user as possible when in the fixed or working position.
- It is dangerous and therefore impermissible for two or more vertical lifelines to be attached to each other, or to the same anchorage point. 29CFR 1926.451 (g) (3) (iv)
- When horizontal lifelines are used, they are to be secured to two or more structural members of the scaffold. 29CFR 1926.451(g)(3)(ii)
- When lanyards are connected to horizontal lifelines or structural members, the scaffold must have additional independent support lines and automatic locking devices capable of stopping the fall of the scaffold in case one or both of the suspension ropes fail. These independent support lines must be equal in number and strength to the suspension ropes. 29CFR 1926.451(g)(3)(iii)
- On suspended scaffolds with horizontal lifelines that may become vertical lifelines, the devices used to connect to the horizontal lifeline must be capable of locking in both directions. 29CFR 1926.502(d)(7)

Guardrail Systems

Guardrails are the primary means of providing fall protection for workers on elevated work platforms. They are required on all open sides or ends and should completely enclose the work platform except for the few exceptions when the work face may be left open. A majority of falls related to scaffolds occur, not because of scaffold collapse, but because of improper planking or lack of complete, adequate guardrails.

- Guardrail systems must be installed along all open sides and ends of platforms, and must be in place before the scaffold is released for use by employees other than erection/dismantling crews. 29CFR 1926.451(G)(4)(i)
- Each toprail or equivalent member of a guardrail system must be able to withstand a force of at least 200 pounds applied in any downward or horizontal direction, at any point along its top edge. 29CFR 1926.451(g)(4)(vii)
- The top edge height of toprails on support scaffolds must be between 36 inches and 45 inches. When conditions warrant, the height of the top edge may exceed the 45–inch height, provided the guardrail system meets all other criteria. (Note: The minimum top edge height on scaffolds manufactured or placed in service after January 1, 2000 is 38 inches). 29CFR 1926451.(g)(4)(ii)
- Midrails, screens, mesh, intermediate vertical members, solid panels, etc., must be able to withstand a force of at least 150 pounds applied in any downward or horizontal direction, at any point along the midrails or other member. 29CFR 1926.451(g)(4)(ix)
- When midrails are used, they must be installed at a height approximately midway between the top edge of the guardrail system and the platform surface. 29CFR 1926.451(g)(4)(iv)
- When screens and mesh are used, they must extend from the top edge of the guardrail system to the scaffold platform, and along the entire opening between the supports. 29CFR 1926.451(g)(4)(v)
- When intermediate members (such as balusters or additional rails) are used, they must be no more than 19 inches apart. 29CFR 1926.451(g)(4)(vi)

- Guardrails must be surfaced to prevent punctures or lacerations to employees, and to prevent snagging of clothing, which may cause employees to lose their balance. 29CFR 1926.451(g)(4)(xi)
- Ends of rails may not extend beyond their terminal posts, unless they do not constitute a projection hazard to employees. 29CFR 1926.451(g)(4)(xii)
- In lieu of guardrails, crossbracing may serve as a toprail or midrails, providing the crossing point is:
- Between 20 and 30 inches above the work platform for a midrails, or
- Between 38 and 48 inches above the work platform for a toprail. 29CFR 1926.451(g)(4)(xv)

Erectors and Dismantlers

The process of erecting and dismantling a scaffold is an extremely dangerous operation. As the components are assembled, workers are required to be in areas that by definition are not complete. Workers engaged in scaffold erection and dismantling require special training in the means and methods of erecting the scaffold. The manufacturer's instructions must be on site and each worker must be familiar with these instructions and understand his role in the erection sequence. Because of the incomplete state of the scaffold during the erection and dismantling procedure, there are special needs for fall protection.

- The fall protection requirements for employees installing suspension scaffold support systems on floors, roofs, and other elevated surfaces, are described in 1926 Subpart M, the Fall Protection standard [note to 29CFR 1926.451(g)(1)].

Competent Person

As with many operations during the use of the scaffold, including daily inspections, there is a need for the presence of a Competent Person during erection. The primary responsibility of the Competent Person during erection and dismantling is the safe and complete assembly and dismantling of the scaffold. During the erection and dismantling of the scaffold, the Competent Person is responsible for the safety of the workers and their fall protection system.

The employer must designate a Competent Person who would be responsible for determining the feasibility and safety of providing fall protection for employees erecting or dismantling supported scaffolds. 29CFR 1926.451(g)(2)

Chapter 14

Single-Point and Two-Point Suspension Scaffolds

Introduction

Two-point suspension scaffolds are frequently used in the construction industry. The most popular of the two-point variety is the swing stage. The overall support requirements of the two-point suspension scaffolds require that they be capable of supporting four times the reaction forces imposed by the load on the scaffold or 1½ times the stall force of the hoisting motor (Figure 14.1).

Methods of Support

A common method of supporting a two-point suspension scaffold is by the use of a cornice hook, a c-shaped device that hangs over a parapet on a building and/or outrigger beams which are positively secured to the roof or counterweighted and extend out over the edge. Cornice hooks must not be used on any cornice or parapet that is not designed to withstand the horizontal load imposed by this type of hook. See Figure 14.2.

It is also imperative that the cornice hook be tied back to a fixed object on the roof of the building and be secured so that it cannot pull forward or pull over the edge of the building. The securement must be at least equal to

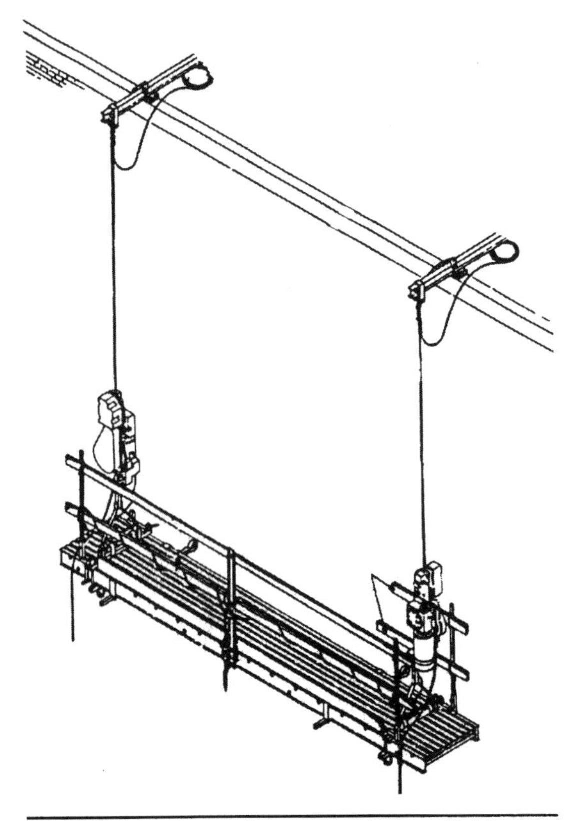

FIGURE 14.1 Two-point suspension scaffold

the load on the scaffold. Pads or blocking placed under the tip of the hook should be adequate and well placed. Hard wood blocking or padding must be used. Soft woods, when loaded with the point of the cornice hook, will split.

Outrigger beams are generally steel beams projecting out from the face of the building. Outrigger beams can be secured by a positive fastener directly to the structure of the building by bolting or clamping. This securement point must be capable of supporting at least four times the intended rated load of the hoist. Alternately these outrigger beams may be of a cantilever type, cantilevered over the edge of the building and counter-weighted sufficiently so that the overturning moment is greater than four times the rated capacity of the hoist. Both the counter-weighted and secured outrigger beams must be tied back to a secured device on the surface of the roof.

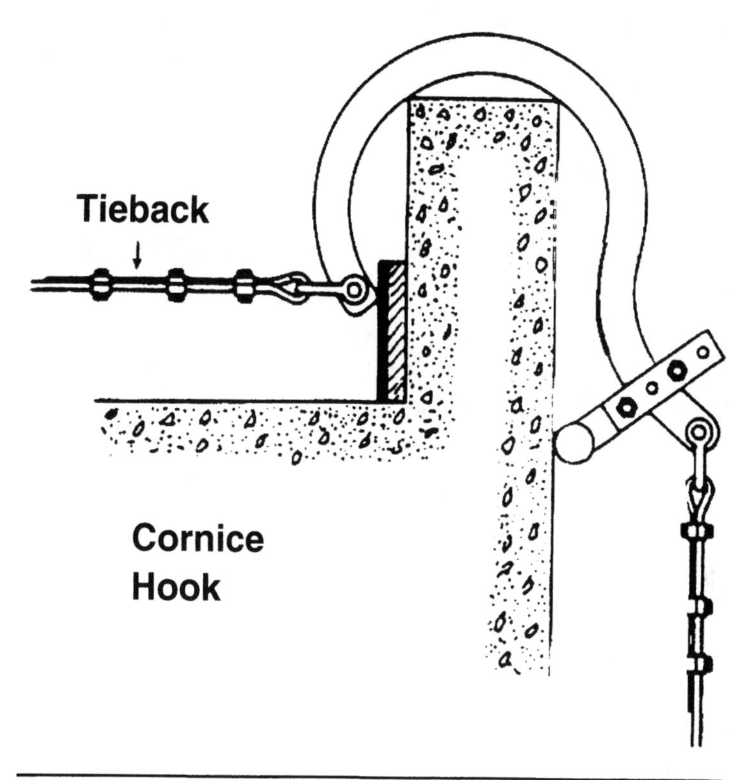

FIGURE 14.2 Cornice hook

The material used as a counter-weight must be secured to the outrigger by a positive means. Loose metal bars or concrete weights may not be used unless secured to the outrigger beam. Flowable materials, such as bags of sand or barrels of water, are not to be used as counter-weights. Their flowability allows them to be easily displaced and the effective counter-weight removed. Water containers may leak or sand may flow through a hole in the bag, and, over time, the effect of the counter-weight is lost.

The suspension ropes must be securely attached to the outrigger beams using shackles or pins as outlined in the section on materials. Shackle pins must be secured by wire to prevent unscrewing. Pins must be provided with a locking clip for securement. The suspension ropes must hang vertically above the support points on the platform.

Great care must be exercised to establish the outriggers at precisely the same spacing as the motors on the scaffold platform so that the load lines hang vertically. When the load lines hang vertically, the load is equal to the weight of the scaffold. If the load lines are not vertical, a horizontal component is induced into the load line. A six-inch difference in spacing may not seem significant when the load lines are one hundred feet long hanging at the lowest work position. This 6 inches results in just 0.5% of the load as a horizontal force. As the scaffold is elevated, the force increases because the load lines are shorter. At ten feet, the horizontal force is 5% of the load on the scaffold. A 1000-pound scaffold would generate 5 pounds of horizontal pull with 100 feet of line extended and 50 pounds of horizontal force at 10 feet of line extended. This is sufficient to cause the cable into the motor to jam. This would also cause a horizontal load on the outrigger and may cause it to buckle, twist, slip off the supports, or otherwise fail.

Outriggers must be suitably secured to maintain this distance and must also be secured in a manner to allow them to remain parallel. If an eyebeam-type outrigger is utilized, the web of the eyebeam must remain in a vertical position. If the load lines are not vertical, horizontal forces are developed in these lines that tend to pull the outriggers together or pull them apart. Conversely, the opposite is occurring at the hoist motor and the suspension scaffold. What may seem like a very small angle at the end of a six or eight story suspended cable becomes very acute and severe as the scaffold is raised, developing significant forces as the scaffold approaches the top. While the angularity, i.e., out of plumbness, of the suspension ropes will tend to add stability longitudinally to the scaffold; it must be avoided because of the adverse affect.

Preventing Scaffold Sway

In addition to providing the support, there needs to be means to prevent scaffold sway. Some of these include angulated roping and static lines. With angulated roping, the upper points of suspension are set closer to the face of the building than to the attachment points on the platform. This causes the platform to press against the face of the building. The major problem with utilizing the angulated roping is the fact that the higher the scaffold rises, the greater the force pressing it into the structure. Static lines may also be used for this purpose.

Static lines are separate ropes secured at the top and at the bottom, closer to the plane of the building face than the outermost edge of the platform. By attaching the platform and drawing the static line tight, the platform is pulled to the face of the building.

The suspension rope and attachment point must also be located at a distance from the face of the building equal to the pickup points on the scaffold platform itself is located. The scaffold will tend to hang directly below the support point. If they are not equidistant, too much overhang will cause the scaffold to hang away from the wall, providing an excessive clearance between the face of the scaffolding and the wall. A slight amount of angle (i.e., angulated rope) is acceptable, but not recommended. Hang the load lines vertical and rely on a static line or ties to the building.

Insufficient clearance will tend to pull the scaffolding into the wall creating drag and excessive force on the rollers of the scaffold. It also greatly increases the tension in the line, which occurs gradually as the scaffold is elevated. This also increases the force on the tie back structure. It is possible while raising a scaffold whose outriggers are not sufficiently extended that, as the scaffold approaches the elevation of the outriggers, sufficient force is generated by the out-of-plumb suspension lines to physically pull the outriggers from the building.

The maximum amount of force allowed to be generated (per OSHA) by this angularity of the suspension lines is 10 pounds. Remember, this force will increase as the scaffold is raised. The 10-pound figure is the maximum allowed at the highest work level. An effective way to provide stability is through the use of a static line secured at the top and the bottom, parallel to the face of the building. See Figure 14.3.

Platforms

Platforms on two-point suspension platforms must be securely connected to the bearers or stirrups. Two-point suspension scaffolds shall not be bridged or otherwise connected to one another during raising and lowering operations unless the bridge connections are articulated and secure.

The danger in bridging between various suspension points is that the scaffolds cannot be kept in a perfectly straight line and even a minute deviation can cause the center or the end supports to accept as much as twice the load to which it is normally subjected, resulting in failure. If the scaffold is hinged and bridged over three support points, the center support point will have twice

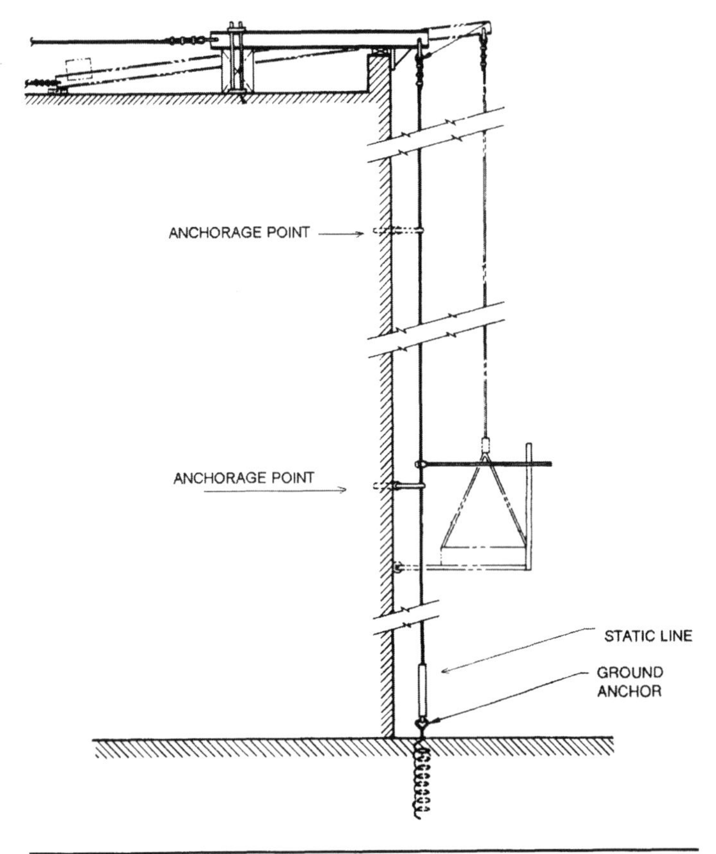

FIGURE 14.3 Static line

the load as the two end support points. Articulation or hinging in this area at the center support allows the loads to distribute in their proper manner and maintains the two scaffolds in contact.

When in use, two-point suspension scaffolds must be level within one inch for every foot of platform length. It is advisable when raising or lowering them to maintain the platforms in a level position. Differences in elevations exceeding one inch per foot of platform length is the maximum that should be tolerated. One inch in a foot is a relatively even slope, however, with a 20- or 24-foot platform, even the 1-inch-per-foot constitutes a significant difference in elevation. Also, the larger the difference in elevation, the more the suspension cables are pulled out of a vertical alignment. It also makes it difficult to walk up the plank. Slipping is extremely probable and materials and tools can slide and fall from the planks.

Multi-level suspended scaffolds must be equipped with additional independent support lines equal in number to the number of points supported and of equivalent strength to the suspension ropes. These independent support lines provide support for the scaffold in the event that the suspension ropes fail. This arrangement keeps the upper scaffold level from coming down onto the lower level and endangering workers on the lower level. Workers tie on to the scaffolds. If done correctly and there is a suspension rope failure, these independent lines carry the weight of the scaffold and workers have the opportunity to exit the scaffold in a safe manner (Figure 14.4).

In order to provide security for these independent support lines, they shall not be attached to the same points of anchorage as the suspension ropes. Tying the independent lines off to anchorage points other than those utilized by the suspension ropes, enforces the integrity of the system and provides safety for the workers. Supports for these scaffold platforms shall be attached directly to the support stirrup and NOT to any other platform.

Single-Point Suspension Scaffolds (Boatswains' Chair)

Single-point suspension scaffolds are used for a variety of maintenance, repair, and construction work. They are designed to provide access for one person and are limited to a maximum of 250 pounds. All of the requirements for the two-point adjustable scaffold for anchorage, tiebacks, and independent lifelines, which apply to the two-point suspended scaffold, also apply to the Boatswains' chair or single-point suspension scaffold (Figure 14.5).

Definition and Requirements

Some of these scaffolds consist of merely a seat suspended from a manila rope, which provides the worker the ability to descend along the face of the work area. Although it is advisable to keep the suspension rope vertical, there are provisions where this type of scaffold can be used with the rope merely supported

FIGURE 14.4 Multi-level suspended scaffold

over the edge of the building or structure and being at a slight angle as the worker proceeds down the face of the building. This provides a significant force holding the worker to the building area. The requirement for this type of suspension must be designed by a Qualified Person, as defined by the Occupational Safety and Health regulations. A $\frac{5}{8}$-inch diameter manila rope, or equivalent, is required for the suspension rope for this type of scaffolding.

Where the rope extends over the edge of the building, provision must be made for adequate padding and/or protection to keep the rope from chafing

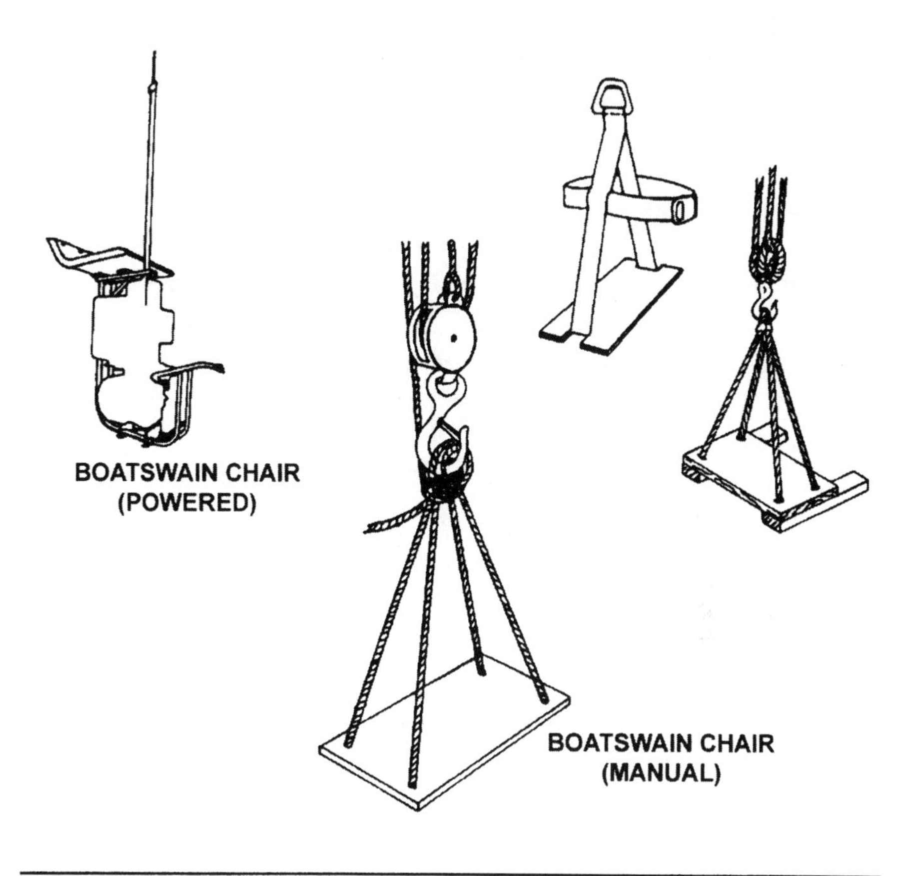

FIGURE 14.5 Boatswains' chair

or wearing. The tieback of the suspension line to a solid support is obviously a critical issue when using a Boatswains' chair or single-point powered suspension scaffold. The securement and proper use of an independent lifeline, and attachment to it, are also critical issues. The blocks, which support the upper portion of the boatswains' chair, must be of a proper size and with adequate bearings as defined in the Occupational Safety and Health regulations.

It is required that the seat of the boatswain's chair be supported by seat slings consisting of a minimum of two, 5/8-inch manila ropes of equal length which cross underneath the seating platform and cross coming up on opposite corners. Having two independent ropes supporting the seat provides for a redundancy should one of the ropes become fatigued or damaged and fail. Where welding or other heat-producing processes are performed, it is required that the seat slings be constructed of 3/8-inch wire rope. The seat itself must be a minimum of a full one-inch thick wood board, or 5/8-inch cross-laminated material such as a marine plywood.

Powered Platform

Many single-point suspension scaffolds are powered platforms, which consist of baskets with guardrails for the worker to stand in and a power-type hoist. The hoist must comply with the standards as defined previously and the suspension system must comply with the regulations as promulgated for the tie-point suspension system outlined previously, except that a qualified person may design a system where the rope is not vertical but is angulated to allow suspension over the edge of the building.

This, to a large extent, limits the accessibility of a powered stage to the top work areas and it is highly recommended that outrigger attachments be provided so that the rope line hangs vertically. As with both single suspension units, the load lines are designed to hang in a vertical position and should not be pulled or swung from the vertical position. This displacement horizontally creates excessive loads on the hoisting mechanism as well as horizontal loads on the support system or the rope where it crosses over the support system. This is not a permitted or recommended procedure.

ANSI A10.8 (2000) also requires in Section 6.4, Electrical Wiring and Equipment:

- All electrical wiring and controls shall be in accordance with ANSI/NFPA 70-1993, ANSI/NFPA 70B-1990. ANSI/NFPA 70E-1988.
- The power supply cable to all hoists shall contain a separate conductor that will serve as a grounding connection for the hoist. All metallic junctions shall also be grounded.
- Strain relief devices or equivalent means shall be provided for power cable connections that may be put under tension during normal operation or while the scaffold is moved from one location to another. Strain relief devices or equivalent means shall also be provided for the power receptacles where the cables are plugged in. The power receptacle connecting to the scaffold equipment shall be fastened to the scaffold so that no strain can be placed on the connection.

Chapter 15

Multiple Point Suspension Scaffolds

Introduction

Multiple point suspension scaffolds come in various configurations. Mostly they are supported by a wire rope mechanism from an overhead structure and are at a fixed elevation.

These scaffolds, because they are suspended from flexible materials, have a tendency to sway when they are in use. There needs to be a means of stabilization to eliminate the sway inherent in this type of platform. Workers moving on the scaffold, tightening bolts, hammering, and performing other work, can generate sufficient force to sway the platform significantly out of a vertical position with the suspension ropes. For those reasons, a means of securing it is highly recommended. These can be flexible ties such as wire rope on opposite sides to an existing structure, or it can be a rigid tie in one direction.

The Interior Hung Scaffold

A popular type is the interior hung scaffold (Figure 15.1). This is generally a large area scaffold that is hung from overhead ceilings or roof supports by fixed length cables, usually wire rope. The roof structure must be carefully analyzed to determine that the structure itself is capable of supporting the intended weight on the scaffold. The support wire ropes are required to be capable of supporting six times the intended load on the scaffold and this number should be used for the analysis of the supporting structure. If the scaffolding is supported by more than three ropes, as previously discussed, without an equalizing system, all of the load of the scaffold may be transferred to two of the four ropes. This reduces the factor of safety by 50%, from a factor of six to a factor of three. Scaffolds supported at more than four points are even more subject to this condition, without some articulation in the platform.

Modifications of the supporting structure, i.e. drilling holes, welding brackets, must be approved by a professional engineer to assure that these modifi-

FIGURE 15.1 Interior hung scaffold

cations do not have a detrimental effect on the permanent structure. In many cases a beam clamp is attached to the bottom of wide flange beams. The beam itself needs to be analyzed by an engineer to determine that the flange of the beam is capable of supporting the additional load imposed by the scaffold. The suspension ropes must remain in a vertical position. Attaching to members of the structure that are farther apart or closer together than the platform width induces a bending or horizontal force on the support structure in addition to the vertical force generated by the weight of the scaffold. The structure may not be capable of resisting this horizontal force, which could cause buckling or failure. This is particularly true of I beam type members and open web steel members such as bar joists. Imposing horizontal loads must not be permitted without a specific analysis by a structural engineer.

The platform shall be supported by a minimum of 2″ by 9″ timber on edge used as a supporting bearer. The wire rope shall be securely fastened to the bearer. It is advisable to provide at least two laps of wire rope around the bearer if it is wrapped and also around the support member. Care should be taken with the support member so that the rope is not kinked or cut by sharp edges and consideration given to the upper flange strength of I beam members when supporting loads. The wrapping of cable around the flange of a beam can cause the flange to roll down under the load, resulting in failure of the beam.

All interior hung scaffolds require standard guardrails on all exposed sides, as well as wire mesh installation from the mid-rail to the platform to protect persons who may be passing below from falling objects.

FLOAT SCAFFOLDS

Float scaffolds are usually work platforms supported by four manila or fiber ropes (Figure 15.2a and Figure 15.2b). They are limited to a maximum of three-person occupancy and have a minimum platform size of 3′ by 6′. Normally the platform has two bearers which strengthen the platform and extend six inches past the platform for the attachment of the ropes. It is customary to utilize four ropes, each attached to a corner of the bearer to suspend the scaffold up to the overhead structure. It is possible, if two manila ropes are available, to rig the scaffolding with the rope so that it attaches to the bearer, crosses under that platform, and attaches to the bearer on the opposite end, leaving two ends to tie up to the support form. With two ropes, there are four ends to attach two support points. The platform is normally ¾-inch plywood or equivalent. Just as with interior hung scaffolding, stability becomes a major concern. These scaffolds must be protected from swaying by either tying or a rigid guide.

Float scaffolds are not normally equipped with guardrails, but it is required that each person on the float scaffold have and utilize fall protection. This requires an independent lifeline for each person working on the scaffold.

CATENARY SCAFFOLDS

Catenary scaffolds are supported by a horizontal wire (Figure 15.3a and Figure 15.3b). This wire is attached to an anchorage at each end and is limited to two

FIGURE 15.2A Float scaffold

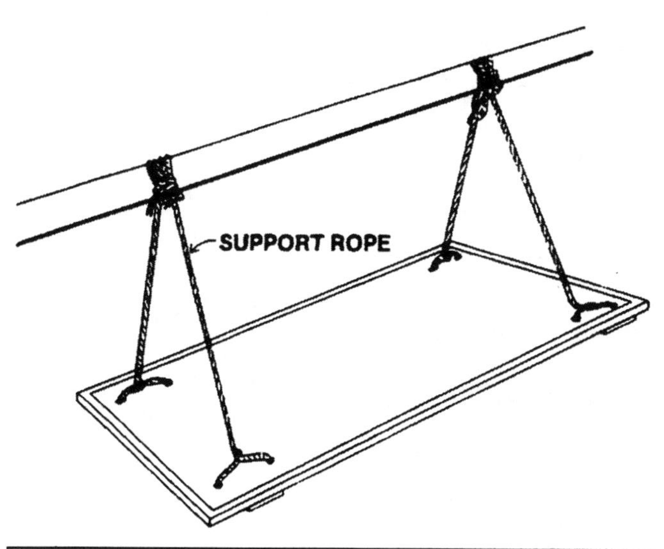

FIGURE 15.2B Float scaffold

persons or 500 pounds. The horizontal support wire rope must not be tensioned too tightly. A sufficient amount of drape or sag must be permitted in order to allow a vertical component to develop in the wire. For a catenary scaffold without intermediate supports, a slope of approximately one in ten is necessary to provide the required vertical component without overstressing the wire rope. If intermediate supports are utilized, they will sustain most of the vertical load and the catenary wire may be much flatter. An intermediate support must be used if more than one platform is attached to the catenary system. A platform on each side of the vertical support is necessary. No more than two platforms may be utilized on any one catenary support.

Because of the nature of the platform, they also require a means of horizontal stabilization, much like the cloak scaffold or interior hind scaffolding. The platforms for a catenary scaffold must have hooks which would secure them to the catenary wire should one wire fail and one end of the platform drop, and also to prevent one end from slipping off the wire. All persons working on a catenary platform must have an independent lifeline and fall prevention device. Wire ropes between the anchorages must be continuous. If a come-along or mechanically tensioning device is utilized, it is recommended that a tension force of not more than 2,000 pounds be placed on the catenary.

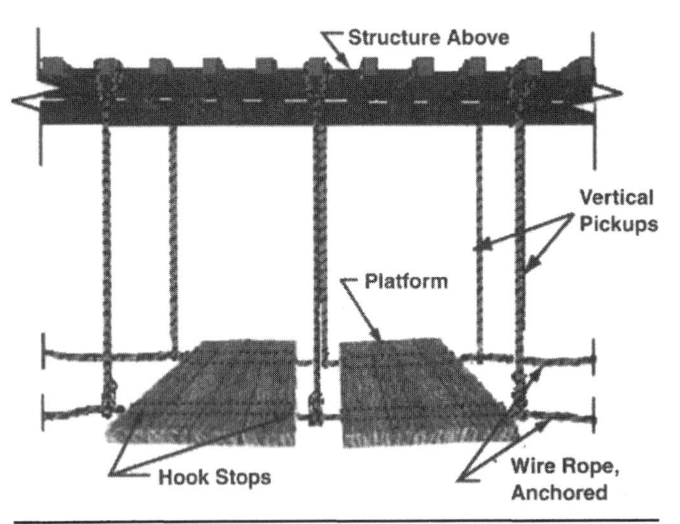

FIGURE 15.3A Catenary scaffold

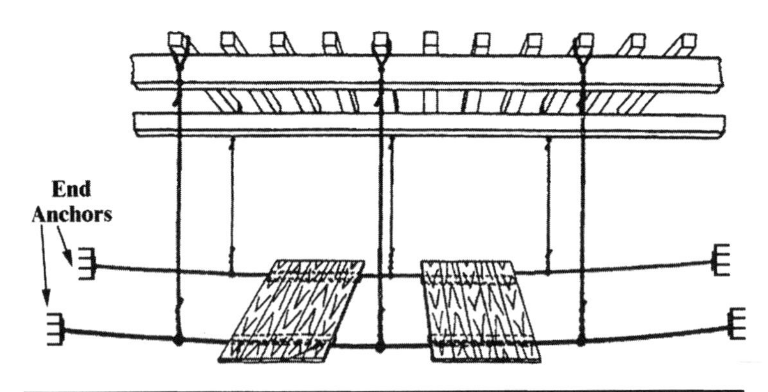

FIGURE 15.3B Catenary scaffold

Chapter 16

Adjustable Multi-Point Suspension Scaffolds (Stone Setters — Mason's)

What is commonly be referred to as a stone setters scaffold is supported by four supporting wire ropes. This type of scaffold consists of a work platform suspended between bearers. The end of each bearer is in turn supported by a wire rope.

Older manually adjusting stone setters scaffolds utilized a system with two support lines coming off of a drum. One support line was threaded, went downward and through a rigging system beneath the work platform, and returned vertically up the inside of the scaffold. The other support cable, from the top of the winding mechanism, went upwards to the overhead support point (Figure 16.1). This four-point suspension system provides greater stability for a suspended scaffold. The newer multi-point suspension scaffolds are equipped with a powered hoist. Most frequently, one hoist is utilized on each corner of the scaffold, attached directly to the beams. This configuration eliminates the rollers and rigging below the work platform.

Mason's Multi-Point Scaffolds

The mason's multi-point scaffold is of the same general configuration as the stone setters. However, it is designed for a minimum 50 pound-per-square-foot load and requires appropriate planking and spacing to support the intended load. In addition, a masons scaffold is required to have a 7/16 inch-diameter support rope (Figure 16.2). The Appendix to the Occupational Safety and Health Act regulations for scaffolding specifically suggests that a 7-inch, 15.3 pound-per-foot I beam be used for an outrigger and that its maximum length be 15 feet, with the maximum extension over the edge of the building of 6 feet 6 inches. Distances that exceed these tabular values would require that the scaffolding be designed by a qualified engineer.

On occasion, both types of scaffolding are utilized with multiple work levels. In these instances, the design should be by a professional engineer. When utilized with multiple working levels, the same criteria apply to providing fall protection to all those working on the scaffold. Independent lines should be supplied to the support points and individuals working on the lower levels should be tied to the scaffold. See the chapter on Fall Protection.

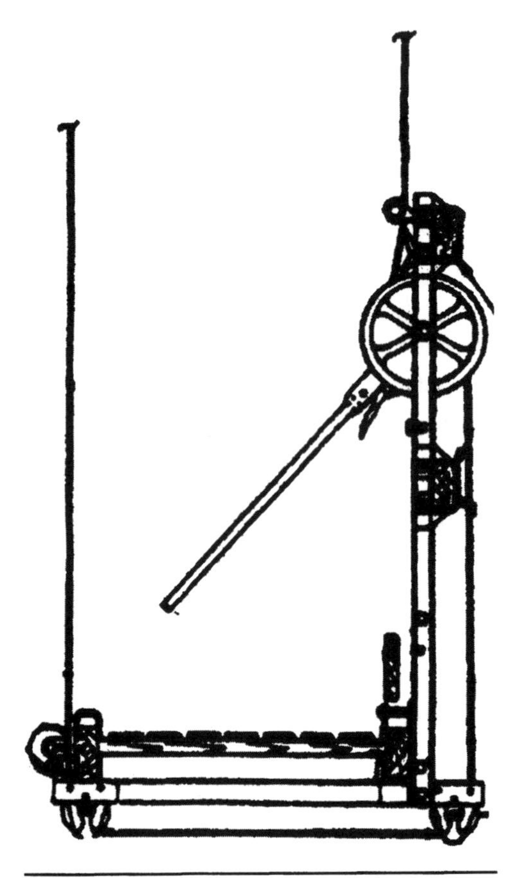

FIGURE 16.1 Manually adjusting stone setters scaffold

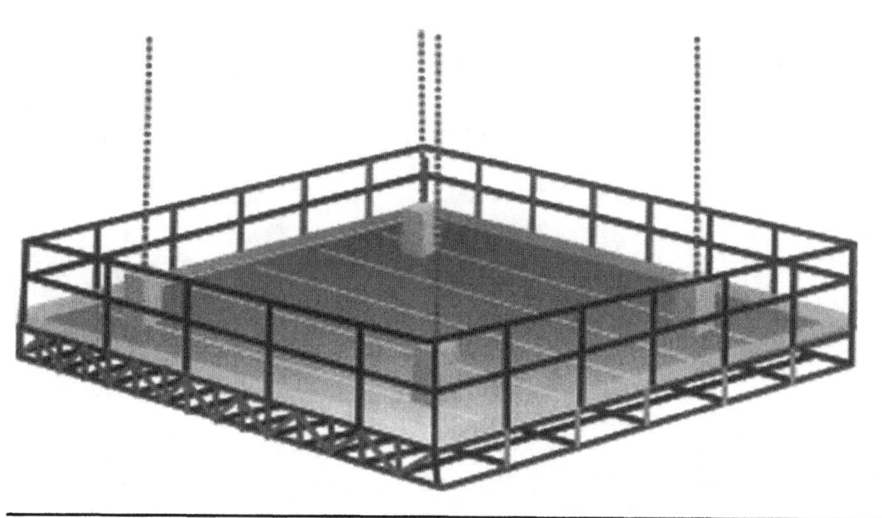

FIGURE 16.2 Masons scaffold

The stone setters arrangement with continuous wire rope provides a means of equalizing the load on each support cable. The multi-point mason's scaffold with individual suspension lines does not have that capability.

These scaffolds, as pointed out earlier, are very susceptible to allowing weight transfer to individual wires. It is possible on these four-point suspension scaffolds to have all of the weight supported by two wires. The overhead outriggers should be designed so that they may support, with reasonable factors of safety, this additional weight.

The hoisting motors should be appropriate for this usage and rated to accept these capacities. All of the requirements for the rated load of the capacity of the hoist and the outrigger structure being able to support the stall load of the hoist with appropriate factors of safety remain as they do for any suspended scaffold.

Chapter 17

Outrigger Type Scaffolding

Introduction

There are numerous configurations of scaffolding that are utilized to work outside of the face of a building or structure that depend upon the structure itself for at least partial support of the scaffolding. The point of support or attachment to the building is always a significant issue in the stability of the scaffolding. A competent person must inspect this area on a regular basis. Because many of these scaffolds are constructed of wood that is a commonly available material on the jobsite, all of the components must be securely fastened to the structure. Inadequately secured components are easily removed by workers who may be looking for materials on the jobsite. This type of scaffolding is usually erected for a specific purpose and mostly for short duration use. It should be removed from the area as soon as work is complete. These types of scaffolds must also be included as a part of the tagging system and inspected daily.

Needle Beam Scaffolds

Needle beam scaffolds consist of two or more bearers. One end of each bearer is supported on the structure, while the other end is supported outside of the structure by a manila or wire rope attached to an overhead structural member (Figure 17.1). A substantial platform is then attached across the bearers and secured. It is not permitted to utilize cleats or other stop devices. The connection between the platform and needle beams must be a secure connection, nailed or bolted. It is most common to use 4" by 6" structural grade wood members. For bearers, the 6" dimension is maintained vertical. On occasion, it is possible to use equivalent metal members, small eye beams, or channel shape sections. Whatever the shape or the material of the members, they must be secured in a vertical position to keep them from rolling over. It is also advisable to a secure the bearers to the permanent structure to prevent movement in either horizontal direction. It is more important that they not be permitted to slide from the support ledge causing the scaffold to fall. The better the securement to the permanent structure the more stable the platform.

Workers on a needle beam scaffold must wear fall arrest devices consisting of an adequately anchored independent lifeline and an adequate harness as described previously.

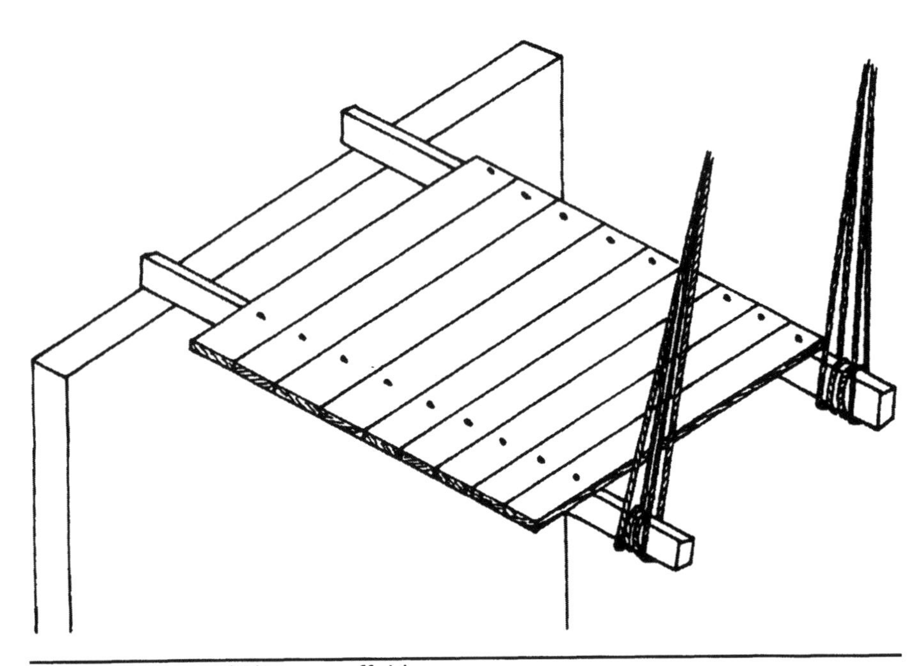

FIGURE 17.1 Needle beam scaffold

Outrigger Beam Scaffolds

Outrigger beam scaffold is a type of scaffold that is composed of a bearer that rests on a supporting member of the permanent structure. A portion of the bearer extends inside the building and a portion outside the building. The bearing area on the structure required by the OSHA regulations is a minimum of 6 square inches (Figure 17.2).

It is also an OSHA requirement that the length of the bearer inside of the structure from the fulcrum or tipping point to the inside end of the bearer be one and a half times the projection outside of the building. It is advisable not to extend these platforms more than 6 feet from the face of the building. A 6-foot extension from the face of the building would require a minimum of one and a half times the 6 feet or a 9-foot projection inside of the building. The inside end of the outrigger bearer is normally braced by struts which act in compression to the underside of the ceiling or other structural member. In many cases, where the ceiling of the structure is a finished ceiling or consists of low strength materials such as drywall, wood paneling etc., it may be necessary to provide a sill across the top of the vertical bracing in order to prevent the bracing from punching through the ceiling. Whenever direct bearing on a structure member is not possible, an evaluation should be performed by a qualified engineer to determine the capacity of such ceilings. These struts are in compression and as such are subject to bending a bracket. Depending upon their cross section (size and length), bracing is required in both directions to prevent bowing.

It is also possible, as an alternative, to secure the inside end of the outrigger beam to the floor in a downward direction. When secured in this manner the securing member is in tension. A wire rope or manila rope responds better in tension than wood members. When attempting to use wood members in tension, it should be considered that they have less strength than they do in compression and the connections between the bearer and the wooden tension member are more difficult to accomplish. The work platform must be secured, i.e. through nailing or bolting to the outrigger beams.

Although industry practice has been to utilize this type of scaffolding and it is commonly used and erected by workers on the jobsite, the Occupational Safety and Health regulations as published in 29 CFR 1926.452 (i), requires that these types of scaffold be designed by a professional engineer and erected

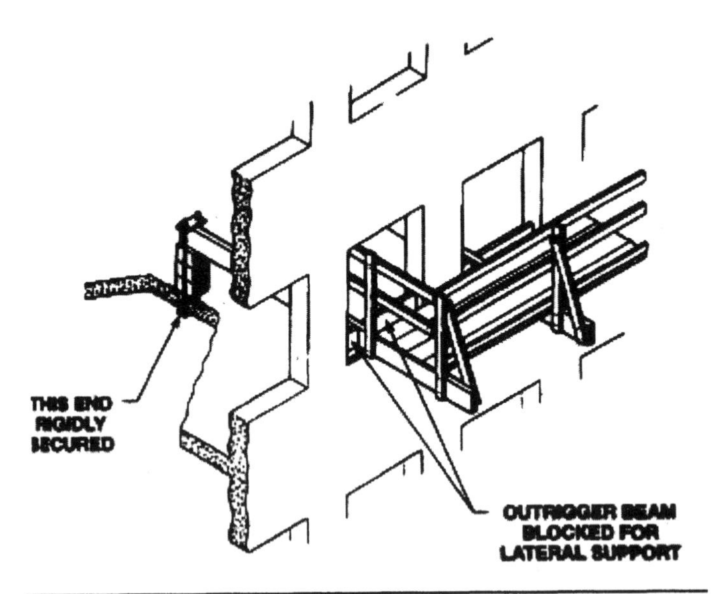

FIGURE 17.2 Outrigger beam scaffold

in accordance with that design. The design must include not only the members, but also the anchorages to the structure, as well as the ability of the structure to resist the loads imposed on it by the scaffold, the workers and their tools, and materials.

This type of scaffolding can be easily constructed with guardrails, thereby eliminating a need for fall protection. However, if guardrails are not provided, a fall arrest system must be provided with the scaffolding. Guardrails must be of a standard type and comply with the standard.

Window-Jack Scaffolds

Window-jack scaffolds are commonly a fabricated platform that extends through a window opening, hooking on to the sides of the window structure, either the wood framing or metal framing they extend outside the window opening providing a platform. Although this type of scaffolding is relatively easy to install, it depends entirely upon the stability of the structure inside the window to prevent the tipping or disengagement of the scaffold. The structure around the window and the window anchorage to the building must be inspected to assure that it is in good condition and securely anchored. Rotted or deteriorated members must be avoided. This type of pre-manufactured platform must be manufactured in accordance with the standards as indicated in the OSHA regulations and should comply with the requirements and recommendations for scaffolds as published in ANSI A10.8. These scaffolds do provide access to the area immediately around a window for maintenance, repairs, and painting.

Although these scaffolding platforms project from the opening of the window, it is not permitted to bridge or plank from one window-jack scaffold to another located in an adjacent window. This type of arrangement is prohibited.

Most of the platforms come equipped with guardrails to provide for fall protection for persons on the platform. If guardrails are not provided, an independent lifeline must be provided and the user of the scaffold must wear appropriate fall arrest equipment.

Chapter 18

Manually Propelled Rolling Towers

Various Uses

Rolling scaffold towers are very popular, particularly on construction sites, for interior finishing. These are narrow scaffolds that may be easily rolled from room to room through doorways and halls. They are also popular for many other uses. Rolling towers provide access to heights and may be easily relocated to provide access to various work areas.

Guidelines for Use

Rolling towers may only be used on smooth, level, and clear floor areas. The floor area must be reasonably level. Block outs or depressions in the floor for such things such as electrical lines, column bases, sewer lines, vertical heat ducts, and water pipes must be covered or protected when rolling scaffolds are in use. The entire work area in which a rolling scaffold is utilized must be free of holes and unevenness. Areas that are not suitable for rolling tower use should be barricaded.

Construction and Safety

Most rolling scaffolds are constructed of welded-frame scaffold sections fitted with casters or specially designed steel frame scaffolds fitted with casters. On occasion, they may be constructed of tube and coupler components fitted with casters. However, this is not the norm.

The primary consideration for safety and the cause of most accidents with a mobile tower is the erected height of the tower. Heights in excess of a height to width ratio of four to one is prohibited. This is the standard criteria for all scaffolds - scaffolds may not be unsupported when the height to width ratio exceeds four. In the case of fixed scaffolds, they are secured to the building or structure that provides external support. Although this is not possible with a mobile scaffold, the same four to one restriction remains. This can be accommodated through the installation of extended base sections or outriggers, which provide additional width. The physical height to the working platform must be measured or calculated using the frame or member heights, the thickness of any connecting pins that fit in the joints, and the height of casters and/or screwjacks if they are used. The total height of all of these components must be analyzed to determine the four to one ratio.

The second most critical issue with mobile scaffolds, which differs from the supported scaffolds discussed earlier, is the requirement for diagonal bracing. These scaffolds must have support in a horizontal diagonal direction between opposite leg supports. If this diagonal bracing is not provided, the scaffold will be distorted out of its rectangular shape into a parallelogram. This shape results in reduced width of the parallel sides and a resultant loss of stability. It is critical that the diagonal bracing be installed to prevent this from occurring.

CASTERS

The casters on the mobile scaffold must be specifically designed for the frame members or tube and coupler on which they are utilized and, according to the manufacturer, and must be specifically rated for mobile scaffolds. There have been instances where casters from other manufacturer's components were utilized on a mobile tower scaffold and, because of the caster, i.e. offset of the wheel centerline from the post centerline, significant bending moments into the support member are induced and can cause failure. The casters must be specifically recommended by the manufacturer for the mobile tower application.

In addition, casters for mobile scaffolds must have a wheel lock, which would prevent horizontal travel, and also a castering lock to prevent the wheel from swiveling. These should be provided on all four corners.

Generally the casters are offset from the centerline of the support by less than one inch and their effect on the base width is negligible. However, when moving a scaffold, particularly if advancing the scaffold with the direction of travel perpendicular to the short side of the scaffold, the casters will turn inward and the support point is then behind the center of the support post. When measured from the location where the wheel makes contact with the concrete surface, the post is in front of this position. Although this is a relatively small distance, it can have a major impact on the stability of the scaffold particularly when the scaffolding is being moved in that direction. The force to move the scaffold is applied in the direction of least resistance to overturning. This is particularly true if obstructions or unevenness are encountered while moving the scaffold that would increase the force required to move the scaffold. When designing a mobile tower, it should be anticipated that the castering effect of the wheels results in the placement of the wheel in the least favorable condition. Designs should provide that the scaffold maintain adequate stability with the casters in that position.

Moving a Scaffold

When relocating a scaffolding, the force necessary to move it should be applied to the scaffolding as close as practical to the base. The greater the distance upward from the floor or supporting structure the force is applied, the more overturning moment that is developed and the more likely the scaffold is to tip over. (Special note: Some states have requirements for a restriction of three times the height to width ratio [3:1], others three and a half times the height [3.5:1]. Check the local codes to see which states have restrictions more stringent than OSHA.)

In order to control this overturning moment, the Occupational Safety and Health act, in its specific requirements for mobile scaffolds, requires that the manual force used to move the scaffold be applied as close to the base as practical and would restrict the distance to five feet as the maximum height above the supporting surface. This allows persons on the ground to propel the scaffold while working in a normal position and still minimize the overturning moment. Persons on top of a scaffold should never pull or "scootch" the scaffold along by repeatedly shifting their weight or grabbing on to components of the structure. Force to move the scaffold applied at the elevation of the work platform is very likely to cause overturning.

In addition, when scaffolds are moved with persons onboard, there are very specific requirements that should be adhered to. The first requirement is that the floor must be level within plus or minus three degrees and free of pits, holes, and obstructions. This is a very minimal slope and the requirement must be adhered to.

Secondly, the height to base width ratio is restricted. It is reduced from four to one (4:1) to two to one (2:1) for scaffolds that are being moved with persons on board. There is an exception that provides that if the scaffold is designed in accordance with the ANSI/SIA A92.5, (Boom-Supported Elevating Work Platforms) and A92.6 (Self-Propelled Elevating Work Platforms) standards for stability as outlined in the non-mandatory Appendix A, the scaffold may be utilized in accordance with those standards.

Third, before movement of the scaffold occurs, each employee should be aware that the scaffold is about to be moved. Special note regarding the casters: the caster and wheel stems must be pinned or secured into the ends of the scaffold legs or the adjustment screws. Failure to do this, if the scaffolding is on slightly uneven floor surfaces, may allow the caster to fall out. There have been numerous accidents where a scaffolding started to tip and, as it is raised off the casters, they fell out. When the tipping was arrested, the scaffolding settled back on the leg that no longer contained the caster and tipped in the opposite direction.

OUTRIGGERS

Outriggers used to extend the base width of the scaffolding or clamped to the uprights can provide additional support. However, as the uprights are round, the clamps may fail to grip the outrigger and may slip or rotate around the post. This effect would change their position from extending the narrow side of the scaffold to extending in the plane of the length of the base section. These outriggers must be firmly and securely attached in the place and direction in which they are providing support. If clamps are used as outlined above, they must be securely attached. A pin or positive attachment to prevent rotation of the outrigger in the plane of the floor is most desirable. Removal or retraction of outriggers to provide access through doorways or other obstructions requires that persons on the scaffold dismount as the four to one ratio required for stability has been reduced.

ACCESS

As with all scaffolds, access to mobile towers is required. It is critical that a well designed access be provided to persons climbing the ends of scaffold frames or support points if they climb on the outside of the scaffold. If they climb on the inside of the scaffold, they are obstructed by work platforms as they approach the work elevation. When providing access to mobile scaffolds, it is recommended that access be provided through the interior of the scaffold, i.e. climbing within the four sides of the frames or ladder-type end frames, and that a platform which has an access way to allow a worker to climb directly through to the work platform be provided. This access way, once the worker has reached the work platform, can either be closed with a trap or enclosed with a standard guardrail or barricade.

Chapter 19

Aerial Work Platforms

Introduction

Aerial work platforms (also called aerial lifts, aerial platform scaffolds, aerial platforms, manlifts, or personnel hoists) consist of vehicle-mounted elevating and rotating aerial platforms (bucket trucks), boom-supported elevating platforms (boom lifts), and elevating work platforms (mostly scissor lifts). This chapter discusses the various types of aerial work platforms, OSHA regulations and industry standards, and associated hazards and precautions. Case studies will also be presented that outline particular hazards and precautions associated with the various types of aerial work platforms.

Types of Aerial Work Platforms

Vehicle-mounted aerial work platforms of the type commonly called "bucket trucks" are intended for use on streets and highways. They are mobile rotating and elevating work platforms that have adjustable position platforms and are supported by a vehicle base. They can have telescoping booms, articulating booms, or a combination of both. See Figure 19.1a and Figure 19.1b.

Boom-supported elevating platforms are self propelled mobile rotating and elevating work platforms that have adjustable position platforms, are supported

FIGURE 19.1A Bucket truck with telescoping boom platform *(Reprinted with permission of the Scaffold Industry Association.)*

from ground level by a structure, and extend away from the base. See Figure 19.2. They are also called boom lifts, cherry pickers, rough terrain booms, articulated booms, up and over booms, straight mast or straight stick booms, narrow aisle booms, slab booms, or by the customary manufacturer's name (e.g., JLG, Condor, Snorkel).

Boom lifts come in two basic types: straight mast or telescoping booms and articulated booms. Telescoping booms until recently made up most of the mar-

FIGURE 19.1B Bucket truck with telescoping and articulating boom platform

ket. Articulating booms have become more popular in the last ten years. Some models have jibs at the end of the boom to allow maneuvering of the boom basket.

Scissor lifts are elevating work platforms with a platform that cannot be positioned horizontally beyond the base. They may be equipped with a deck/platform extension to extend reach. Self-propelled scissor lifts are capable of travel at full extension unless outriggers or stabilizers are required. See Figure 19.3.

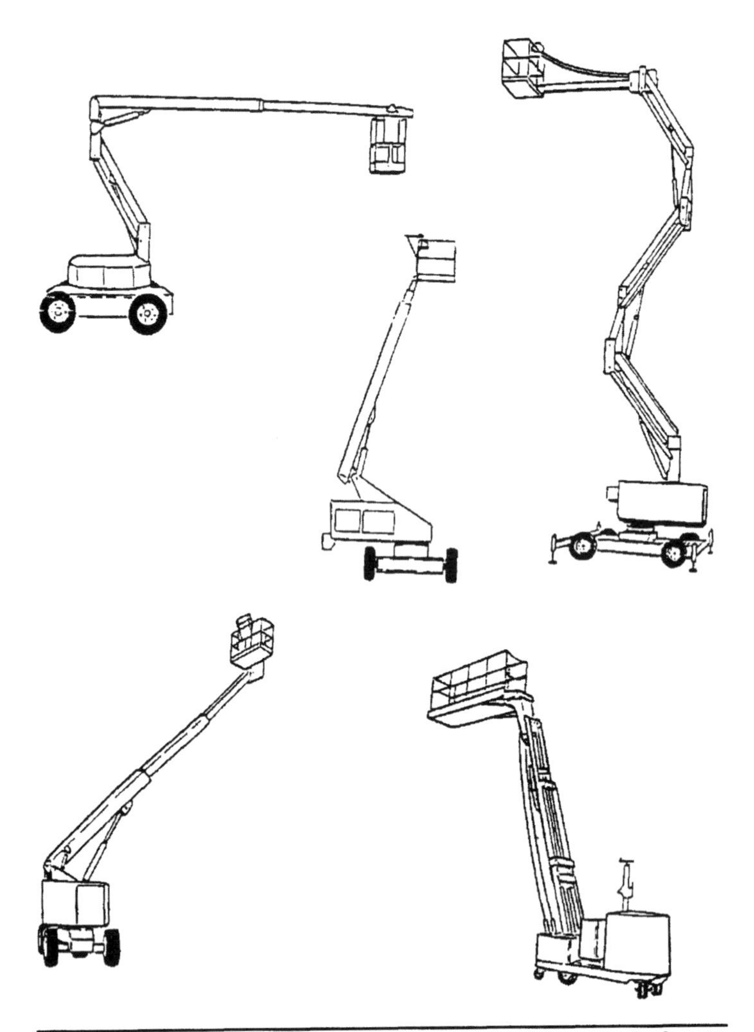

FIGURE 19.2 Examples of boom-supported aerial work platforms. *(Reprinted with permission of the Scaffold Industry Association.)*

OSHA Regulations

Vehicle-mounted aerial work platforms (called aerial lifts by OSHA) are regulated by OSHA under 1926.453 in Subpart L on scaffolds. These involve vehicle-mounted aerial lifts conforming to ANSI A92.2-1969 Vehicle Mounted Elevating and Rotating Work Platforms. These include extensible (telescoping)

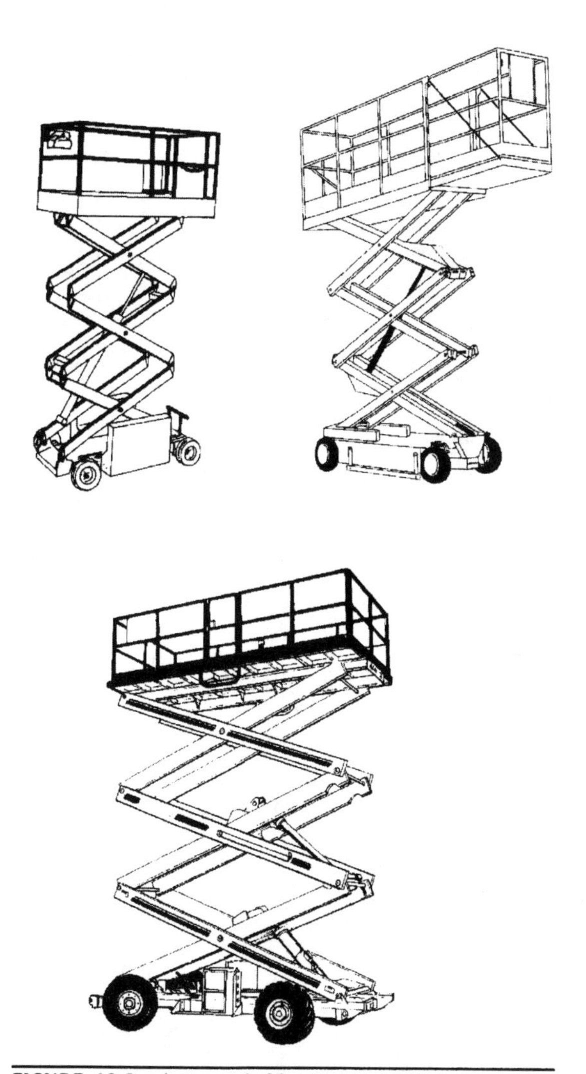

FIGURE 19.3 (upper left) Narrow-aisle scissor lift (upper right) Narrow-aisle scissor lift with deck extension (bottom) Rough terrain scissor lift *(Reprinted with permission of the Scaffold Industry Association.)*

and articulating boom platforms (bucket trucks), aerial ladders, vertical towers, and any combination of them. OSHA also regulates self-propelled boom-supported elevating work platforms under 1926.453 (1). Scissor lifts and other aerial work platforms are regulated by OSHA as mobile scaffolds (29 CFR 1926.452 (w)) and by the general requirements for scaffolds (1926.451) (2).

OSHA requirements under 1926.453 for bucket trucks (and boom lifts) include:

- Lift controls shall be tested each day prior to use. (1926.453 (b) (2) (i))
- Only authorized persons shall operate an aerial lift. (1926.453 (b) (2) (ii))
- Belting off to adjacent poles or other structures while working from a lift shall not be permitted. (1926.453 (b) (2) (iii))
- Employee shall always stand on the basket floor and not climb on the basket edge or use planks, ladders or other devices for a work position. (1926.453 (b) (2) (iv))
- Personal fall protection is required and the lanyard must be attached to the boom or basket. (1926.453 (b) (2) (v)) Personal fall arrest systems must conform to 1926.502 (e) of Subpart M. If a body belt is used as a tethering or restraint device, it must conform to 1926.502 (e), and the manufacturer must be notified if it is not already specified in the operations manual.
- Boom and basket load limits specified by the manufacturer shall not be exceeded. (1926.453 (b) (2) (vi))
- Brakes shall be set (on non-hydraulic brake models); when outriggers are used, they shall be positioned on pads or a solid surface. Wheel chocks shall be installed before using an aerial lift on an incline, providing they can be safely installed. (1926.453 (b) (2) (vii))
- An aerial lift truck shall not be moved when the boom is elevated and occupied unless designed for that type of operation. (1926.453 (b) (2) (viii))
- Aerial lifts shall have easily accessible upper controls and lower controls that can override the upper controls. (1926.453 (b) (2) (ix))
- Electrical tests of bucket insulation shall conform to the requirements of ANSI A92.2-1969 section 5 or equivalent d.c. tests. (1926.453 (b) (3))

As stated earlier, scissor lifts are regulated by OSHA as mobile scaffolds and by the general requirements for scaffolds. (1) The most important OSHA requirements for scaffolds that would apply to scissor lifts include:

- The platform must be designed to be capable of supporting its own weight and at least 4 times the maximum intended load. (1926.451 (a))
- Scaffolds (including scissor lifts) must not be loaded in excess of their maximum intended loads or rated capacities, whichever is less. (1926.451 (f) (1))
- A Competent Person must inspect the scaffold (scissor lift) before each work shift, and after any occurrence that could affect its structural integrity. (1926.451 (f) (3))
- A minimum safe approach distance must be maintained between the scaffold (scissor lift) or any conductive object on the scaffold and energized power lines. This clearance distance is 3 feet under 300 volts; 10 feet for 300 volts to 50 kV; and 10 feet plus 0.4 inches for each kV over 50 KV. (1926.451 (f) (6))
- Makeshift devices such as boxes and ladders must not be used on top of scaffold (scissor lift) platforms to increase working height. (1926.451 (f) (14) and (15))
- Either guardrails or personal fall arrest systems must be provided to protect employees. (1926.451 (g) (1) (vii)) Personal fall arrest systems must meet the requirements of 1926.451 (g) (3) and 1926.502 (d). Guardrails must comply with 1926.451 (g) (4).

Employees shall not be allowed to ride on mobile scaffolds unless: (i) the surface is within 3 degrees of level, and free of pits, holes and obstructions; (ii) it meets the stability test requirements of ANSI/SIA A92.6; (iii) outriggers, if used, are installed on both sides of the scaffold; (iv) for self-propelled scaffolds, the maximum speed is one foot per second; and (v) no employee is on any part of the scaffold extending beyond the wheels of the scaffold. (1926.452 (w) (6)) The operating manual for ANSI/SIA A92.6 clearly defines the work surface, slope, and ability to operate on a grade. (See discussion of industry standards below.)

Industry Standards

The American National Standards Institute (ANSI) and Scaffold Industry Association (SIA) have a variety of industry standards for aerial work platforms. The standard for bucket trucks is ANSI/SIA A92.2-2001 Vehicle-Mounted Elevating and Rotating Aerial Devices (3); for boom-type aerial work platforms, ANSI/SIA 92.5-1992 Boom-Supported Elevating Work Platforms (4); and for scissor lifts, ANSI/SIA A92.3-1990 Manually Propelled Elevating Aerial Platforms (5), and ANSI/SIA 92.6-1999 Self-Propelled Elevating Aerial Work Platforms (6). The ANSI A92.2 standard referenced by OSHA in 1926.453 is the 1969 version, not the most recent. The most recent version should be used. These ANSI standards cover the following topics:

- Scope, purpose, requirements, and application
- Referenced and related American National Standards
- Definitions
- Design requirements
- Electrical systems & devices (ANSI 92.2 only)
- Responsibilities of manufacturers
- Responsibilities of dealers and installers
- Responsibilities of the owners and users
- Responsibilities of renters and lessors
- Responsibilities of operators

Specifications for Aerial Work Platforms

Aerial work platforms in general can be classified by power source, tire type, maximum platform height, wheel base width, and load capacity. Aerial work platforms also have a variety of safety features, many of which are specific to the type of aerial work platform. Some scissor lifts also have deck extensions.

Power Source

Aerial work platforms can be powered by gasoline, diesel fuel, propane gas, electricity, or a combination of these. Propane gas and electric aerial work platforms are intended for indoor use, and gasoline and diesel for outdoor use. Some boom lifts and scissor lifts can be manually powered. Bucket trucks, as on-the-road vehicles, are usually gasoline- or diesel- powered.

Tire Type

Tire type can determine the application. Traction tread tires are used for rough terrain work, balloon-type tires for outdoor use, and solid rubber slab application tires (e.g., hard rubber tires) for use on asphalt or concrete (e.g., indoors). Aerial work platforms with solid rubber tires or slab model tires have low clearance, meaning the bottom of the machine is closer to the ground. On uneven surfaces, this can create tipping over problems. Pneumatic or air-filled tires are commoner for outdoor use. Liquid-filled or foam-filled tires are intended for outdoor use, or in work environments that may deflate a pneumatic tire (e.g., where drywall contractors may drop screws).

Maximum Platform Height

The maximum platform height is the maximum height above the surface to which the aerial work platform can be raised. The maximum working height is usually considered as the maximum platform height plus 6 feet, although this does not take into account the differences in heights of workers.

Telescoping boom lifts commonly reach a platform height of 80 feet with 50-60 feet of horizontal reach. Articulating boom lifts commonly reach a platform height of 80 feet with 34 feet of horizontal reach. Boom lifts are available that can go as high as 120 feet for telescopic lifts and 150 feet for articulating lifts (with 80 feet of horizontal reach).

The most popular types of scissor lifts have a maximum platform height of about 20-25 feet, but some can reach 60 feet or more.

Wheel Base Width

Stability is the main problem with aerial work platforms, especially as vertical height and horizontal reach increases. Wider wheelbases increase the stability but decrease the ability to be used indoors with narrow doorways or in other tight spaces.

Wheelbases for boom lifts vary from 4 feet for smaller ones to over 10 feet for large boom lifts.

The problem with scissor lifts that can reach high distances is the height to base ratio. As this ratio increases, the stability of the scissor lift when extended decreases. Therefore scissor lifts with high reaches have larger wheelbases to make them more stable, sometimes as much as 10 feet. Accessibility is also an important issue. Scissor lifts with narrow wheelbases can go through doors, enter elevators, or get into tight spaces. For this reason, scissor lifts with a 20-foot platform height and a 32-inch wheelbase are popular. Scissor lifts with a 19-foot platform height and a 27- or 28-inch wheelbase can go through elevator doors but are more restricted in platform size and load capacity.

Load Capacity

The *unrestricted rated load capacity* of an aerial work platform is the maximum allowable load the aerial work platform is designed to carry — including personnel, tools and materials (and bucket liner, in case of bucket trucks) – when spread evenly over the platform. Load placement requirements will be described in the manufacturer's operator's manual. It should also be on a decal on the aerial work platform.

Some manufacturers instead supply the *restricted load capacity*, along with a restricted load capacity chart. This gives multiple load capacity ratings for different configurations of the aerial work platform, for example, with outriggers extended or not, with platform deck extensions in use, and for different work platform locations. Some aerial work platforms also have restricted load capacity rating indicators.

Unrestricted rated load capacities for telescoping boom lifts are now up to 300 pounds and for articulating boom lifts up to 1000 pounds.

The unrestricted rated load capacity of scissor lifts can vary from 500 to 750 pounds or more for 20-foot electric scissors, and from 1500 pounds or more for 33-foot diesel rough terrain scissors. Scissor lifts with extension platforms have separate platform capacities for the deck extension. Some scissor lifts with the platform extending beyond the scissors would require that the load be centered and evenly distributed.

Safety Features

Aerial work platforms have a variety of safety features, some of which are mandatory and others optional depending on the type and model of aerial work platform.

Every aerial work platform has emergency controls at ground level so that the platform can be lowered in case of incapacitation of the operator, failure of upper platform controls or equipment malfunction, running out of gas, or other problem. These emergency controls override the upper controls. Some aerial work platforms come equipped with outriggers or extendible axles that must be extended when the platform is elevated, and bucket levelers to level the bucket when the chassis is on a slope. Other safety features that can be found on some aerial work platforms include pothole guards, motion alarms, and tilt alarms.

Some vehicle-mounted buckets are insulated, meaning the bucket is insulated from the rest of the truck. This does not prevent phase-to-ground or phase-to-phase electrical contact if the operator or bucket contacts power lines.

Uses

Aerial work platforms are used to position personnel, along with their tools and materials, at elevated work locations. They can also be used to lift personnel to elevated locations. Bucket trucks are used for street and highway work where an aerial work platform is needed that can travel on the streets and highways. Examples of users include electric, telecommunication and telephone utilities and utility contractors, street light contractors, sign contractors, and public works departments.

Boom lifts are used on construction, industrial, and commercial sites where reach and height are more important than platform work area and lifting capacity. They can be used both indoors and outdoors. Typical uses include structural steel erection, glazing, painting, electrical installation, and general maintenance.

Scissor lifts are used instead of boom lifts where platform work area and lifting capacity are more important than reach and height. Scissor lifts are increasingly replacing ladders as work platforms and can reach vertical locations that ladders can't. Scissor lifts can also maneuver to reach the best working position, both horizontal and vertical.

Rental/Lease and Purchase

Aerial work platforms can be purchased from a dealer or rented or leased from a rental company (lessor). Utilities and public works departments usually buy their bucket trucks and have the responsibility for maintaining them. In construction, rental or lease is the normal way contractors get aerial work platforms such as boom-supported lifts and scissor lifts. They can rent or lease them for a few days or for extended periods of time. A major disadvantage of the rental process is that a contractor and his employees rarely get the same model from job to job. This puts an extra burden on training to ensure the operator is qualified to operate that particular model.

Another disadvantage is that the contractor is never sure what condition the aerial work platform is in when he receives it. Lessors are supposed to perform a pre-delivery inspection, service and adjustments by a qualified mechanic before the contractor accepts delivery with the contractor getting a copy of the inspection report. However, this frequently does not take place. In addition, the lessor is supposed to familiarize the person who accepts delivery with particular features of that model (review of ground and platform controls, location of all manuals and safety features). This is often very abbreviated. Lessors are responsible for the frequent and annual inspections and for maintenance of the aerial work platform.

Training Requirements

There are two basic types of training: training of operators and training of mechanics. All training shall be performed by a qualified person who is experienced with the particular model of aerial work platform being used. Training records, including name of trainer and trainee, date of training, and training materials should be retained for 4 years.

Proof of training should be issued to all trained personnel. Required information for record retention should include: 1) the trainee's name; 2) date of training; 3) aerial work platform type (e.g., bucket truck, boom-supported, scissor, self-propelled, manually propelled); 4) name of trainer; and 5) name of company or entity providing training.

Operator Training

Before being authorized to operate a particular model of aerial work platform, users shall ensure that all operators are formally trained and qualified to operate that aerial work platform. Training should include the following topics:

- Regulations and standards, including relevant OSHA regulations, ANSI standards, manufacturer's requirements, and responsibilities of Dealers, Owners, Users, Lessees, Lessors, and Operators
- General training on aerial work platforms with similar characteristics and purposes (e.g., bucket truck, boom-type, scissor lifts, manually propelled aerial work platforms, etc.)
- Physical characteristics such as fuel sources, tire types, heights and sizes, axle types
- Center of gravity for different types of aerial work platforms
- Model-specific training for aerial work platforms with unusual characteristics, e.g., EE-rated (electrically encapsulated to withstand a certain amount of electrical current without exploding), models with excessively rated platform capacities
- Operator responsibilities, including knowing and understanding the Operations Manuals that comes with the aerial work platform (e.g., equipment limitations, intended uses, rated capacities, etc.)
- Familiarization with the individual aerial work platform prior to being authorized to operate it; location of the weather-resistant compartment containing the operation manuals; understanding control functions, placards and warnings; auxiliary power sources and other safety devices

- Review of Frequent and Annual Inspection and Maintenance Requirements logs
- Worksite inspections required before operating an aerial work platform at a site
- Modification requirements that must be approved in writing
- Safety devices (e.g., auxiliary power down, manual bleed down valve, outriggers, bucket levelers, etc.)
- Identification of worksite hazardous locations, such as the potential presence of flammable or explosive gases, and overhead obstructions and powerlines
- Tie-off requirements. Operator should be trained in the use of personal fall protection systems where required. This should include a discussion of lanyard length, fall arrest vs. fall restraint, fall distances, anchorage points, etc.
- External load requirements. Overhanging of objects outside the platform must be approved by the manufacturer.
- Shutdown of aerial work platforms

Training of Mechanics

Aerial work platform mechanics shall be qualified persons, who are formally trained to maintain aerial work platforms. Training shall include:

- General training on aerial work platforms with similar characteristics and purposes, e.g., boom-type, scissor lifts, manually propelled aerial work platforms, etc.
- Training on specific models with unusual characteristics, e.g., EE-rated models, models with excessively rated platform capacities
- Familiarization with documentation of individual aerial work platform upon delivery, transfer of care, rentals, lease or loan (date, time), executing party (delivery person, owner's representative), authorizing party (person being familiarized and authorizing the use of the aerial work platform), equipment model, serial number, unit rental number, manufacturer name
- Annual and frequent inspection requirements

Inspection Requirements

There are three types of inspections required by the ANSI standards: annual inspections; frequent inspections; and pre-operation inspections. These inspections must be done by the owner. When aerial work platforms are rented or leased, annual inspections and frequent inspections should be done by the rental company (lessor). There should also be a pre-delivery inspection before an aerial work platform is turned over to a renter, lessee, or new owner. This should involve inspection, servicing, and adjustment to the manufacturer's requirements. The maintenance log should be located on the aerial work platform or be easily accessible. This log should contain proof of frequent inspections and pre-delivery inspections. Dealers are responsible for inspections of rentals in the field whose inspections expire while on long-term rental/lease. See "Chapter 19 Checklists on Aerial Work Platforms" in the Appendix for forms and checklists that should prove useful for the three types of aerial work platforms.

Annual Inspections

Annual inspections are required to be performed no later than 13 months from the date of the previous annual inspection, as required by the manufacturer and relevant ANSI standard. Inspection records should be maintained for a period of four years.

Frequent Inspections

Frequent inspections shall be conducted on an aerial work platform: (1) that has been in service for 3 months or 150 hours, whichever comes first; (2) that has been out of service for longer than 3 months; or (3) that was purchased used. Frequent inspections shall be documented and available upon delivery to the receiving party, as specified in the relevant ANSI standard.

Pre-Operation Inspection

A pre-operation inspection should be performed prior to use and operation of aerial work platforms, or after another operator has operated the aerial work platform, as required by the manufacturer and relevant ANSI standard.

It is the responsibility of the operator to inspect the machine before the start of each workday. It is recommended that each operator inspect the machine before operation, even if it has already been put into service under another operator. Also check for the manuals. The manufacturer's operations and ANSI manuals are required to be on the aerial work platform or it is considered inoperable.

Common Hazards and Precautions

According to U.S. Bureau of Labor Statistics data, there were 247 deaths of construction workers due to aerial work platforms from 1992-1999 (7). Of these, 70% were due to bucket trucks and other boom-supported aerial work platforms and 30% to vertical lifts, mostly scissor lifts. The major causes of death were: electrocutions (33%), mostly due to bucket trucks; falls (31%); collapses/tipovers (25%), mostly scissor lifts and boom-supported aerial work platforms; struck by incidents (6%); caught in/between incidents (5%); and other (4%).

Electrocution Hazards

About half the electrocutions in bucket trucks involved direct contact of part of the operator's body with energized overhead power lines, mostly involving electricians or electrical power installers. About one-third involved the bucket or boom contacting an energized overhead power line.

Many bucket trucks have insulated buckets. Contrary to common belief, this does not prevent electrocution. The insulation is at the point of attachment of the bucket to the boom and is intended to isolate the bucket and occupant from a ground path through the bucket. Buckets can be insulated against different voltages. If that voltage rating is exceeded, then the insulation protection is overcome. However, a more common scenario is where the hand, arm, or shoulder of a bucket operator or part of the bucket contacts an energized power line while the bucket operator is holding a grounded de-energized line, creating a path from phase to ground and electrocuting the operator. Another scenario is where the contact is from one energized line to another creating a phase-to-phase contact.

If the bucket is not insulated, as is common when it is used by non-electrical trades, then a part of the operator's body contacting an energized power line can create an electrical path through the body to the bucket, to the boom, and through the truck body to ground. In addition contact of the bucket or boom can result in electrocution of the operator or someone on the ground in physical contact with the bucket truck.

Most of the electrocutions involving boom lifts involved workers accidentally contacting overhead power lines, rather than electrical workers working on the power lines. Boom lifts are generally not insulated. Similarly, half of the electrocutions occurring while in a scissor lift were due to the uninsulated lift or operator contacting an overhead power line.

Electrocution Precautions

The key to protection is knowing the height of the equipment in use and the voltages in your work environment and ALWAYS LOOK IN THE DIRECTION YOU ARE TRAVELING — LOOK UP.

- Inspect the work area for nearby overhead power lines.
- Workers who are not qualified electrical workers must stay at least 10 feet away from overhead power lines. This includes any part of the lift and any tools the workers may be holding. Read the ANSI Manual of Responsibilities for MSAD (Minimum Safe Approach Distance), and contact your supervisor for overhead hazards and voltages. See also Table 19.1.
- Qualified electrical workers must de-energize/insulate live power lines or use proper PPE/equipment.
- According to OSHA, insulated buckets must meet electrical tests conforming to ANSI A92.2-1969 section 5 or equivalent d.c. tests. (1926.453 (b) (3)) The more recent A92.2-2001 should be used instead.

Fall Hazards

Causes of falls from bucket trucks or boom lifts include catapulting out of the aerial work platform after being struck by vehicles or other objects by sudden movement of the boom, transferring to or from the bucket or platform at a height, falls while the aerial work platform was moving with the bucket or boom extended, standing on or leaning over the edge, standing on boxes or other objects to reach higher, and not fastening the entrance door. In many cases the

TABLE 19.1 Minimum Safe Approach Distances (MSAD)

Voltage (phase to phase)	MSAD
0 to 300 volts	Avoid contact
over 300 v to 50 kV	10 feet
over 50 kV to 200 kV	15 feet
over 200 kV to 350 kV	20 feet
over 350 kV to 500 kV	25 feet
over 500 kV to 750 kV	35 feet
over 750 kV to 1000 kV	45 feet

causes were unknown. However, in all cases, personal fall protection systems – which could have prevented these fatal falls — were not worn or used.

Fall Precautions for Bucket Trucks and Boom Lifts

- Always use personal fall protection, even when transferring to or from the bucket or boom platform at a height. See the Fall Protections Section.
- Be careful around possible falling objects such as tree branches, telephone poles, and crane loads.
- Do not stand on or lean over the bucket edge – always keep feet on the platform.
- Do not stand on boxes, ladders or other devices to achieve additional height.
- Make sure the bucket door or entrance chain is securely fastened.
- Do not move a bucket truck with the bucket extended and occupied unless the bucket truck is designed for that purpose.
- When working near traffic with bucket trucks, set up a proper work zone with cones or other warning signs.

Fall Precautions for Scissor Lifts

- Make sure the gate chain is fastened and guardrails are present.

- Do not stand on or lean over the guardrails – always keep feet on the platform.
- Do not stand on boxes, ladders or other devices to achieve additional height.
- Use personal fall protection if the scissor lift does not have guardrails or they have been removed. A personal fall protection system may also be used with guardrails attached as an extra precaution if an adequate anchor is present.

Collapses/Tipovers

Aerial work platforms are intended to be used on solid, flat level surfaces. If the aerial work platform is on a surface with a sideslope or on a front to back grade (see Figure 19.4), then the center of gravity shifts. Depending on the position of the platform and the ground support (e.g., soft earth), this shift might be enough to tip over the aerial work platform. The operations and maintenance manual lists limits for your aerial work platform. This can also be found on caution placards on your machine.

Hazards of Boom Lifts and Bucket Trucks

Collapses of the boom (often due to hydraulic failure), overturning of the bucket truck (mostly due to improper extension of outriggers or defective outrigger

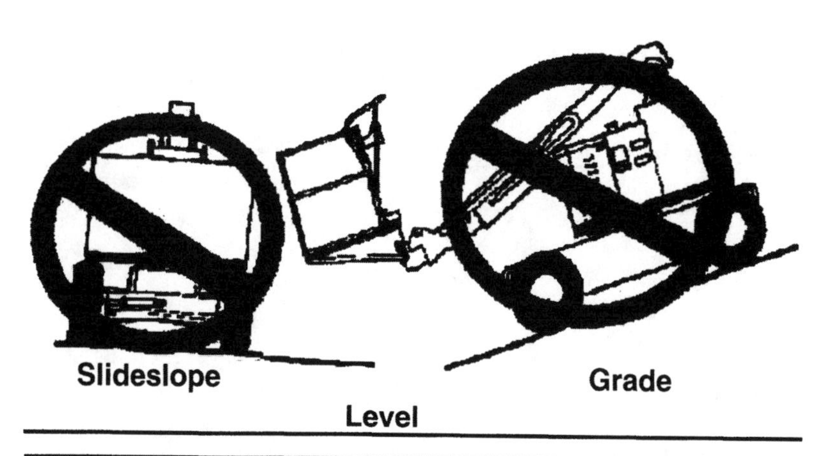

FIGURE 19.4 Sideslope and grade. *(Reprinted with permission of Genie Industries/Terex)*

controls) and collapses of the bucket (mostly due to struck by incidents) were the main causes of death. Other causes included poor maintenance, overloading, overextending the boom, being struck by vehicles or other objects, and unstable or uneven surfaces.

The stability of a boom-supported lift or bucket truck depends on its fulcrum point, the center of gravity, the workload, and sideslope and grade. A boom lift works on the principle of loads supported over a set of two wheels (the fulcrum) (Figure 19.5). When the boom platform is extended over its drive wheels, these wheels becomes the fulcrum. If the boom platform is rotated 90 degrees to one side, then the set of wheels on the side over which the boom is extended become the fulcrum.

The center of gravity of the boom lift is the combination of the center of gravity of the load and the center of gravity of the boom lift counterweight system. This combined center of gravity changes as you move the platform horizontally and vertically. It also shifts if you increase the load. If the combined center of gravity extends beyond the set of wheels acting as the fulcrum, the boom lift will tip over.

Precautions with Bucket Trucks and Boom Lifts

- Always carry out pre-operation inspections. Check the maintenance log. If there are any questions, do not operate the bucket truck or boom lift.
- Check the stability of the surface.
- Do not exceed the manufacturer's recommended sideslope and grade limits.
- Make sure outriggers, stabilizers, extendible axles, and any other stability-enhancing features (if available) are properly extended and supported.
- Be careful around possible falling objects such as telephone poles.
- Never disable safety features.
- Check the load chart before extending the boom.
- Do not raise platform in windy or gusty conditions.

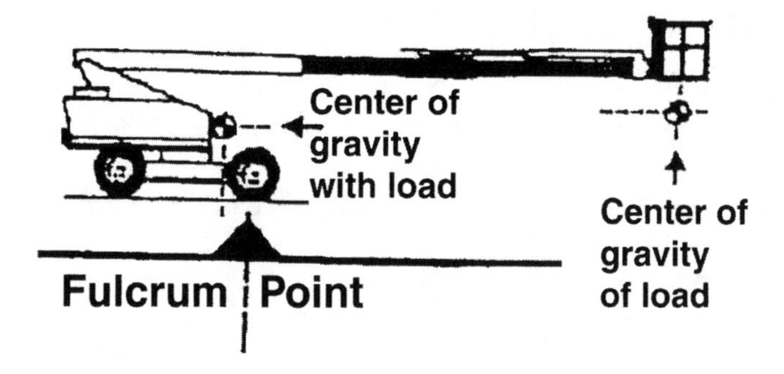

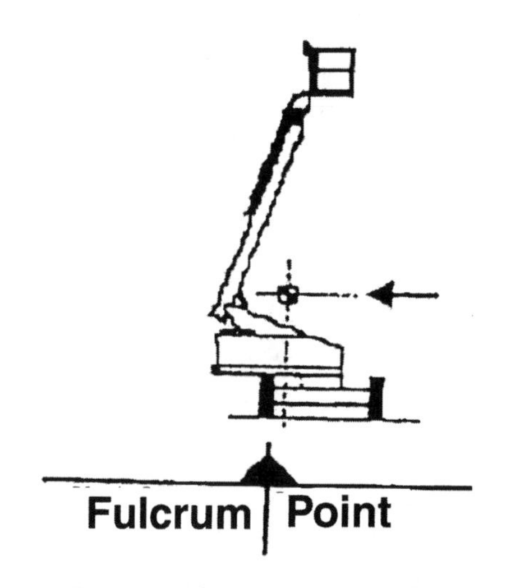

FIGURE 19.5 Fulcrum point and center of gravity of boom-supported lifts. *(Reprinted with permission of Genie Industries/Terex.)*

- Do not exceed platform capacities. Know your worksite hazards, e.g., overhead powerlines, overhead obstructions.

Hazards with Scissor Lifts

Most tipovers occurred with the scissor lift elevated over 15 feet. Causes included hitting a curb or hole while driving the scissor lift with the platform elevated (causing one-quarter of tipovers), the scissor lift being struck by an object, unequal distribution of platform load, overloading, excessive sideways force due to pulling or pushing work operations, failure to extend outriggers, and so forth.

The stability of a scissor lift depends on its center of gravity, the load, and slideslope and grade. The center of gravity of the scissor lift is the combination of the center of gravity of the platform and load (including any deck extension) and the center of gravity of the base. The center of gravity must be over the base or it will tip over. This is especially important when deck extensions are used.

Precautions with Scissor Lifts

- Never disable safety features.
- Do not drive long distances with the platform extended.
- Make sure that any pathway over which you drive is level, clear of holes, curbs, and other obstructions.
- Do not raise the platform on uneven or soft surfaces.
- Do not drive onto uneven or soft surfaces when elevated.
- When repositioning, limit travel speed.
- Do not raise the platform on a slope exceeding manufacturer's recommendations or drive onto the slope when elevated.
- Do not raise platform in windy or gusty conditions.
- Avoid excessive horizontal forces when working on elevated scissor lifts
- Do not exceed platform capacities.
- Do not lower platform with deck extension out.
- Know your worksite hazards including overhead powerlines and other obstructions

Other Hazards and Precautions

Struck-by deaths occurred when falling materials struck the operator and overhead obstacles were struck while moving the aerial work platform. Caught between deaths typically involved operators getting caught between the edge of the bucket, lift platform, or the scissor lift guardrail and a beam or other obstruction while moving the aerial work platform. Getting pinned between the scissor lift platform and base while doing maintenance is also a hazard. Take the following precautions:

- When working near traffic, set up a proper work zone with cones or other warning signs.
- Inspect work area for possible overhead hazards, including possible falling materials and obstructions at body level.
- Limit travel speed according to surface conditions.
- Use lockout/tagout procedures when doing maintenance.
- Always look in the path of travel; this includes looking up.

Fall Protection

Guardrails or bucket sides are the primary means of fall protection for aerial work platforms. If guardrails are used, there must be a top rail 42 inches above the platform floor, plus or minus 3 inches. This top rail must be able to withstand 300 pounds of force (4,6). There must also be a mid rail and toeboards at least 4 inches high. Chains may be used across access opening less than 30 inches wide. Guardrails are the primary means of fall protection for scissor lifts. OSHA only requires guardrails or personal fall protection since they regulate scissor lifts as scaffolds.

There are two types of personal fall protection: fall arrest and fall restraint. Fall arrest systems are intended to stop your fall and prevent you from striking a lower level. In order to accomplish this, you must be tied off to an anchorage that can withstand a force of 5000 pounds. Personal fall arrest systems must conform to 1926.502 (d) of Subpart M, which requires a full body harness.

Fall restraint systems, on the other hand, are intended to keep you inside the platform and prevent you from falling. The lanyard should be short enough to prevent an operator from climbing onto the guardrails but still able to maneuver around the platform. OSHA allows a body belt to be used as a tethering or restraint device, but it must conform to 1926.502 (e). A full body harness, however, is preferred since it provides better protection.

Bucket Trucks

Bucket trucks have such anchorages built in. Other boom-supported aerial work platforms and scissor lifts don't necessarily. You must tie off to the designated anchor point, not to the guardrails or somewhere else.

Boom Lifts

If the boom lift has an anchor point, you must tie off to the designated anchor point, not to the guardrails or somewhere else. If it does not have a designated anchor point, one must be designed by a Professional Engineer or other qualified person.

Scissor Lifts

The problem with fall arrest systems in a scissor lift is the anchor, which must be able to withstand a load of 5000 pounds for one person. If the lift cannot take this load, then the scissor lift would tip over on top of you. One alternative that can work in some cases is to use an overhead anchor or tie off to an overhead horizontal lifeline, which would still give some freedom to move the scissor lift.

Fall restraint systems are intended to keep you inside the platform and prevent you from falling. This would prevent the scissor lift from tipping over since there is no sideways force on it. Some manufacturers provide anchor points in scissor lifts. Before using it, determine whether it is for fall arrest or fall restraint and that it can meet your individual needs. The lanyard should be short enough to prevent an operator from climbing onto the guardrails but still able to maneuver around the platform.

Typical Malfunctions and Injuries

Following are true case studies that illustrate the hazards, precautions, and safety principles discussed in this chapter on aerial work platforms. Bucket truck case studies are given first, followed by boom lift case studies, and then scissor lift case studies.

Bucket Truck Case Study 1: Electrocution (8)

A lineman about 29 feet off the ground in the bucket of an aerial lift was grounding overhead guy wires with bare copper wire. He was electrocuted when he contacted the bracket holding the energized wire (7,200 volts) while still holding the grounded guy wire.

The lineman should have been wearing insulated gloves and insulated sleeves rated for 7200 volts.

Bucket Truck Case Study 2: Electrocution (9)

A lineman was working in an insulated aerial bucket near energized 7,200-volt overhead power lines. He was installing new lines that were de-energized and in contact with the ground. The energized lines were temporary with pigtails protruding from the temporary connections. The bucket upper controls were malfunctioning, resulting in unpredictable bucket movements. He was electrocuted when he moved the bucket too close to the energized pigtail, and contacted it with his hand while his other hand was holding the grounded, de-energized wire.

The lineman should have done a pre-operation check on the controls and not operated it until they were fixed. He also should have been wearing insulated gloves and/or sleeves rated for the voltage.

Bucket Truck Case Study 3: Electrocution (10)

A utility construction crew was using an uninsulated aerial lift truck to trim trees from the right-of-way of a 7,620-volt power line. A worker was leaning against the side of the truck. Another worker was in the bucket over the high voltage line trimming trees. When repositioning, the boom touched the high voltage line, energizing the truck. The worker on the ground was electrocuted.

The bucket truck operator failed to maintain sufficient clearance between the boom and the energized truck. The worker on the ground should not have been in contact with the truck.

Bucket Truck Case Study 4: Fall (11)

Two construction employees were ascending from ground level to an elevated highway construction level in the bucket of an articulating aerial lift during a night shift. An eighteen-wheel truck struck the platform as it swung over the unlighted and poorly marked interstate highway throwing the unbelted employees onto the highway. A second vehicle ran over one employee, resulting in death. The second employee was severely injured by the fall and hospitalized.

The work area was not lit and no traffic work zone protection had been set up. The employees were not wearing fall protection and were untrained.

Bucket Truck Case Study 5: Tipover (12)

Employees were engaged in tree trimming operations using a bucket truck. After arriving at the site, one of the employees set up the truck by placing the outriggers to stabilize it. While in the bucket, he began to unfold the boom. When the upper boom was approximately 20 degrees from level, the truck overturned causing the boom and bucket to fall 50 feet, striking the ground. The employee in the bucket was killed when he was ejected head first onto the street.

It was later determined that the left outrigger was not extended. Either it hadn't been extended at the beginning or it had been retracted later.

Boom Lift Case Study 1: Electrocution/Fall (13)

A painter was killed when he fell from the bucket of a boom lift after coming in contact with an overhead power line. He had been given a fall protection harness. The painter raised the bucket about 35 feet where it contacted an energized 26,000-volt power line. He fell from the bucket and struck the ground.

The painter was new on the job, had never operated a lift before, and received no training. He had not fastened his harness and did not stay at least 10 feet away from live overhead power lines. The job had not been evaluated for possible electrocution hazards before starting work.

Boom Lift Case Study 2: Fall (14)

A painter fell to his death while painting a roller coaster from a boom lift at a Virginia amusement park. He was about 90 feet high in a cherry picker, a type of boom lift, when the lift started rolling down a slight asphalt incline onto soft ground. The front wheels sank in the dirt, causing the boom basket to jerk, catapulting him out of the basket.

The surface should have been checked beforehand to determine if the incline was within manufacturer's recommendations. The wheels should have been chocked. The operator should have been wearing fall protection.

Boom Lift Case Study 3: Boom Collapse (15)

Two construction workers were using a boom lift while erecting metal towers. They were about 22 feet above the ground when a loud pop was heard and the boom and bucket fell to the ground. Their safety lines held the two workers in the bucket. One worker bled to death when he was forcefully jerked against the guardrails. The other injured worker suffered broken legs and a broken foot.

The cause of the collapse was faulty equipment. The end of the main hydraulic cylinder had come apart, causing the boom to fall.

Boom Lift Case Study 4: Caught Between (16)

Two ironworkers were working on structural steel from the bucket of an aerial lift. They repositioned the bucket, but extended the boom horizontally instead of elevating it. One ironworker was killed when his head, neck, and chest was pinned between a steel rafter and the control pulpit of the bucket. The other ironworker was unhurt.

The excessive forward motion of the aerial lift was caused by over-activation of the control. The controls were not clearly marked as to function.

Scissor Lift Case Study 1: Electrocution (17)

A construction laborer and two ironworkers were electrocuted, and three other construction workers seriously burned, when their scissor lift contacted a 69,000-volt overhead power line. The crew was installing aluminum siding to one side of a 25-foot warehouse under construction. The three-phase power line, which ran parallel to the warehouse, was 7 feet lower on the north end of the warehouse than on the south end. The six construction workers were working from the south end of the warehouse, where adequate clearance for the 25-foot 6-inch platform existed, toward the north end. They moved the scissor lift under the lowest part of the power line at a point where the ground sloped upward to meet the existing roadway. The platform's top guardrail contacted the bottom phase of the power line, and current passed to ground through the platform and the workers who were touching it.

The job had not been evaluated for possible overhead electrocution hazards before starting work. The platform of the scissor lift should have been lowered to ensure that adequate clearance existed.

Scissor Lift Case Study 2: Fall (18)

A painter died when he fell from a scissor lift while painting the ceiling of a newly constructed fitness center. He had positioned the lift near the wall of the building, mounted it, and raised it without fastening the safety chains. The guardrails had been lowered. The painter stepped or fell off the open end of the platform, falling about 8 feet.

The guardrails should have been correctly installed and the safety chains secured.

Scissor Lift Case Study 3: Tipover (19)

A supervisor died when the scissor lift he was operating tipped over and he and the machine fell 15 feet to the floor of a newly constructed department store. He had been assigned to stringing electrical wiring above the drop ceiling. He mounted the scissor lift, elevated the platform to 15 feet above the floor, and then drove the lift in reverse to the working area. On the way, one wheel dropped into an unguarded hole and the scissor lift tipped over.

The area should have been checked for holes and other obstructions. He should not have driven the scissor lift to the work area with the platform elevated.

Scissor Lift Case Study 4: Tipover (20)

An electrician died of head and neck injuries when his scissor lift tipped over as he was preparing to install a light under a building soffit. Due to a large awning below the soffit, a boom lift could not be used, so a battery-powered scissor lift with a 20-foot-high reach and 8-foot-long platform with a 3-foot deck extension was used. A large piece of plywood was placed on the dirt as a base for the scissor lift. The worker was observed sitting on the top guardrail of the scissor lift with his legs wrapped around the bottom rails a few minutes before the scissor lift toppled. He was also wearing stereo headphones.

The scissor lift was designed for indoor use with a safety buzzer when the lift tilted beyond 3 degrees side to side or 5 degrees front to back. The buzzer was working. The OSHA investigator measured the pitch of the plywood at 8 degrees side to side. The dirt under the plywood probably compacted unevenly under the weight of the lift. Sitting on the rails could have thrown the lift even further off balance. There is also the possibility that the electrician exerted horizontal force as he worked. The worker might not have heard the buzzer due to his headphones.

Scissor Lift Case Study 5: Caught Between (21)

An electrician apprentice was killed when his neck was crushed between the top guardrail of his scissor lift and the top of a door opening in a concrete wall. He was supposed to clean out holes 10 feet high in the wall through which conduit would be run and had to work on both sides of the wall. He was given a ladder but borrowed a scissor lift from another contractor without permission. A door opening in the concrete wall had a piece of plywood covering bent over rebar. In driving the scissor lift over the plywood through the door opening, somehow the lift vertical controls became activated and his head was trapped between the top guardrail and the top of the door opening, crushing his neck. The controls might have become activated when the scissor lift hit some unprotected rebar and lurched.

Operator error was a major cause of the fatality. The scissor lift had a remote control, which could have been used to move it through the opening. The scissor lift also had placard warning about operating it without proper instruction. Employers should ensure that unauthorized persons cannot borrow and use their equipment. The working surface over which the scissor lift is to be driven must be stable. The operating controls of the scissor lift should have been protected from inadvertent activation, e.g., such as recessed buttons instead of the exposed toggle switches

References

1. OSHA. (5/11/2001). OSHA Standards Interpretation and Compliance Letters: Standards applicable to aerial lifts; acceptable uses of body belts as restraint systems and positioning devices. Washington, DC: OSHA.

2. OSHA. (2/23/2000). OSHA Standards Interpretation and Compliance Letters: Fall protection, training, inspection and design requirements of aerial lifts and scissor lifts/scaffolds. Washington, DC: OSHA.

3. American National Standards Institute (2001). ANSI/SIA A92.2-2001. Vehicle-Mounted Elevating and Rotating Aerial Devices. New York, NY: ANSI.

4 American National Standards Institute (1992). ANSI/SIA A92.5-1992. Boom Supported Elevating Work Platforms. New York, NY: ANSI.

5. American National Standards Institute. (1990). ANSI/SIA A92.3-1990. Manually Propelled Elevating Aerial Platforms. New York, NY: ANSI.

6. American National Standards Institute (1999). ANSI/SIA A92.6-1999. Self-Propelled Elevating Aerial Work Platforms. New York, NY: ANSI.

7. McCann M. (2003). Deaths in Construction Related to Personnel Lifts, 1992-99, J. Safety Research, 34(s): 507-514.

8. OSHA Office of Statistical Studies and Analysis. (1991). Selected Occupational Fatalities Related to Vehicle-Mounted Elevating and Rotating Work Platforms as Found in Reports of OSHA Fatality/Catastrophe Investigations. Washington, DC: Occupational Safety and Health Administration. Case 3.

9. ibid. Case 27.

10. ibid. Case 14.

11. ibid. Case 16.

12. ibid. Case 11.

13. Washington State Safety and Health Assessment and Research for Prevention Program. (1/9/2003). Painter electrocuted when elevating boom manlift contacts power line. SHARP Report No 71-7-2003. Seattle, WA: SHARP.

14. Munz M. (3/2/03). Painter with St. Louis firm falls to death in Virginia. St. Louis Post-Dispatch.

15. OSHA Office of Statistical Studies and Analysis. (1991). Selected Occupational Fatalities Related to Vehicle-Mounted Elevating and Rotating Work Platforms as Found in Reports of OSHA Fatality/Catastrophe Investigations. Washington, DC: Occupational Safety and Health Administration. Case 23.

16. ibid. Case 19.

17. National Institute for Occupational Safety and Health. (1990). One laborer and two steel workers electrocuted when an elevated work platform contacts 69,000-volt powerline in Ohio. FACE Report: (Report Number FACE-90-01). Morgantown, WV: Author.

18. National Institute for Occupational Safety and Health. (9/16/1996). Painter dies after falling from aerial platform – Virginia. FACE Report: (Report Number FACE-96-20). Morgantown, WV: Author.

19. National Institute for Occupational Safety and Health. (1996). Supervisor dies as a result of injuries sustained in fall with powered vertical aerial platform – Virginia. FACE Report: (Report Number FACE-96-18). Morgantown, WV: Author.

20. Occupational Health Service, New Jersey Department of Health and Senior Services. (7/19/2001). Electrician killed when scissor lift tip over in New Jersey. NJ FACE Investigation # FACE-00-NJ-073-01. Morgantown, WV: Author.

21. California Department of Health and Health Services. (4/27/1998). Electrician apprentice dies when his neck is crushed by a scissor lift in California. California FACE Investigation 98CA003. Morgantown, WV: Author.

Chapter 19 Useful Forms and Checklists

The remaining pages of this chapter include forms to be used in conjunction with the information discussed in this chapter.

CHECKLIST 1
SAMPLE PRE-OPERATION CHECKLIST FOR BUCKET TRUCKS

This is just a sample checklist; each manufacturer has pre-operations checklists for their bucket trucks. It is the responsibility of bucket truck operators to read all manuals associated with the machine, and to inspect the machine before the start of each workday. It is recommended that each operator inspect the machine before operation, even if it has already been put into service under another operator. The operator should be looking for anything that may seem out of place, broken, leaking, loose or tampered with, or that may interfere with the safe use and operation of the equipment.

Conduct a Walk-Around Inspection

- ❑ Be sure all decals and placards are in place and legible.
- ❑ Be sure operators and ANSI manuals are legible and on the unit.
- ❑ Check pins.
- ❑ Check all bolts and other fasteners for signs of loosening.
- ❑ Check for hydraulic fluid leaks.
- ❑ Check all fluid levels: engine oil, coolant, fuel, hydraulic transmission, etc.
- ❑ Check batteries: cell levels, cables, corrosion, connections, etc.
- ❑ Inspect booms for trash and debris.
- ❑ Check that covers and guards are installed and free of damage.
- ❑ Check the following for damage, abnormal wear, or missing parts:
  - ❑ Boom and accessories (welds and cracks)
  - ❑ Fiberglass platform (cracks in mounting ribs, floor, and flanges around top)
  - ❑ Platform entry gate, liners, guards
  - ❑ Electrical components, wiring, and electrical cables
  - ❑ Hydraulic hoses, fittings, cylinders, and manifolds
  - ❑ Fuel and hydraulic tanks
  - ❑ Hydraulic lines and fittings

- ❑ Drive and turntable motors and torque hubs
- ❑ Dents and damage to any part of equipment
- ❑ Leveling system (master cylinder, slave cylinder, hoses, fittings)
- ❑ Tires and wheels (hubs, lugs, wheel rims, worn or damaged tires, tire pressure)
- ❑ Engine and related components
- ❑ Nuts, bolts, and other fasteners
- ❑ Cracks and welds on structural components
- ❑ Synthetic load line
- ❑ Material handling attachments (if any)
- ❑ Fall protection harness, lanyard, anchorage
- ❑ Fire extinguisher and safety equipment

Conduct a Function Test

This test should be conducted in an area that is firm, level and free from overhead and ground level obstructions. Brakes should be set and wheels chocked.

- ❑ Test emergency stop functions.
- ❑ Test lights, flashers and other signaling devices.
- ❑ Test outriggers/stabilizers (when equipped).
- ❑ Check lower controls, all functions.
- ❑ Check upper controls, all functions.
- ❑ Check upper control override at lower control panel.
- ❑ Check operation of DC pump and engine shut-off (if applicable).
- ❑ Verify boom holding valve operation.
- ❑ Verify covers and guards are installed and free of damage.
- ❑ Test the steering.
- ❑ Test the drive and braking system.
- ❑ Test all drive systems, high-low torque/gear.
- ❑ Test the foot switch and/or enabler.

CHECKLIST 2
SAMPLE PRE-OPERATION CHECKLIST FOR BOOM LIFTS

It is the responsibility of bucket truck or boom lift operators to read all manuals associated with the machine, and inspect the machine before the start of each workday. It is recommended that each operator inspect the machine before operation, even if it has already been put into service under another operator. The operator should be looking for anything that may seem out of place, broken, leaking, loose or tampered with, or that may interfere with the safe use and operation of the equipment.

Conduct a Walk-Around Inspection

- ❑ Be sure all details and placards are in place and legible.
- ❑ Be sure operators and ANSI manuals are legible and in the weather-resistant compartment.
- ❑ Check all fluid levels: engine oil, coolant, fuel, hydraulic transmission, etc.
- ❑ Lubricate all fittings, as required by the manufacturer.
- ❑ Check batteries: cell levels, cables, connections, etc.
- ❑ Check the following for damage, abnormal wear, or missing parts:
  - ❑ Boom and accessories (welds and cracks)
  - ❑ Electrical components, wiring, and electrical cables
  - ❑ Hydraulic hoses, fittings, cylinders, and manifolds
  - ❑ Fuel and hydraulic tanks
  - ❑ Hydraulic lines and fittings
  - ❑ Drive and turntable motors and torque hubs
  - ❑ Dents and damage to any part of equipment
  - ❑ Tires and wheels (hubs, lugs, wheel rims, worn or damaged tires)
  - ❑ Engine and related components
  - ❑ Limit switches, horns, alarms, and beacons

- ❑ Nuts, bolts, and other fasteners
- ❑ Platform entry gate
- ❑ Cracks and welds on structural components
- ❑ Fire extinguisher and safety equipment

Conduct a Function Test

This test should be conducted in an area that is firm, level and free from overhead and ground level obstructions.

- ❑ Ground control tests
  - ❑ Test emergency stop.
  - ❑ Test machine functions.
  - ❑ While raising and extending the boom, look for damage in newly exposed parts and body.
  - ❑ Test outriggers/stabilizers (when equipped).
  - ❑ Test tilt sensor.
  - ❑ Test auxiliary ground controls.
  - ❑ Test manual bleed down controls and/or auxiliary power unit (APU) power down when available).
- ❑ Platform control tests
  - ❑ Test emergency stop.
  - ❑ Test service horn.
  - ❑ Test the foot switch and/or enabler.
  - ❑ Test the platform functions.
  - ❑ Test the steering.
  - ❑ Test the drive and braking system.
  - ❑ Test all drive systems, high-low torque/gear.
  - ❑ Test auxiliary controls.

Checklist 3
Sample Pre-Operation Checklist for Scissor Lifts

It is the responsibility of scissor lift operators to read all manuals associated with the machine, and inspect the machine before the start of each workday. It is recommended that each operator inspect the machine before operation, even if it has already been put into service under another operator. The operator should be looking for anything that may seem out of place, broken, leaking, loose or tampered with, or that may interfere with the safe use and operation of the equipment.

Conduct a Walk-Around Inspection

- ❑ Be sure all decals and placards are in place and legible.
- ❑ Be sure operators and ANSI manuals are legible and in the weather-resistant compartment.
- ❑ Check the engine oil, coolant and fuel levels on internal combustion engines, and battery levels on electric models.
- ❑ Check the following for damage or missing parts:
  - ❑ Electrical components, wiring and electrical cables
  - ❑ Hydraulic hoses, fittings, cylinders and manifolds
  - ❑ Fuel and hydraulic tanks
  - ❑ Drive and turntable motors and torque hubs
  - ❑ Scissor extension cables and wear pads
  - ❑ Dents and damage to any part of equipment
  - ❑ Tires and wheels
  - ❑ Engine and related components
  - ❑ Limit switches, horns, alarms and beacons
  - ❑ Nuts, bolts and other fasteners
  - ❑ Platform entry midrails/gate and guardrails
  - ❑ Cracks and welds on structural components

Conduct a Function Test

This test should be conducted in an area that is firm, level and free from overhead and ground level obstructions.

- ❑ Ground control tests:
  - ❑ Test emergency stop.
  - ❑ Test machine functions.
  - ❑ Test outriggers/stabilizers (when equipped).
  - ❑ Test tilt sensor.
  - ❑ Test auxiliary ground controls.
  - ❑ Test manual bleed down controls.
- ❑ Platform control tests.
- ❑ Test emergency stop.
- ❑ Test service horn.
- ❑ Test the foot switch.
- ❑ Test the scissor and platform functions.
- ❑ Test the steering.
- ❑ Test the drive and braking system.
- ❑ Test all drive systems, high-low torque/gear.
- ❑ Test the limited drive speed.
- ❑ Test auxiliary controls.

Checklist 4
Sample Pre-Delivery and Frequent Inspection Form for Boom Lifts

Indicate for each item:
OK / CR–Corrective Action Required / C –Corrected / NA–Not Applicable

Turntable and Chassis

_______ Wheel rim nuts torqued properly; tires inspected for proper inflation, damage and wear

_______ Steer, drive and axle components secure and undamaged

_______ Fluid levels correct: hydraulic tank, engine oil, hubs, coolant and batteries

_______ Grease and lube per Service Manual

_______ Hydraulic and air filters clean

_______ Fuel and hydraulic tank caps tight and vents open

_______ Exhaust and hydraulic system free of leaks

_______ Hood doors open and latch properly

_______ Turntable bearing, swing drive and gear secure, undamaged and properly lubricated. No missing bearing bolts, or signs of looseness

_______ Pump (s) and motor (s) secure, undamaged, and free of leaks

_______ Turntable lock secure, undamaged and operates properly

_______ Engine idle, throttle and RPMs set properly

Platform and Boom

_______ Platform installed and secure; gate closes and latches properly

_______ Boom(s), pins, wear pads and attaching hardware secure and undamaged

_______ Load capacity markings installed and legible, capacity indicator operational

_______ Hydraulic lines and electrical cables secure and undamaged. No leakage

______ Cylinder pins, pivot pins and attaching hardware secure and undamaged

______ Check cylinders for leaks and damage

______ Boom chains/cables for proper adjustment and attaching hardware secure and undamaged

Functional

______ Control levers, switches, gauges and instruments operate properly, including options (horn, lights, etc.)

______ Footswitch operates properly (shuts off function when released).

______ Emergency stop switch operates properly (shuts off controls and engine).

______ Machine functions operate properly at both ground and platform controls (lift, telescope, manual descent, etc.).

______ Auxiliary power system operates properly.

______ Drive and operate unit to test all systems.

______ Brakes operate properly (Swing and Drive).

______ Axle and outrigger interlocks operate properly.

______ All function and speed cutouts operate properly.

Comments:____

____

____

____

____

The undersigned certifies that this machine has been inspected, serviced and adjusted according to the applicable manufacturer's operating, service and maintenance manuals, procedures, and criteria and all discrepancies have been corrected.

______________________________ ______________________

Authorized Signature Date

Source: Adapted from JLG Boom Lift Aerial Work Platform Pre-Delivery and Frequent Inspection Form.

Checklist 5
Sample Frequent Inspection Checklist for Scissor Lifts

Frequent Inspections are individual to each manufacturer. Although equipment is universal in nature, they contain individual parts that may need to be replaced more or less often. Always refer to the manufacturer's recommendations in their manuals for these inspections. These inspections shall be performed by a formally trained mechanic and all repairs and corrective measures shall be recorded.

Indicate for each item:
OK / CR–Corrective Action Required / C –Corrected / NA–Not Applicable

Battery System

______ Electrolyte level correct

______ Battery cables secure and undamaged

______ Batteries charged

______ Terminals secure and undamaged

Engine Oil

______ Fluid level correct

______ Engine oil free of leaks

Engine Fuel System

______ Fuel level correct

______ Fuel system free of leaks

______ Air cleaner satisfactory

______ Sediment bowl clean

______ Fan belts secure and undamaged

Engine Coolant

______ Coolant level correct

______ Antifreeze percentage correct

Service

______ Engine oil changed

______ Oil filter changed

______ Fuel filter changed

______ Hydraulic filter changed as required by manufacturer's manual

______ Pressure checks (psi)

Hydraulic Oil

______ Fluid level correct

______ Hydraulic system free of leaks

______ Pipes and fittings secure and undamaged

Emergency System and Other Safety Features

______ Emergency lowering valve works (manual bleed down control)

______ Emergency Stop switch operates properly (shuts off controls and engine)

______ Exterior cable secure and undamaged

______ Outriggers/stabilizers operate properly (when equipped)

______ Tilt sensor and alarms (audible or flashing) operate properly

Platform

______ Gate/chain closes and latches properly

______ Deck and railings secure and undamaged (check guard rail attachment, points-nuts, bolts, screws, and fasteners)

______ Welds secure and undamaged

______ Extension deck operates properly (extends and retracts)

______ Slew ring/scissor pack greased

______ Slew ring locks

Tires and Rims (inside and out)

______ Tires and rims undamaged
______ Air pressure satisfactory
______ Wheel nuts secure

General

______ Safety decals and placards in place and legible
______ Manuals/key in place
______ Full check on operations
______ Lift controls operate properly
______ Foot switch operates properly
______ All drive systems operate properly (high-low torque/gear, limited drive speed)
______ Steering controls operate properly
______ Braking system operates properly
______ Service horn operates properly
______ All fitted lights operate
______ Auxiliary ground controls operate properly
______ Emergency power
______ Engine RPM
______ Limit switches operate properly

Comments:____

____

____

____

____

The undersigned certifies that this machine has been inspected, serviced and adjusted according to the applicable manufacturer's operating, service, and maintenance manuals, procedures, and criteria and all discrepancies have been corrected.

______________________________ ____________________
Authorized Signature Date

Checklist 6
Sample Annual Inspection Checklist for Bucket Trucks

Visual Inspection

The visual inspection shall include at least the following, and shall include the removal of inspection covers:

- ❑ Outriggers: pads, structure/welds, bolts, hoses, fittings, cylinders, check valves, pins and retainers.
- ❑ Chassis: truck frame, aerial subframe/mounting, suspension, PTO, brake hoses, brake (microbrake) lock assembly, pintle hook, electrical system, hydraulic/electrical components, steering components, exhaust system and cooling lines.
- ❑ Pedestal: mounting bolts/welds, pedestal structure, diagonal brace, attachment welds/pins, hydraulic swivel joint, hydraulic components, swing drive gear box/mounting bolts, backlash between swing pinion/bullgear and electric collector ring/brushes.
- ❑ Rotation bearing: upper/lower bearing attachment weld/bolts, vertical movement of bearing, and proper torque on accessible bearing bolts.
- ❑ Turntable: turntable structure, bucket leveling cables, leveling cylinder, compensating chain/sprocket, hydraulic components and lower control operation.
- ❑ Lower boom: boom structure, welds, lift cylinders/attachment, hydraulic components/lines, leveling cables/rods, upper/extend cylinder/attachment, push links, boom rest supports, tie down straps, lower boom insulator/mounting, boom extension roller assembly, and wear pads.
- ❑ Elbow area: elbow structure, hydraulic hoses and leveling cables.
- ❑ Upper boom (extension): boom structure, welds, leveling cables/rods, wear pads, upper boom insulator/mounting, hydraulic lines/components, jib structure/mounting, tool circuit hoses/fittings and pole claw arms/mounting brackets.

- ❑ Platform (bucket): mounting bracket/bolts, leveling system, exterior condition, control operation and hydraulic lines/components.
- ❑ Winch: mounting brackets/bolts/pins, gearbox, hydraulic motor/lines load line condition and control operation.
- ❑ Placards: Load rating chart, electrical hazard placards, MADDDC placards and upper/lower control operation placards.
- ❑ Pin joints and retainers: on outriggers, boom pivot, lower/lift cylinder, upper/extend cylinder, pushlinks, elbow pivot, platform (bucket), leveling cylinders.
- ❑ Ultrasonic test (pins). *
- ❑ Dielectric test: upper boom insulator, lower boom insulator, baskets/liners, jib (for insulated lifts only). *
- ❑ Acoustic emission test: upper boom insulator, lower boom insulator, structural components. *

* *The results of these tests shall be recorded for review of paperwork.*

Structural Inspection

The structural inspection shall be performed by a 'qualified' mechanic, as required by the manufacturer, government and local regulations, and shall include at least the following:

- ❑ Magnetic particle inspection shall be performed on all critical welds, plates and castings.
- ❑ Dye penetrate inspection shall be performed on all critical welds, plates and castings made of nonferrous material and any area requiring verification of the magnetic particle inspection.
- ❑ Ultrasonic inspection shall be performed on all accessible pins including outriggers, pedestal, boom and platform.

Testing

- ❑ An acoustic emission (and load) test shall be performed on the fiberglass structures and metal components as specified by ASTM (f914_85)/ANSI (A92.2) standards.

- ❑ Functional and operational tests are to be performed to check the operation of controls, bearings, pins, bushings, cylinders, holding valves, bucket leveling mechanism, etc. According to the ASTM/ANSI standards.
- ❑ Dielectric test to 100,000 volts to verify the electrical insulating in the FRP upper/lower booms, double buckets and upper controls (for insulated lifts only).
- ❑ A dielectric test on the hydraulic oil ASTM D1816 standards shall be performed.

Comments:____

____

____

____

____

The undersigned certifies that this machine has been inspected, serviced and adjusted according to the applicable manufacturer's operating, service, and maintenance manuals, procedures, and criteria and all discrepancies have been corrected.

______________________________	____________________
Authorized Signature	Date

Source: Aerial Lift Vehicles: Inspection & Testing for New Jersey, Department of Transportation (T-1503).

Checklist 7
Sample Annual Inspection Form for Boom Lifts

Indicate for each item:
OK / CR–Corrective Action Required / C –Corrected / NA–Not Applicable

Inspection procedure codes:

(1) Weld cracks, dents and/or rust

(2) Installation

(3) Leaks

(4) Operation

(5) Condition

(6) Tightness

(7) Residue buildup

(8) See placards and decals inspection chart

A. Chassis

______ Structural (1)

______ Steering cylinder pin caps (2,6)

______ Steering linkage (4)

______ Tires (5)

______ Hydraulic tubes and hoses (3, 5)

______ Decals and placards (8)

______ Drive motor brake (4)

______ Torque drive wheel lugs per placard on wheel (6)

______ Torque steel wheel lugs per placard on wheel (6)

______ Right drive motor (3)

______ Left drive motor (3)

______ Lubrication points

B. Turntable

______ Structural (1)

______ Torque turntable top bolts @ 180 ft. lbs (6)

______ Torque turntable bottom bolts @ 180 ft. lbs (6)

______ Hydraulic hoses and tubes (3, 5)

______ Centerpost (3)

______ Centerpost securing bolts (6)

______ Lift the cylinder pin caps and tie wire (2,6)

______ Emergency bleed down valve (4)

______ Cowling (5,6)

______ Wire harness (2,5)

______ Lift cylinder and holding valves (3,4)

______ Decals and placards (2,8)

______ Master level cylinder pin caps and tie wire (2,6)

______ Master level cylinder and holding valves (3,4)

______ Rotation brake (4)

______ Rotation backlash (5)

______ Correct operations manual in document holder

______ System pressure

______ Lubrication points

C. Ground Control Station [NL1]

______ Station selector switch — ground controls

______ Platform controls DO NOT work

______ Station selector switch — platform controls ground controls

______ Ground controls DO NOT work

______ Ground controls switches (4)

______ Turntable rotation (4)

______ Boom UP-DOWN (4)

______ Boom extension IN-OUT (4)

______ Platform leveling (4)

______ Emergency power, all functions (4)

______ Platform rotation (4)

D. Booms

______ Structural (1)

______ Lift cylinder pin caps and tie wire (2,8)

______ Hydraulic hoses and tubes (3,5)

______ Extension cylinder and holding valve (3,4)

______ Hose carrier tube (7)

______ Hose carrier support (7)

______ Plastic hose track (5)

______ Electrical wires (5)

______ Extend/retract wire ropes and sheaves (2,5,6)

E. Platform

______ Structural (1)

______ Decals and placards (2,8)

______ Gravity gate (4)

______ Hydraulic tubes and hoses (3,5)

______ Slave level cylinder pin caps and tie wire (2,6)

______ Slave level cylinder and holding valve (3,4)

F. Platform Control Station

______ Foot switch, unit will not start when engaged (4)

______ Foot switch, functions operate when engaged (4)

______ Engineering START-STOP (4)

______ Electric throttle, if equipped (4)

______ Turntable rotation (4)

______ Boom UP-DOWN (4)

______ Boom extension IN-OUT (4)

______ Platform leveling (4)

______ Platform rotation (4)

______ Drive speed, HIGH-LOW (4)

______ Drive range, HIGH-LOW (4)

______ Engine choke (gasoline unit only) (4)

______ Emergency power, all functions (4)

______ Emergency shutdown control (4)

______ Gradual start to stop lift/swing/drive controls (4)

______ Speed limit switch (4)

______ 5 ° tilt alarm (4)

______ 110 AC power to platform (2,4)

G. Hydraulic Reservoir

______ Filler/breather cap and tank (8)

______ Hydraulic fluid level (5)

H. Engine

______ Charging system (4)

______ Air filter (5)

______ Hour meter (4)

______ Belts and hoses (5)

______ Hydraulic pump (4)

______ Battery terminals (6)

______ Indicator gauges (4,5)

______ Engine START-STOP (4)

______ Electric throttle (4)

______ Electric choke, if equipped (4)

______ Engine governor setting (4)

I. Platform Rotator

______ Structural (1)

______ Hydraulic rotator actuator (3)

______ Platform rotation holding valve (4)

______ Mounting bolts and pin caps (2,6)

J. Optional Equipment

______ 110 volt onboard AC generator (2,4)

______ Dual fuel system (2,4)

______ LP gas system and control (2,4)

______ Horn (2,4)

______ Motion warning alarm (2,4)

______ Operation warning light (2,4)

______ Key type master switch (2,4)

______ Cold weather start kit — diesel only (2,4)

______ Hydraulic system, cold weather warm up kit (2,4)

______ Extension cylinder bleed down (2,4)

______ Driving lights (2,4)

______ Platform work lights (2,4)

______ Sandblast protection kit (2,4)

______ Airline to platform (2,4)

______ Tow package (2,4)

______ Optional platform (2,4,6)

______ Manual rotator (2,4,6)

______ Hydraulic rotator — self contained (2,4,6)

______ Four-wheel-drive (2,4)

______ Listing lugs (2,5)

K. Placards and Decals

______ 4 decals, Snorkel brand logo

______ 8 decals, danger electrical hazard

______ 1 placard, emergency bleed down valve

______ 1 decal, danger you must not operate

______ 1 placard, pre-start instructions

______ 1 placard, platform capacity (consult factory)

______ 3 decals, yellow arrow

______ 3 decals, blue arrow

______ 1 placard, caution serial number

______ 12 pop rivets

______ 1 placard, platform identification

______ 1 decal, hydraulic oil

______ 2 decals, made in USA

______ 1 literature compartment

______ 1 decal, safe operation information

______ 1 placard, ANSI standard

______ 1 decal, notice manual reorder

______ 2 placards, lug nut torque

______ 4 bumpers

______ 4 decals, lift

______ 1 decal, lube recommendations

______ 2 placards, 125 volts 15 amps

______ 1 decal, operating manual enclosed

______ 1 crankcase oil tag (used for shipping only)

______ 5 decals, danger cylinder failure (1 per cylinder)

______ 2 caps

______ 1 decal, danger do not reach

______ 1 decal, attach fall restraint

______ 1 decal, rotate while greasing

______ 1 record pouch

______ 1 placard, engine RPM

______ 2 decals, inspect wire ropes

______ 1 decal, danger electrical/tipping hazard

______ 2 nuts 1/4-20 self locking

______ AR plugged, during a half-inch diameter

______ 4 rubber bumpers

______ 2 placards, lug nut torque

______ 2 capscrews, ¼-20 x ½ inch long, hex head grade 5

______ 4 decals, tied down symbol

______ 1 decal lift/tie down

______ 1 decal, TB60 logo

______ 1 decal, gasoline fuel (or diesel fuel)

______ 3 decals, danger rotating engine parts

______ 1 decal, engine control module (Ford only)

______ 1 decal, caution governor damage (Ford only)

______ 2 decals, 4 x 4 logo

______ 1 decal, 60 logo

Options Placards and Decals

______ 1 decal, danger do not ride

______ 1 decal, engine block heater

______ 2 placards, danger foam filled tires

______ 1 placard, caution liquid withdrawal

______ 1 decal, hydraulic warmup instructions

______ 1 decal, dual fuel instructions

______ 4 placards, tire pressure

______ 2 decals, towing instructions

______ 1 decal, danger crashing hazard

______ 1 decal, danger tow package

______ 1 decal, LPS logo (LP fuel only)

Corrective Action Required: ______________________________

____

____

____

____

All items have been properly checked and tested and found to be operating satisfactorily or necessary corrective action has been completed.

______________________________ ____________________

Inspected By Date

Source: Adapted from Snorkel Model TB 60 Pre Delivery and Inspection Report

CHECKLIST 8
SAMPLE ANNUAL INSPECTION FORM FOR SCISSOR LIFTS

Indicate for each item:
OK / CR–Corrective Action Required / C –Corrected / NA–Not Applicable

Inspection procedure codes:

(1) Weld cracks, dents and/or rust

(2) Installation

(3) Leaks

(4) Operation

(5) Condition

(6) Tightness

(7) Residue buildup

(8) See placards and decals inspection chart

A. Chassis

______ Structural (1)

______ Steering cylinder (2, 3, 4)

______ Steering cylinder fasteners and linkage (1, 2)

______ Tires and wheels (5)

______ Hydraulic tubes and hoses (3, 5)

______ Decals and placards (2, 8)

______ Cowling covers (5,6)

______ Torque wheel lugs nuts/bolts 70-80 ft. lbs. (2, 6)

______ Right drive motor (3,6)

______ Left drive motor (3,6)

______ Parking brakes (3,4)

______ Free wheeling valve (4,6)

______ Cords and wire assemblies (2,5)

______ Pothole protector interlock and alarm (1,2,4)

______ Level sensor interlock and alarm (4,5)

______ Hour meter (4)

______ Swing-out trays (2,4,5)

______ Lowering cushion valve (3,4)

______ Hydraulic oil reservoir (3,5)

______ Filler cap (6)

______ Hydraulic fluid level (2,5)

______ Return filter (2,6)

______ Oil level (3,4)

B. Scissor Arm Assembly

______ Structural (1)

______ Scissor arm pivot pins, snap rings and roll pins (2,5)

______ Lift cylinder and valves (2,3,4)

______ Safety bar (2,4,5)

______ Manual lowering valve (4)

______ Hydraulic tubes and hoses (3,5)

______ Wires and electrical cables (2,5)

______ Decals and placards (2,8)

C. Base Control Station

______ Control select switch "BASE"

______ PLATFORM CONTROLS DO NOT WORK (4)

______ Control select switch "PLATFORM"

______ BASE CONTROLS DO NOT WORK (4)

______ Toggle switch, lift — UP/DOWN (4)

______ Lowering alarm (4)

______ Battery disconnects switch – OFF

______ all controls DO NOT WORK (4)

______ Emergency stop switch — OFF

______ all controls DO NOT WORK (4)

D. Platform

______ Structural (1)

______ Guardrails properly installed and fastened (2,5)

______ Platform entrance chain or gate (2,5)

______ Wires and electrical cables (2,5)

______ Decals and placards (2,8)

______ Correct operator's manual in document holder (2)

______ Roll-out deck extension (2,4,5,8)

______ Swing-down rails, detents, and pins (2,4,5)

F. Platform Controls

______ Control select switch (4)

______ DRIVE/STEER (4)

______ Control select switch activated on "DRIVE"

______ Speed switch, speed 1 — SLOW (4)

______ Speed switch, speed 2 — MEDIUM (4)

______ Speed switch, speed 3 — FAST (4)

______ When elevated, drives slow ONLY (4)

______ When lowered, drives fast, medium or slow (4)

______ Platform UP/DOWN switch — UP (4)

______ Control select switch activated on "PLATFORM"

______ Platform UP/DOWN switch — DOWN (4)

______ Control select switch activated on "PLATFORM"

______ Emergency stop – OFF

______ NO CONTROLS WORK (4)

F. Electrical

______ Motor and hydraulic pump assemblies (2,4)

______ Batteries (2,4)

______ Battery terminals (5,6,7)

______ Battery electrolyte level (5)

______ Battery specific gravity reading (5)

______ Wires and cables (2,5,6)

______ Battery charger (2,4,5)

G. Optional Equipment

______ Entry gate — spring-loaded (2,4,5)

______ Battery charge indicator (2,4)

______ AC 110 volt outlet on platform (2,4)

______ AC generator — battery-operated (2,4,5)

______ Operator horn (2,4)

______ Motion warning alarm (2,4)

______ Backup horn (2,4)

______ Flashing light (2,4)

H. Placards and Decals

_______ 2 platform ratings

_______ 1 ground control schematic

_______ 1 battery charger, 115 volt only

_______ 2 danger towing instructions

_______ 2 danger tip over/electrical hazard

_______ 2 danger tip over hazard

_______ 1 danger do not alter switch

_______ 1 ANSI standards

_______ 1 emergency bleed down valve

_______ 2 caution cylinder disassembly

_______ 1 hydraulic oil level

_______ 1 danger safety prop

_______ 2 SL 25 logos

_______ 4 Snorkel logos

_______ 1 check battery before charging

_______ 1 serial number

_______ 1 emergency stop

_______ 3 danger cylinder failure

_______ 1 hydraulic fluid level

_______ 1 hydraulic schematic

_______ 1 danger shearing/crushing hazard

_______ 2 made in USA

_______ 6 forklift

_______ 1 ground controls

_______ 1 warning stripes

______ 1 platform control box top

______ 1 lube recommendations

______ 1 caution no step

Additional for Fixed Rail Platforms

______ Operator's checklist

______ 1 operating manual enclosed

Additional for Retractable Rail Platforms

______ 1 operators checklist

______ 7 danger pinch point

______ 10 pinch point keep clear

______ 3 danger swing down rails

______ 1 warning stripes

______ 1 operating manual enclosed

Optional Placards and Decals

______ 1 EE logo yellow

______ 2 factory mutual system EE

______ 1 horn

______ 1 danger fire hazard NFPA505

______ 2 notice – use OEM parts

______ 1 125 volt 15 amp power to platform

______ 2 EE logos 5 inch white

Corrective Action Required: ____________________________________

____

____

____

____

All items have been properly checked and tested and found to be operating satisfactorily or necessary corrective action has been completed.

______________________________ ______________________

Inspected By Date

Source: Adapted from Snorkel Model SL 25 Pre Delivery and Inspection Report

Chapter 20

Stairways

Stairways are a connected series of steps used to access and exit other levels. They are also a primary means of emergency escape from a building. Figure 20.1 shows stair definitions.

Types of Stairways

The basic types of stairways are straight stairways with landings for rest platforms and to change direction, and spiral stairways which are a series of steps arranged in a cylindrical fashion around a central pole. Temporary stairways used in construction have landings every 12 feet. Temporary spiral stairways are not allowed in construction. Stairways can have handrails on unprotected edges or sides or stairrail systems and vertical barriers along the unprotected edges. Landings with unprotected sides and edges must have standard guardrail systems. Figure 20.2 shows a typical staircase with landing.

OSHA Regulations

OSHA regulates stairways under 1926.1051, 1926.1052, and 1926.1060 in Subpart X of the construction industry standards. Stairway openings (before stairs are built) are regulated under 1926.500(b)(1) and 1926.501. OSHA also

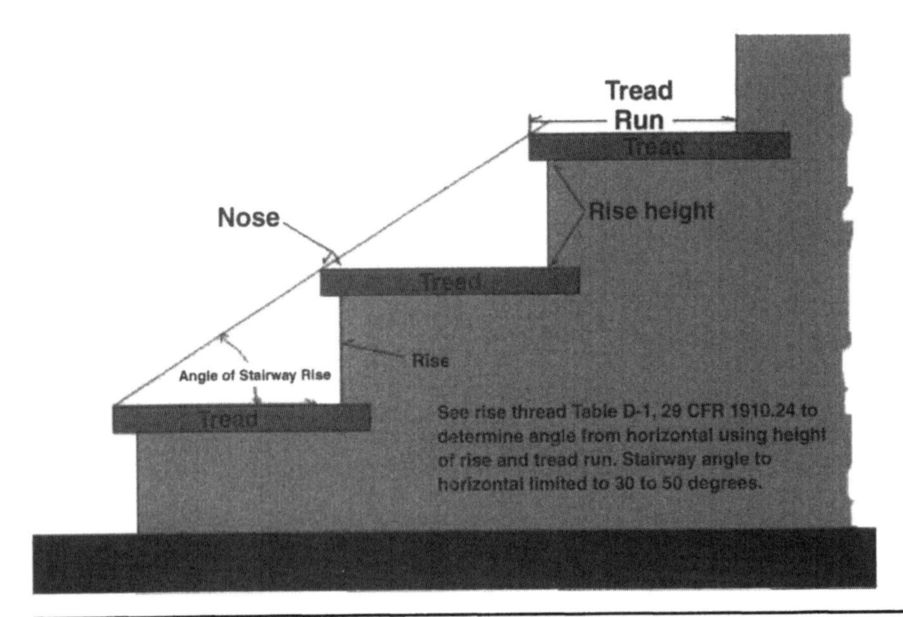

FIGURE 20.1 Stair definitions

regulates stairways under 1910.23(d) and 1910.24 of the general industry standards. The OSHA standards can be found at www.osha.gov. The most important construction requirements for stairways include:

- Requirements for presence of stairways and ladders (1926.1051)
- Landing requirements for temporary stairways (1926.1052(a)(1))
- Angle of stairways (1926.1052(a)(2))
- Riser height and tread depth (1926.1052(a)(3))
- Door platforms (1926.1052(a)(4))
- Filling of metal pan landings and steps (1926.1052(a)(5))
- Hazardous and slippery conditions (1926.1052(a)(6-7))
- Temporary treads and landings (1926.1052(b)(1-3))
- Requirements for handrails and stairrail systems (1926.1052(c)(1-2))
- Height of stairrails (1926.1052(c)(3))

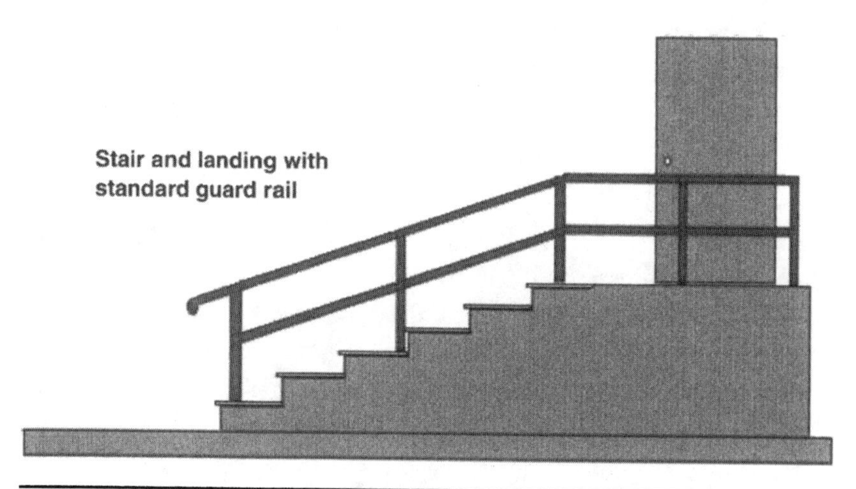

FIGURE 20.2 Staircase with landing

- Midrails, screens, mesh, etc. (1926.1052(c)(4))
- Strength of handrail and top rails of stairway systems (1926.1052(c)(5))
- Height of handrails (1926.1052(c)(6-7))
- Other requirements of handrails and stairrail systems (1926.1052(c)(8-11))
- Guardrail systems for landings (1926.1052(c)(12))
- Training requirements (1926.1060)
- Load-bearing requirements for fixed stairways (1910.24): "Fixed stairways shall be designed and constructed to carry a load of five times the normal live load anticipated but never of less strength than to carry safely a moving concentrated load of 1000 pounds."

Industry Standards

The National Fire Protection Association code NFPA 101B, Code for Means of Egress for Buildings and Structures, 2002 edition, is a national combined building code with requirements for stairs used as emergency exitways. (1) These requirements include minimum stair widths, tread depth, riser heights, requirements for handrails, direction of door openings, fire ratings, and more. This code is a successor to NFPA 101 Life Safety Code (2), and is updated every three to five years.

The American Society of Civil Engineers standard ASCE 7-02, Minimum Design Load for Buildings and Other Structures, and SEI/ASCE 37-02, Design Loads on Structures During Construction, describes load requirements for permanent and temporary stairways. (3,4) Besides the live loads on the stairways from traffic, these standards include other loading factors such as earthquakes, wind, snow, ice, etc. ASCE 37 covers unusual loads such as construction materials) during the construction process.

INSTALLATION

The predominant hazard related to the construction of stairs is falling through a stairway opening. The hole in the floor for the stairs is designed into the plans and either filled with lightweight material (knockouts) or left open. If the floor opening is more than 6 feet above a lower level, then workers in the area — including the workers installing the stairs — have to be protected by personal fall protection systems, a guardrail system, or a cover over the hole. When a cover is used, it must be identified as a "cover" or that there is a "hole" under the covering. Fall protection must remain in place until construction is completed.

All stair railings and landing guardrails should be in place before installers are permitted to use the stairways. Open-sided landings need standard guardrails (that are 42 inches plus or minus 3 inches for a toprail, a midrail half-way between, and a toeboard if objects could fall to lower levels.

Other workers are not allowed to use the metal pan stairways until the treads and/or landings have been filled in, either temporarily with wood or other materials, or permanently with concrete or other final materials. Similarly, skeleton metal frame structures and steps cannot be used unless they are fitted with secured temporary or permanent treads and landings. Temporary treads must be made of wood or other solid material and installed the full width and depth of the stair. If permanent lighting has not been installed, temporary lighting (at least 5-foot candles) should be installed in stairwells.

Training Requirements

OSHA requires that a Competent Person train employees using stairways. OSHA defines a Competent Person as one who "is...capable of identifying existing and predictable hazards...and has authorization to take prompt measures to eliminate them." Areas in which workers are to be trained include:

- Nature of fall hazards
- Proper construction, use, placement, and care in handling of stairways
- OSHA stairway regulations

Retraining must be done as needed to maintain employee understanding and knowledge about stairway safety. Figure 20.3 is a checklist for stairs and stairways.

Common Hazards

In 2001, falls down stairs or steps accounted for 2,134 injuries in construction, 5% of all construction injuries. (6) An analysis of the causes of 58 construction deaths from 1992-2000 related to stairs was done using data from the Bureau of Labor Statistics database, the Census of Fatal Occupational Injuries. (7) The results are shown in Table 20.1.

This study shows that falls through stairway openings and falls while carrying loads on stairs accounted for half of all stair fall deaths.

One study of stairway risk factors found that the risk of stairway falls was decreased by using riser heights under 7 inches, tread depths over 11 inches, and noser projections (overhang on the tread) less than 11/16 inches while descending. (8) Other factors associated with safer stair usage included tread materials (with concrete or stone being best), unobstructed straight ahead views, and not using handrails as a climbing aid. Other research has found that handrail use increases the number of falls, probably due to falls while pulling oneself up by the handrail. (9, 10)

Other factors increasing the risk of stair falls include slippery surfaces, improper rail height, short flights of stairs (1-3 risers), poor contrast between

- ❑ Do temporary stairways have landings at least 30 inches deep and 22 inches wide for every 12 feet or less of vertical rise?
- ❑ Is the angle of the stairway at least 30 and not more than 50 degrees from the horizontal?
- ❑ Are variations in riser height and stair tread depth ¼ inch or less?
- ❑ Can the stairs support 5 times the normal anticipated live load, or a minimum of 1000 pounds of a moving concentrated load?
- ❑ Do doors or gates opening into the stairway have platforms extending at least 20 inches beyond the swing of the door?
- ❑ Are metal pan treads and landings secured in place before filling?
- ❑ Are the stairs free of dangerous projections such as nails?
- ❑ Are the stairs kept free of slippery conditions?
- ❑ Are spiral stairways only used by workers if they are a permanent part of the structure?
- ❑ Are metal pan stairways used by workers during the construction process only if the metal pans have been filled in (except for workers constructing the stairway)? (The filling must be concrete or other permanent materials, or temporary wood or other materials.)
- ❑ Are skeleton metal frame stairways used by workers during the construction process only if the stairs are fitted with secured temporary treads and landings (except for workers constructing the stairway)? (The treads must be made of wood or other solid material and installed the whole width and depth of the stair.)
- ❑ Do stairways with 4 or more risers or rising more than 30 inches (whichever is less) have at least one handrail?

FIGURE 20.3 Checklist for stairs and stairways *(continued on next page)*

- ❑ Do winding or spiral staircases have handrails preventing workers from using interior edges of stairs where the tread depth is less than 6 inches?
- ❑ Are stair rails installed after March 15, 1991 at least 36 inches high?
- ❑ Are any mid rails located halfway between the top of the stairway system and the steps?
- ❑ Do any meshes or screens extend from the top rail to the step and along opening to the top rail supports?
- ❑ Are any vertical members (e.g., balusters) a maximum of 19 inches apart, or do other intermediate structural members have maximum openings 19 inches wide?
- ❑ Can handrails or top rails of stair rail systems support 200 pounds in any downward or outward direction?
- ❑ Are handrails 30-37 inches from the top of the rail to the step?
- ❑ Are top rails used as handrails 36-37 inches from the top of the rail to the step?
- ❑ Are stair rail systems, handrails, and their ends surfaced to prevent punctures, lacerations, or snagging of clothes?
- ❑ Do handrails have an adequate handhold to enable them to be grasped to prevent falls?
- ❑ Do temporary handrails have a minimum clearance of 3 inches between the handrail and walls or other objects?
- ❑ Are unprotected sides and edges of the stairway landing provided with standard 42-inch guardrail systems?

FIGURE 20.3 *(Continued from previous page)* Checklist for stairs and stairways

TABLE 20.1 Analysis of 58 construction deaths related to stairs, 1992-2000

Cause of Fall	Number of Deaths	%
Falls through unguarded stairway openings	21	36
Slips/trips carrying load on stairs	8	14
Fall from unguarded stairs/landings	6	10
Fall while working on landings	6	10
Falls while installing stairs	5	9
Unknown cause/other	12	21

floor ending and stair beginning, poor lighting, putting only part of the foot on the tread, and visual distractions.

Common Precautions

Precautions in the construction and use of stairways include:

- When the knockout is removed to expose the stairway opening, place guardrails or a secured cover over the hole until the stairway is installed and ready to use.
- Wear personal fall protection equipment when installing stairs and working more than 6 feet above a lower level.
- Make sure the stairways are constructed and guarded properly.
- Mark the top and bottom of the stairs if the stair ends are not clearly visible.
- Make sure stairways are lit properly.
- Keep stairways clean and free of oil, grease, and obstructions that could cause slips or trips.
- Do not carry objects on stairs that could interfere with your visibility or balance.
- Do not work from ladders on stairs or landings.

Fall Protection

Fall protection systems for stairway openings and stairways can consist of:

- Guardrail System: A standard guardrail system consists of a top rail, 38-45 inches high, a mid rail (or mesh, screening, or panels), and a toeboard. The top rail must be able to hold 200 pounds, and the mid rail 150 pounds. (29 CFR 1926.502(b))
- Cover: A floor opening cover must be able to support at least twice the weight of workers, equipment, and materials that may be imposed on it at any one time. Covers should be color-coded or marked "HOLE" or "COVER". (29 CFR 1926.502(i))
- Personal Fall Protection System: A standard personal fall protection system consists of a full body harness, lanyard, and 5000-pound anchorage to which to attach the lanyard. There may also be other components such as shock-absorbing lanyard, horizontal lifelines, or vertical lifeline. Personal fall protection systems should be designed by a professional engineer or other qualified person. (29 CFR 1926.502(c))
- Stairrail Systems and Handrails: A stairrail system consists of a vertical barrier at least 36 inches high erected along unprotected sides and edges of the stairway. The upper edge may be a handrail, whose height must be from 30-37 inches from its upper surface to the surface of the tread, or the top edge of a stairrail system, which must be between 36-37 inches from its upper edge to the tread. Handrails must be able to support at least 200 pounds within 2 inches of the top. Midrails, screens, mesh, intermediate members or the equivalent must be provided between the top rail of the stairrail system and the stairway steps.

Typical Malfunctions and Injuries

Following are case studies of typical malfunctions and injuries involving stairways.

Case History No. 1: Fall through Stairway Opening (11)

A 56-year old carpenter was killed when he fell through a second floor stairway opening of a house under construction. He was cutting plywood to hand to men on the roof. He was close to a stairway opening that did not have guardrails nor was it covered. He attempted to hand the 8 foot by 43-inch piece of plywood to the men on the roof through the trusses, and fell through the opening with the plywood while trying to reposition it.

The stairway opening should have been covered or protected with guardrails.

Case Study No. 2: Fall while Installing Staircase (12)

A 44-year old carpenter was killed after falling into a stairwell while trying to install a wooden staircase. The staircase was a large 20' by 20' stairwell from the basement to the second floor. It was installed in several pre-made sections, each half a floor high with a small landing in between. The victim was working on the top second floor landing with another carpenter on the landing below to install the last section. While repositioning the section, they lost control of it. The victim tried to grab the stairs as he fell, lost his balance, and fell 21 feet to the basement.

The carpenters should have been wearing fall protection harnesses attached to vertical lifelines.

Case Study No. 3: Fall from Stairway Landing (13)

A 20-year old drywall installer was killed when he fell from an unguarded second floor landing. He was sanding a sheetrock ceiling on the landing when he stepped or fell off the open side of the landing, unaware of how close he was to the edge.

Open-sided stairway landings must be provided with standard guardrail systems.

References

1. National Fire Protection Association. (2002). NFPA 101B, Code for Means of Egress for Buildings and Structures, 2002 edition. Quincy, MA: NFPA.

2. National Fire Protection Association. (2000). NFPA 101, Life Safety Code, 2000 edition. Quincy, MA: NFPA.

3. American Society of Civil Engineers. (1978). ASCE 7-02, Minimum Design Load for Buildings and Other Structures. Reston, VA: ASCE.

4. American Society of Civil Engineers. (2002). SEI/ASCE 37-02 Design Loads on Structures During Construction. Reston, VA: ASCE.

5. Occupational Safety and Health Administration. (1997). Stairways and Ladders. OSHA Publication 3124, revised ed. Washington, DC: OSHA.

6. Bureau of Labor Statistics. (2001). Table R64. Number of nonfatal occupational injuries and illnesses involving days away from work by event or exposure leading to injury or illness and industry division, 2001. Available at: www.bls.gov/iif/oshcdnew.htm, Table R64, 2001.

7. McCann, M. (2003). Unpublished data.

8. Templer J, Archea J, Cohen HH. (1985). Study of factors associated with risk of work-related stairway falls. J. Safety Research 16(4): 183-196.

9. National Bureau of Standards. (1979). Guidelines for Stair Safety. Washington, DC: NBS.

10. Johnson D, Pain K. (1987). Occupational Falls: A Research Review. Edmonton, Alberta: Occupational Safety and Health Division, Alberta Community and Occupational Health.

11. Occupational Health Services, New Jersey Department of Health. (1991). New Jersey FACE Report No. 91NJ01501: Carpenter dies after a 18 foot 9 inch fall through a floor opening. Morgantown, WV. Author.

12. Occupational Health Services, New Jersey Department of Health. (1998). New Jersey FACE Report No. 98NJ05501: Carpenter dies after falling into a stairwell while installing a staircase. Morgantown, WV: Author.

13. National Institute for Occupational Safety and Health. (1994). FACE Report No. 9413: Drywall mechanic dies after 10-foot fall from an open-sided floor B South Carolina. Morgantown, WV: Author.

Chapter 21

Ladders

Ladders are widely used in construction. They are one of the most common tools used by construction workers, and everyone thinks they know how to use a ladder. The severity of the risk from ladder use is not generally recognized. There is a need for ladder safety training.

TYPES OF LADDERS

Basically, ladders are upright climbing devices used either for gaining access to another level or as a work platform. There are two basic types of ladders: portable ladders and fixed ladders. Following are three ladder classifications:

1. Portable (ladders are rated based on maximum working load)

   a. Non-self-supporting (used as work platforms or to access a higher level)

      – Extension ladders (maximum length – 60 feet)

      – Straight ladders (maximum length – 30 feet)

   b. Self-supporting (used as work platforms, available in heights up to 20 feet)

      – Step ladders (sometimes referred to as an "A" frame)

  - Platform ladders (stable working platform at the highest standing level)
  - Trestle and extension trestle ladders (used to support planks or staging)
  - Ladder stands (mobile, self-supporting ladder in the form of stairs)

c. Job-made ladders

- Handrails must extend at least 36 inches above landing

2. Fixed (used to access a higher level, fixed ladders set at 90 degrees or less).

- Less than 24 feet
  - Handrails must extend at least 42 inches above the landing.
- Greater than 24 feet
  - Handrails must extend at least 42 inches above the landing.
  - Cages, ladder safety device, or self-retracting lifelines are required.

Portable Ladders

Portable ladders are the common work ladders that can be carried from one location to another.

Non-Self-Supporting Ladders

Non-self-supporting portable ladders include straight ladders — the workhorse of portable ladders — and extension ladders (see Figure 21.1). Both types of ladders have feet on them. Straight ladders commonly go up to 20 feet in length. Extension ladders consist of two or more ladder sections that travel in guides or brackets that can be assembled to give adjustable lengths. The maximum length of an extension ladder is 60 feet.

Job-Made ladders

Job-made ladders are portable wood ladders that are assembled by employees on the job to provide access to other levels. Sometimes, they will be double-

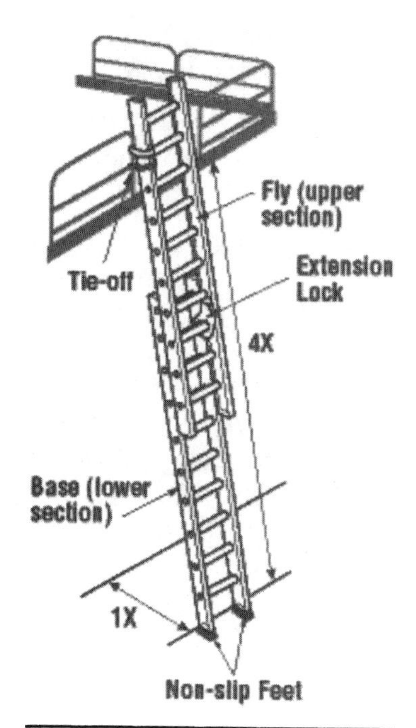

FIGURE 21.1 Extension ladder *(Reprinted with permission of the Canadian Centre for Occupational Health and Safety (CCOHS), 250 Main Street East, Hamilton, Ontario L8N 1H6; Phone (905) 572-4400; Toll-free: 1-800-263-8466; Fax: (905) 572-4500; E-mail: inquiries@ccohs.ca.)*

cleat ladders that are double width and have a center rail to allow simultaneous two-way traffic up and down.

Self-Supporting Portable Ladders

Self-supporting portable ladders include common stepladders, two-way stepladders, platform ladders, ladder stands, trestle ladders, and extension trestle ladders (Figure 21.2). These ladders are usually used as work platforms. When used properly, self-supporting ladders have metal spreaders or locking devices to keep them stable.

Stepladders have flat, slip-resistant steps, and a hinged back that is for support only. Only one person at a time can be on a stepladder. The two-way stepladder has steps on both sides, and models made by some manufacturers can be used by two people at a time. Platform ladders have a stable working platform at the highest standing level. Ladder stands are mobile, fixed-size, self-supporting ladders consisting of a wide flat-tread ladder in the form of stairs. They may have handrails. Trestle ladders are hinged at the top and form equal angles with the base. Extension trestle ladders have a vertically adjustable straight ladder that can be locked in place. The vertical ladder must extend at least three feet into the base. Extension trestle ladders are used in pairs to support planks or staging. Self-supporting ladders are available in heights up to 20 feet.

Straight ladders and extension ladders are used to gain access to other work levels or as a work platform. Almost all trades use them. Self-supporting ladders are used mostly as work platforms. Painters, electricians, sheet metal workers, and other trades working on or above ceilings commonly use them.

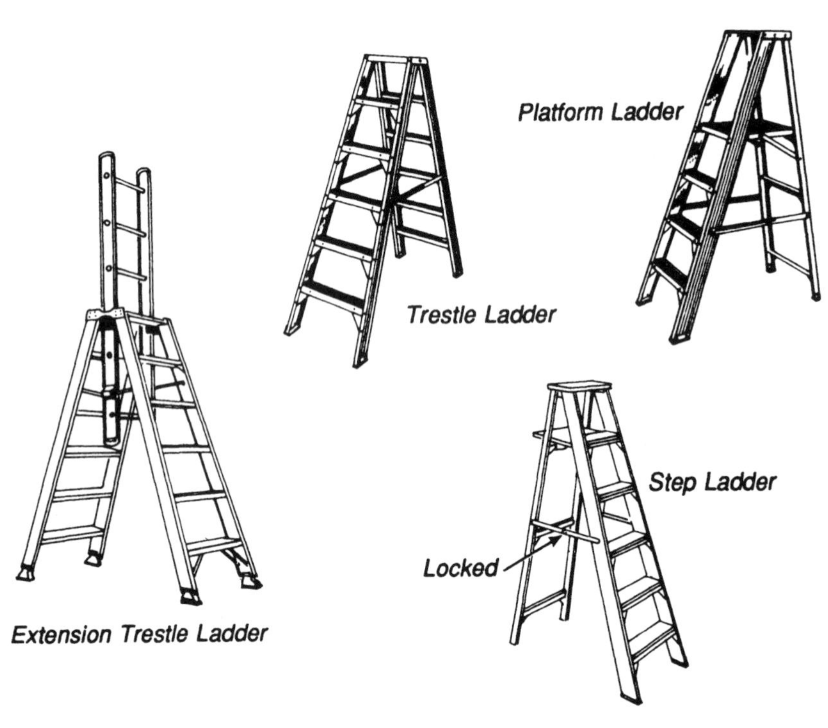

FIGURE 21.2 Self-supported ladders *(Courtesy Construction Association of Ontario)*

Fixed ladders are used for access to another level only. They are commonly used where access is infrequent and stairs are not practical.

Both self-supporting ladders and non-self-supporting ladders can be made out of wood, metal (usually aluminum), or reinforced fiberglass. Fiberglass ladders are classified as nonconductive and are used where electrical contact is possible, although these ladders can conduct electricity if they are dirty, contaminated with road salt, mud, or some latex paints.

Portable ladders are rated based on their maximum working load, the maximum weight they can bear safely, including the worker and his or her tools. The ratings are:

- Light duty (Type III) for up to 200 pounds
- Medium duty (Type II) for up to 225 pounds
- Heavy duty (Type I) for up to 250 pounds
- Extra heavy duty (Type IA) for up to 300 pounds
- Special duty (Type IAA) for up to 375 pounds.

Only Type I, IA, and IAA ladders should be used in construction. The duty ratings of a ladder are listed on a label on the ladder (Figure 21.3).

Fixed Ladders

Permanent fixed ladders are made out of metal and are permanently fastened to a structure such as a building, water tank, and so forth. They can either have individual rungs attached to the vertical surface or have the rungs attached to side rails which are bolted to the vertical surface. See Figure 21.4. Fixed ladders are used to gain access from one level to another.

OSHA Regulations

OSHA regulates ladders under 1926.1050, 1926.1051, 1926.1053, and 1926.1060 in Subpart X of the construction industry standards and in Suppart D of 20 CFR 1910 of the General Industry Standards. The General Industry standards apply to construction jobs, for example, when a fixed ladder is used during the course of construction. Section 1926.1053 deals with ladder safety.

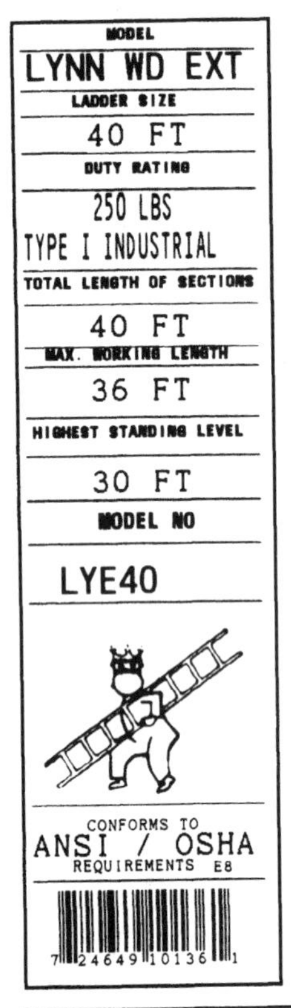

FIGURE 21.3 Duty rating label (Courtesy Lynn Ladder & Scaffolding Co., Inc.)

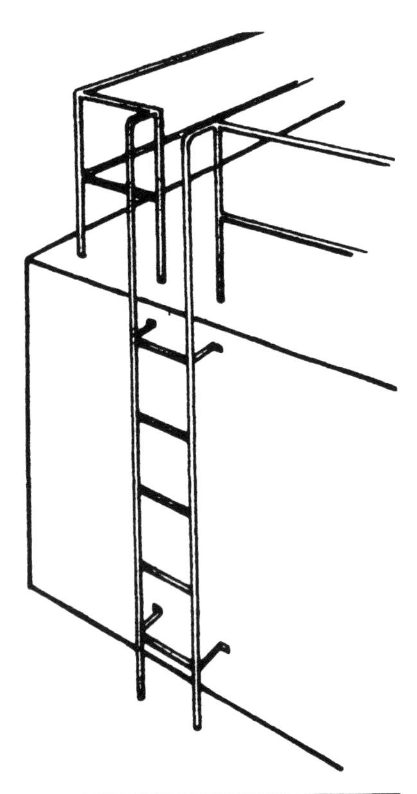

FIGURE 21.4 Fixed ladder (Courtesy Construction Association of Ontario)

Ladder Construction

The most important general requirements for ladder construction include:

- Load capacity and design (1926.1053(a)(1)) and Appendix A to 1926 Subpart X — Ladders)
- Spacing of rungs, cleats, and steps (1926.1053(a)(2-3))
- Spacing of side rails (1926.1053(a)(4))
- Structural requirements for ladders (1926.1053(a)(5-6, 11))

- Do not tie or fasten ladders together unless made to be used that way. (1926.1053(a)(7))
- Requirements for platforms and landings between ladders (1926.1053(a)(10))
- Painting of wood ladders (1926.1053(a)(12))
- Clearance requirements for fixed ladders (1926.1053(a)(13-17))
- Requirements for cages, wells, ladder safety devices or self-retracting lifelines with fixed ladders (1926.1053(a)(18-23))
- Extension of side rails of fixed ladders above upper level (1926.1053(a)(24))

Ladder Use

OSHA requirements for ladder use, including job-made ladders, include:

- Extension and supporting of side rails of portable ladders at upper level (1926.1053(b)(1, 10))
- Slipping hazards on ladders (1926.1053(b)(2))
- Maximum ladder loads (1926.1053(b)(3))
- Use ladders only for designed purpose (1926.1053(b)(4))
- Setting up straight and extension ladders (1926.1053(b)(5-9))
- Do not move or extend ladders when occupied. (1926.1053(b)(11))
- Nonconductive side rails (1926.1053(b)(12))
- Forbidden to climb on the cross-braces or backs of regular stepladders (1926.1053(b)(14))
- Inspection of ladders and tagging of defective ladders (1926.1053(b)(15-17))
- Forbidden use of single-rail ladders (1926.1053(b)(19))
- Climbing and descending ladders (1926.1053(b)(20-22))
- Ladder training (1926.1060)

Industry Standards

The American National Standards Institute (ANSI) has a variety of industry standards for ladders (1-7). These standards set rules and establish minimum requirements for the construction, testing, selection, care, and use of the common types of ladders. They also set the rules and minimum requirements for labeling. The ANSI ladder standards include:

- ANSI A14.1-2000 American National Standard for Ladders — Wood Safety Requirements
- ANSI A14.2-2000 American National Standard for Ladders — Portable Metal — Safety Requirements
- ANSI A14.3-2002 American National Standard for Ladders — Fixed — Safety Requirements
- ANSI A14.4-2002 American National Standard Safety Requirements for Job-Made Wooden Ladders
- ANSI A14.5-2000 American National Standard for Ladders — Portable Reinforced Plastic — Safety Requirements
- ANSI A14.7-2000 American National Standard for Mobile Ladder Stands And Mobile Ladder Stand Platforms
- ANSI A14.10-2000 American National Standard for Ladders — Portable Special Duty Ladders

In addition, a proposed ANSI standard, A14.8-2003 American National Standard for Ladders — Portable Ladder Accessories – Safety Requirements, is near completion. In Appendix A to 29 CFR 1926 Subpart X, OSHA references the following ANSI standards as nonmandatory guidelines: ANSI A14.1-1982; ANSI A14.2-1982; ANSI A14.3-1984; ANSI A14.4-1979; and ANSI A14.5-1982. However, the most recently published standards should be used.

Installation and Removal of Portable Ladders

Use of portable ladders involves several preliminary steps: selecting the right ladder for the job; inspecting it; getting it to the work location; and setting it up. After use, they have to be removed and stored properly.

Ladder Selection

Choose the right ladder for the job. This includes the type of ladder, length of ladder, the duty rating of the ladder, and the ladder material. Straight ladders or extension ladders are used for accessing another level or as a work platform on a vertical surface. For ladder lengths over 30 feet, extension ladders should be used. The maximum working height of a straight or extension ladder is its highest standing level as listed on the duty rating label on a ladder plus 5 feet. For all practical purposes, the highest standing level is the fourth rung. When using ladders for accessing a higher level, the top of the ladder should extend 3 feet above the upper surface. For accessing heights over 60 feet, two or more extension ladders would be needed, with landing platforms in between. Never splice two straight ladders together to make a longer ladder.

Stepladders are commonly used indoors for working at heights under 20 feet, especially for overhead work. They should only be used on a firm, level surface such as a floor to prevent them from tipping over sideways. The most comfortable working height is at shoulder level, about 5 feet above standing level for the average worker. Since the maximum standing height for a worker on a stepladder is on the second step from the top – about 2 feet down from the top of the ladder – then the maximum working height of a stepladder is its height plus 3 feet. Stepladders over 20 feet in height should never be used.

The maximum intended load on the ladder – the weight of the worker plus tools and materials – determines the required duty rating or maximum load capacity for the ladder. The ANSI-required duty-rating sticker on a commercial ladder tells you its maximum load capacity. According to ANSI, only Type I, IA, and IAA ladders should be used in construction.

The working environment determines the material from which the ladder is manufactured. If there is a chance of contact with energized electrical current, then a nonconductive fiberglass ladder should be used. If wooden ladders are not completely dry, they can conduct electricity. If ladder weight is a concern, for example, having to carry a ladder for long distances, then the lighter metal ladder is best, followed by a fiberglass ladder. Make sure the steps of metal ladders are roughened or grooved to prevent slipping. All domestic ladders are.

After selecting a ladder, it must be inspected. Defects and contamination could compromise its safety. The inspection checklists in the appendix at the end of this chapter will help in this important step.

Moving Ladders

Lifting ladders off the roofs of trucks or vans and carrying them to a work location can result in slips and trips or back injuries. Ladder racks for trucks are available with hydraulic devices so they can be rotated down to shoulder level, making it easier to lift off the ladder. Two workers should carry long and heavy ladders when possible. If a ladder being carried starts to fall, do not try to recover it. Dump it and get out of the way. Reinspect the ladder for damage. Never move or reposition extension ladders without collapsing the sections. Make sure to keep fingers away from moving or sliding parts.

Setting Up All Portable Ladders

The first step is to check the area where the ladder will be set up for hazards. This includes overhead power lines, other overhead obstructions, clutter and traffic around the base of the ladder, a level and firm surface for the ladder base, and a firm and secure surface for the top of the ladder. If necessary, rethink the ladder location. If the ladder will be located near traffic, either pedestrian or vehicular, then the area around the base of the ladder should be barricaded off. Lock doors that would allow people to exit near the ladder. Also make sure that the weather conditions (wind, snow, ice) are suitable for setting up and using a ladder.

Next, inspect the ladder again for possible damage in transport before setting it up. Make sure it is clean. Mud, oil, and grease can make a nonconductive ladder conduct electricity.

Setting Up Straight and Extension Ladders

If possible, use two people to erect straight and extension ladders. The ladder should be at an angle of 75.5 degrees. At a lesser angle, the base of the ladder might slide away; at greater angles, the ladder might fall over.

Follow these steps to erect a straight or extension ladder:

- Place the bottom of the ladder at the base of the structure

- Walk the ladder up hand-over-hand. Extend the ladder 3 feet above resting point on roof if used for access to higher level.
- Pull the base of the ladder slightly away from the building. For extensions ladders, go around to the back of the ladder, lift the fly section with the rope and pulley, and lock it.
- Pull the base of the ladder out until it is at an angle of 75.5 degrees. The base should be 1 foot away for every 4 feet of ladder length to its support position (4:1 rule). See Figure 21.5. The length of the ladder is best estimated by counting the number of rungs to the resting place since rungs are about 12 inches apart.

The common Fireman's rule method for estimating the angle is not as accurate as the 4:1 rule (8). It involves the worker standing in front of the ladder with toes against the base and arms stretched directly in front. The ladder is at the proper angle when the worker's palms touch the rung without having to bend forward or backward. The angle varies with ladder length and worker height and arm length.

Leveling and Stabilizing the Ladder

The ladder feet should be level. Ladder levelers can be attached to most straight and extension ladders to help level the ladder on uneven ground or for use on

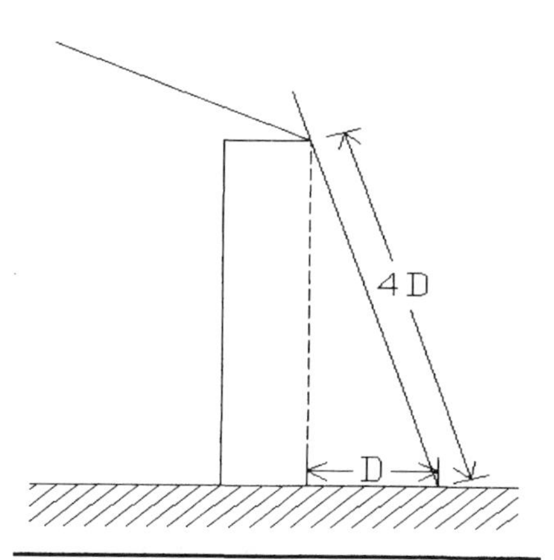

FIGURE 21.5 The 4:1 rule for erecting ladders

steps (Figure 21.6). Note that they are not designed for use on soft ground or slippery surfaces.

The base of the ladder should be secured, for example, by tying to stakes or placing a secured board against the feet. If the surface is soft, uncompacted, or rough, use a mud sill. See Figure 21.7. Slip-resistant safety feet, although useful, are not a substitute for securing the ladder base.

The side rails at the top of the ladder should be placed so they have equal support. Ladder stabilizers that attach near to the top rung provide additional width stability for the tops of ladders (Figure 21.8). Stabilizers can also be used for corners (Figure 21.9). Ladder hooks can also be used to stabilize the tops of ladders where there is a surface to hook on to (Figure 21.10).

When it is needed to set up ladders against poles or similar objects, Fixed V-Rung devices or pole grips, steel bands with a V indent in the middle which replaces the top rung, Figure 21.11) and Pole Straps, similar to V-Rung devices but made

FIGURE 21.6 Ladder levelers (Courtesy Lynn Ladder & Scaffolding Co., Inc.)

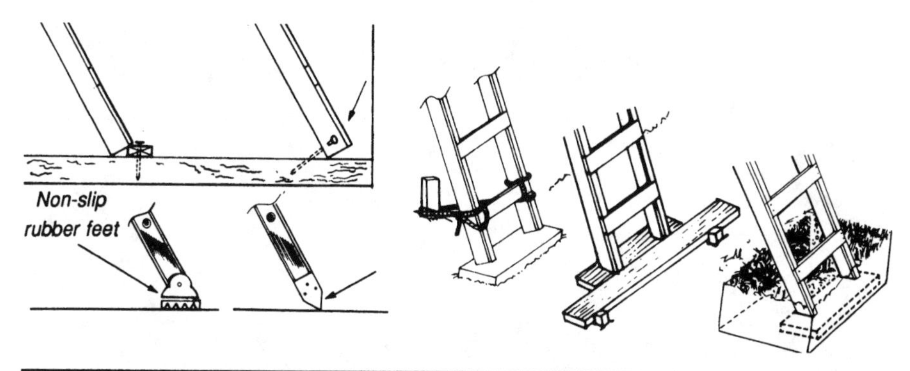

FIGURE 21.7 Securing the ladder base. (Courtesy Construction Association of Ontario)

of adjustable nylon, (Figure 21.12) can be used. Pole Lashes that attach to top of ladder and wrap totally around the pole and Ladder-Cinches, long-adjustable straps, are other devices for attaching ladders to poles and round objects.

For ladders used for access to an upper level, the top of the ladder should extend at least 3 feet above the landing level. If this is not possible due to ladder length, the top must be secured to prevent the ladder from slipping sideways. This is recommended for all ladders. A grabrail or similar device provided to help workers get on and off the ladder. For maximum safety, have another worker hold the ladder.

Each new section of an extension ladder must overlap the lower section by a minimum distance that depends on the overall length, and must lock in place. Always check to make sure the locks have properly engaged. See the Table 21.1.

After setting up the ladder, check it over before climbing. Check for levelness of the ladder base, stability of the base, security of rung locks for extension ladders, and so forth. See the Appendix for a checklist on setting up extension ladders.

Setting Up Self-Supporting Ladders

Stepladders and other self-supporting ladders should be placed so that all four legs are on solid, level ground. Unevenness in leg positioning can result in the ladder falling over sideways. The spreader brace must be locked in the fully open position. Never use a self-supporting ladder as a straight ladder by leaning it against a wall.

FIGURE 21.8 Ladder stabilizer. (Courtesy Lynn Ladder & Scaffolding Co., Inc.)

FIGURE 21.9 Corner ladder stabilizer. *(Courtesy Lynn Ladder & Scaffolding Co., Inc.)*

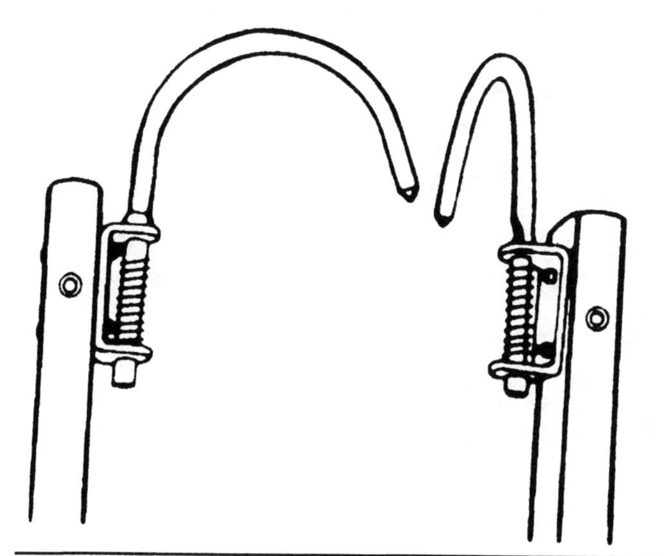

FIGURE 21.10 Ladder hooks. *(Courtesy Lynn Ladder & Scaffolding Co., Inc.)*

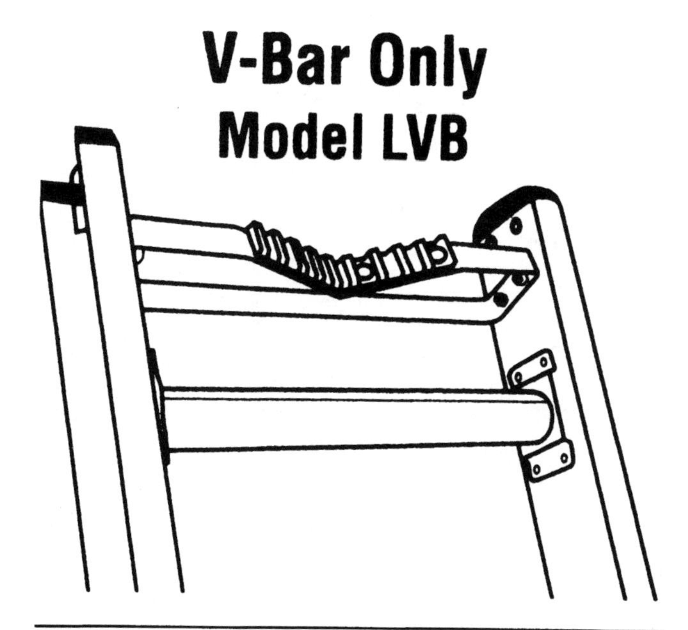

FIGURE 21.11 V-Rung device. *(Courtesy Lynn Ladder & Scaffolding Co., Inc.)*

FIGURE 21.12 Pole Strap. *(Courtesy Lynn Ladder & Scaffolding Co., Inc.)*

TABLE 21.1 Minimum overlap for different lengths of extension ladders.

Ladder Length	Overlap
Up to and including 36 feet	3 feet
Over 36 feet through 48 feet	4 feet
Over 48 feet though 60 feet	5 feet

Removing Ladders

To remove ladders, reverse the steps above for setting them up. Since you will be walking backwards when lowering a straight or extension ladder, check for tripping hazards first. Lower it slowly.

Ladder Care

- Keep all ladder components, including safety shoes, in good condition.

- Wood ladders to be used outside should be treated with a clear finish or transparent penetrating preservative to prevent weather damage.
- Never paint a wood ladder. It can cover dangerous cracks or fill or hide them.
- Lubricate metal parts periodically, including extension ladder rung locks and pulleys, to prevent rusting and to keep them working properly.
- Discard metal or fiberglass ladders that have been exposed to fire (such as welding) or exposed to strong chemicals that can damage them.
- Store wood ladders away from heat or dampness; store fiberglass ladders away from sunlight or other ultraviolet sources such as arc welding.
- Store ladders on racks that can support them.
- Never sit on ladder side rails.

Installation of Fixed Ladders

Fixed ladders built and tested according to ANSI A14.3-1984 will satisfy the OSHA requirements for fixed ladders. However, good practice says to use the most recently published version (ANSI A143.-2002). Fixed ladders must be set at pitches of not more than 90 degrees. Handrails for fixed ladders must extend at least 42 inches above the landings. Ladders with a climb length of more than 24 feet or whose top is more than 24 feet above a lower level must have cages, ladder safety devices or self-retracting lifelines. See the fall protection section for information on cages, ladder safety devices, and self-retracting lifelines.

TRAINING REQUIREMENTS

The OSHA construction standard requires that a Competent Person train employees using ladders. OSHA defines a Competent Person as one who "is...capable of identifying existing and predictable hazards...and has authorization to take prompt measures to eliminate them."

Areas in which workers are to be trained include:

- Nature of fall hazards
- Correct procedures for erecting, maintaining, and disassembling fall protection systems
- Proper construction, use, placement, and care in handling of ladders
- Maximum intended load capacity of ladders used
- OSHA ladder regulations

Retraining must be done as needed to maintain employee understanding and knowledge about ladder safety.

Inspection Checklists

A competent person must inspect ladders for damage and contamination. If the ladder is damaged, it must be tagged "DO NOT USE", or similar language, and taken out of service until repaired. The appendix at the end of this chapter contains inspection checklists for ladders.

Common Hazards

Falls are the leading cause of death in construction, resulting in an average of 301 deaths per year, based on statistics from the Census of Fatal Occupational Injuries (CFOI) for 1992-99 (10). CFOI is a U.S. Bureau of Labor Statistics database. Of these 301 falls per year, 42 (14%) were due to falls from ladders (11).

An analysis of 137 fall deaths and injuries from extension ladders showed that 47% occurred while working from ladders, 31% while climbing or descending ladders, and 10% while in transition to and from ladders at upper levels (12). Ladder collapses caused 35% of the extension ladder deaths and injuries. Another cause of ladder falls can be electrical shock. A study of 61 electrical injuries resulting in emergency room treatment found that one-quarter of the electrical exposures resulted in ladder falls (13).

A study of 123 occupational injuries caused by falls from portable ladders (14) found that 60% occurred while working from a ladder, 26% while descending, and 14% while ascending. Over half (57%) occurred with stepladders, 30% with extension ladders, 10% with other straight ladders, and the rest with other types including job-made ladders. The types of events resulting in the ladder accidents are found in Table 21.2.

A review of the literature on ladder falls commissioned by NIOSH (15) concluded that more research is needed on falls due to mechanical failure of ladders, stepladders tipping sideways, straight ladders slipping out at the base, straight ladders tipping at the top, workers tripping/slipping on ladders, and transitions to and from elevated platforms.

These studies show that factors that can contribute to falls include slippery or oily ladder rungs or steps, carrying objects in hands, reaching, loss of balance (especially when transiting to and from ladders at upper levels), electrical shock, and so forth. Factors that can result in the ladder collapsing include sliding of the base of straight or extension ladders, tipping sideways of stepladders, sliding of straight and extension ladders at the top, and insecure ladder components. Often these ladder collapses can result from pushing or pulling activities while working from the ladder.

Table 21.3 is a summary of causes of wooden portable ladder accidents taken from ANSI A14.1-2000 (1).

TABLE 21.2 Events resulting in ladder accidents

Overreaching	23	19%
Slip on rungs	17	14%
Misstep on rungs	12	10%
Structural failure	11	9%
Struck by/attempted to catch/ avoid falling object	10	8%
Applying excessive force	9	7%
Leaning stepladder against a structure	9	7%
Transition on to or off of ladder	8	6%
Standing on top rung	7	6%
Other miscellaneous	17	14%
Total	123	

Common Precautions

Some of the most important precautions for ladders are involved in setting up the ladder. Refer to the guidelines in the appendix at the end of the chapter. The precautions for ladder use are adapted from the guidelines for extension ladders (109):

- Recheck the ladder setup during climb (levelness, stability of base, tie-offs, rung locks on extension ladders, spreader braces on stepladders.
- Recheck your shoes for mud, paint, oil, or other slick/sticky substances.
- Recheck the work area for clutter and make sure that any traffic barriers needed are in place. Make sure the weather is suitable for climbing.
- Use a spotter whenever possible to help stabilize the ladder.
- Maintain 3-point contact with ladder (two feet and one hand or two hands and one foot) at all times. Never carry tools or other materials in your hands.
- Face the ladder when climbing, descending, or working from a ladder
- Keep your body centered within the frame of the ladder
- Recheck the ground for clutter so you don't trip when you get off a ladder when descending.
- For access to upper level with straight/extension ladders, recheck that the ladder extends 3 feet above resting point

Working from a Ladder

Working from a ladder can be hazardous. In many instances, another means of reaching an elevated workstation – such as using an aerial lift – is safer. If you work from a ladder, the following procedures can help minimize the risk:

- Keep both feet on the same rung/step while working.
- Keep your body centered inside the frame of the ladder between the side rails.

TABLE 21.3 Summary of significant injury causes from portable ladders.

Ladder Design	Cause of Accident	Possible Factors Involved in Accident				
		Base Support or Surface	Top Support	Set-Up	User Location	Ladder Selection and Condition
Self-supporting (stepladder)	Stability	Soft Uneven Unstable surface Slipper surface Differences in firmness of surface Slope of surface	Not applicable	Unlocked spreaders One or more feet unsupported Used unstable or insufficient supports Ladder not close enough to work location User's physical condition	Above "highest standing level" label Reaching out too far, particularly laterally Climbing onto ladder from above Handling heavy loads or unstable objects Applying side load	Too short (size) Incorrect ladder type (IA, I, II, or III) Incorrect style ladder (for example, platform or single versus stepladder) Use of improper equipment (for example, ladder versus scaffold)
	Sliding	Uneven surface Low friction Unstable surface Ice snow or wet surface Slippery surface	Not applicable	Used as a non-self-supporting ladder Ladder not close enough to work location Worn, missing or contaminated feet	Reaching out too far Stepping off ladder Applying side load	Use of improper equipment (for example, ladder versus scaffold) Removal or deterioration of slip-resistant ladder feet
Non-self-supporting (extension and single)	Human slip	Not applicable	Not applicable	Pitch (angle) too steep (especially with flat rungs) User's physical condition or shoes	Descending	Dirty, oily, or icy step surfaces Use of improper equipment (for example, ladder versus scaffold)
	Lateral sliding (left or right) at top support	Uneven surface Differences in firmness of surface Unstable surface	Uneven surface Slippery surface Unstable surface (pole, tree, corner of building, and the like) Ice, snow, or wet surface Insufficient top support	Not tied off Not held at base Inadequate or excessive extension above top support Ladder not close enough to work location Pitch (angle) too steep Feet unsupported or unstable Extension locks not engaged	Getting on or off ladder to roof Reaching out too far laterally Applying side load	Too short or too long (size) Not extended far enough (too steep) Use of improper equipment (for example, ladder versus scaffold)
	Outward sliding at lower base support	Unstable surface Low friction Loose surface Ice, snow, or wet surface Slippery surface	Overextension above top support	Used unstable or insufficient supports Pitch (angle) too flat Ladder not footed Ladder not tied off or blocked Worn, missing or slippery feet Improper selection of feet or slip-resistant bearing surfaces Extension locks not engaged	Above "highest standing level" label Sliding tendency increases above ladder working length midpoint Careless climbing onto or off of ladder (from or to roof, and the like) Applying side load	Wrong foot or slip-resistant bearing surface Too long (size) or extended too far Use of improper equipment (for example, ladder versus scaffold)

This chart is provided courtesy of the American Ladder Institute, the Accredited Standards Developer for Ladder Safety Standards under the American National Standards Institute. For more information on ladder safety standards and a comprehensive safety resource list, please visit www.americanladderinstitute.org.

- Use a rope or lift to raise or lower tools or materials. Attach tools to your belt. Don't carry loads up and down the ladder in your hand unless there is no other option and it won't affect your balance. While climbing or descending, place both feet on each rung before repositioning a hand.
- Recheck the area below in case you have to drop the load.
- Tieoff the top of straight or extension ladders when possible.
- For pushing or pulling activities, use a body harness and separately anchored lanyard when possible.
- Maintain 3-point contact with the ladder at all times.

Stepping On/Off a Ladder at Height

Many ladder injuries occur while transferring to or from a ladder at a height. The following procedures can minimize that risk for straight, extension ladders and fixed ladders:

- Free your hands of all materials before stepping on/off the ladder.
- Maintain 3-point contact at all times.
- Recheck the tie-off at the top of the ladder.
- Check for hazards before stepping on/off the ladder.

Descending Ladders

More injuries occur while descending a ladder than while climbing one. The following procedures can minimize the risk:

- Maintain 3-point contact with the ladder at all times.
- Face the ladder as you walk down.
- Put both feet on each rung/step as you descend.
- Check the ground for clutter before stepping off the ladder.

Fall Protection

OSHA does not require fall protection for ladders, with the exception of certain fixed ladders.

Fixed Ladders

OSHA only requires fall protection on fixed ladders whose tops are 24 feet or more above a lower level (1926.1053(a)(18)) or which are more than 24 feet long. (1926.1053(a)(19) Such fixed ladders must be provided with cages, wells, ladder safety devices, or self-retracting lifelines to prevent falls (see Figure 21.13).

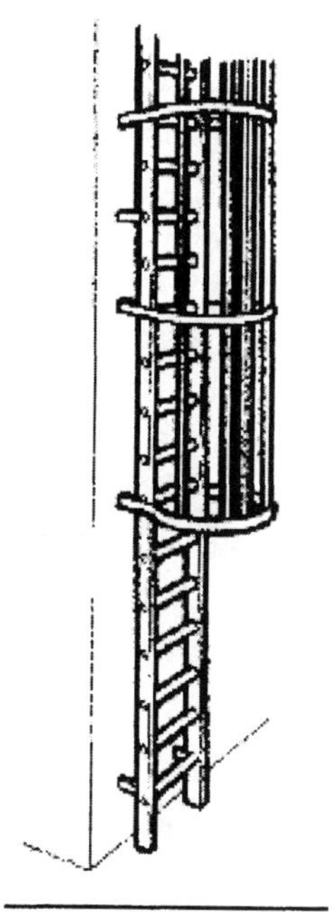

FIGURE 21.13 Fixed ladder cage

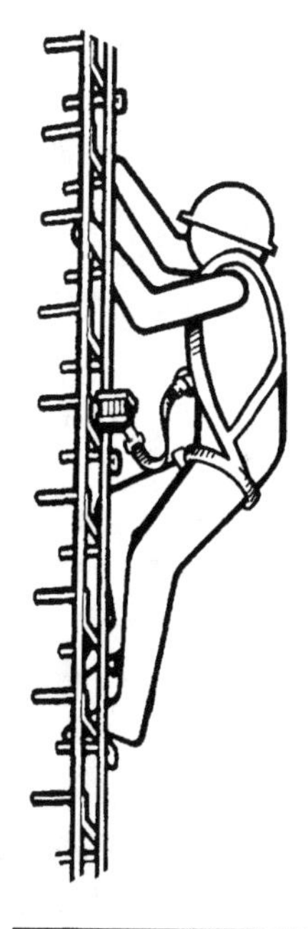

FIGURE 21.14 Ladder safety device for fixed ladder

Figures 21.13 and 21.14 reprinted with permission of the Canadian Centre for Occupational Health and Safety (CCOHS), 250 Main Street East, Hamilton, Ontario L8N 1H6; Phone (905) 572-4400; Toll-free: 1-800-263-8466; Fax: (905) 572-4500; E-mail: inquiries@ccohs.ca.

Cages or wells must be designed by a professional engineer or otherwise qualified person. The cage must start seven to eight feet above the lower level and consist of horizontal bands fastened to vertical bars and either to the side rails or the fixed stairs or the building. The insides of cages shall be free of projections. The bottom of the cage must be seven to eight feet above the point of access to the fixed ladder. There must be a landing every 50 feet. Wells have similar requirements to cages except they completely surround the fixed ladder.

Ladder safety devices for fixed ladders are intended to catch a climber by activating within two feet after a fall occurs. The climber wears a fall protection harness that attaches to a sleeve or collar and travels on a rail or wire rope attached to the fixed ladder. This connection between the harness and collar must be nine inches or less. As the person climbs, the collar must move along without any help from the climber. If he or she falls, the device activates and clamps down on the cable or rail within two feet. The ladder safety device must be capable of withstanding an 18-inch drop of a 500-pound weight. Self-retracting lifelines must have rest platforms at least every 150 feet. They must meet the requirements of 29 CFR 1926 Subpart M for personal fall arrest systems.

Fall Protection on Portable Ladders

OSHA does not have any requirements for fall protection on portable ladders since, in OSHA's opinion, the ladders were intended to be used for access to an upper level rather than as a work platform. Sometimes, personal fall protection is advisable, especially when performing strenuous work from a ladder. Do not tie off to portable ladders, however, since they are not stable enough to be an anchor, especially if there are pushing or pulling work activities that could displace the ladder. An exception is, when the ladder is tied at the top and bottom, the fall personal fall protection system is a positioning device, and the anchors are adequate. If possible, when working from a ladder, tie off to a self-retracting lanyard or lifeline that is anchored to a stable structure above shoulder level (for example on the roof). It should be engineered for a maximum free fall distance of two feet. Such fall protection systems should be designed by a professional engineer or other qualified person.

Typical Malfunctions and Injuries

Case History No. 1: Fall from Extension Ladder (16)

A healthy 73-year-old handyman fell 25 feet to his death from an extended aluminum ladder positioned against the side of a house. The ladder was held in position by a helper. The handyman was washing the side of the house with a scrub brush in his right hand. To reach a section to his right beyond his reach, he removed his right foot from the ladder, leaving only his left hand and foot for a grip. He apparently slipped, falling 15 feet to a roof overhang, and then another 10 feet onto the driveway.

A basic rule on ladders is always to have 3-point contact, either two hands and one foot or two feet and one hand. Violation of this rule with the overreaching probably caused this death.

Case History No. 2: Collapse of Extension Ladder Section (17)

A construction laborer died when the extension ladder he was climbing collapsed. He was instructed to carry a pail of mortar up one level. He placed one end of the top section of an aluminum extension ladder (without safety feet) on the wet concrete floor and leaned the other end against the wall. He then began to climb. The ladder apparently slipped on the wet floor causing him to fall about 12 feet.

Upper sections of extension ladders should never be used as single ladders. Employees should be trained in the proper use of ladders.

Case History No. 3: Collapse of Extension Ladder (18)

A 37-year old plumber was killed when he fell about 20 feet from an extension ladder which collapsed. The plumber had climbed the ladder, which they had leaned against the roof, onto the roof to examine pipes. He then stepped back onto the ladder to descend to the ground. The ladder collapsed, with one section sliding into the other. The victim fell to the pavement.

Examination of the extension ladder showed that the rung locks would occasionally get hung up on braces on the second to top rung of the lower section. The rung locks are designed to grip over two rungs, one from each ladder section. In this case, the locks would not cover completely the upper section rung. When weight was put on the upper section of the ladder, the upper section slid down. It is crucial that ladders are thoroughly inspected before use, and that the rung locks are checked to ensure that they are properly locked in place. In this case the ladder should have been taken out of service.

Case History No. 4: Movement of a Step Ladder (19)

A 46-year-old sheet metal worker was killed when he fell off an 8-foot stepladder, which moved. The worker was using an 8-foot stepladder to reach a 9-foot, 4-inch sheet metal duct. He was moving up and down several steps of the ladder in an attempt to connect the duct to a fire damper fastened to the wall. He had his right foot on the fifth ladder step, his left foot on the step above, and was reaching when the ladder spun counterclockwise tangling his feet in the steps. He fell to the concrete floor.

When working from a ladder, always have both feet on the same step. By standing on two steps, most of his weight was on the lower step, creating an unbalanced load on one side of the ladder. Reaching and pushing created a further instability. A side load on a stepladder can cause it to lift the other foot off the ground and spin around.

Case History No. 5: Fall from a Fixed Ladder (20)

A 23-year-old elevator mechanic helper was killed when he fell 20 feet while climbing a fixed ladder on the side of a warehouse. He was carrying a 15-pound box on his shoulder and fell after placing it on the roof at the right side of the ladder. He might have had one foot on the roof when he fell.

Investigation showed that the fixed ladder was in fair to poor condition with visible rust. It had been welded together from several ladders and had different distances (11 inches and 14 inches) between rungs at different parts of the ladder. The distance between side rails was 15 inches (16 inches minimum is required). The fixed ladder didn't extend more than 3 feet above the landing (42 inches required). At the step-off area on the roof, there was a 3-foot by 2-foot rubber mat. The helper should not have been carrying the box; it could have been lifted to the roof by a line.

Case History No. 6: Electrocution While Carrying Metal Ladder (21)

A 21-year-old painter was killed when the metal ladder he was carrying contacted an overhead power line. The victim and two co-workers had been painting the exterior of a two story residence, using a 40-foot extension ladder extended to 26 feet. A 19,500-volt overhead power line was located about 10 feet away from the residence at a height of 24 feet. After a 12-hour day, the painters were in the process of cleaning up. The victim stood at the base of the extension ladder and pulled it away from the side of the building, standing it upright without collapsing the sections. The rope for the pulley system to lower the top section was missing. The top section contacted the overhead power line. The electric current passed through the ladder, his hands, body, and feet to ground, electrocuting him.

A nonconductive ladder should have been used, not a metal one. If possible, the utility company should have been contacted to de-energize or insulate the line. The extension ladder should have been inspected and the missing pulley rope replaced. The victim could then have collapsed the extension ladder to a safe height before moving it. The effect of a 12-hour day and evening darkness might also have been a contributory factor.

Case Study No. 7: Roofer Falls from Ladder After Contacting Overhead Power Line (22)

A 36-year-old roofer was killed when he fell 30 feet from a wooden ladder after the 6-foot metal pole of his mop contacted a 3600-volt overhead power line, which was 24-36 inches from the building. He was climbing the ladder to the roof of a house while carrying the mop. As he was stepping off the ladder onto the roof, the aluminum handle of his mop contacted the power line, shocking him, and causing him to fall to the ground.

Before setting up, an environmental survey should have been done to look for overhead hazards. The ladder should not have been located so close to the overhead power line. It should have been at least 10 feet away. If possible, the utility company should have been contacted to de-energize or insulate the line. He should not have been carrying the mop, but should have lifted it with the pulley wheel attached to the ladder.

REFERENCES

1. American National Standards Institute/American Ladder Institute (ALI). (2000). ANSI A14.1-2000 – American National Standard for Ladders – Wood Safety Requirements. Chicago: ALI. Available at: http://www.americanladderinstitute.org/ali/faq.asp.

2. American National Standards Institute/American Ladder Institute. (2000). ANSI A14.2-2000 – American National Standard for Ladders – Portable Metal —Safety Requirements. Chicago: ALI. Available at: http://www.americanladderinstitute.org/ali/faq.asp.

3. American National Standards Institute/American Ladder Institute. (2002). ANSI A14.3-2002 – American National Standard for Ladders – Fixed — Safety Requirements. Chicago: ALI. Available at: http://www.americanladderinstitute.org/ali/faq.asp.

4. American National Standards Institute/American Ladder Institute. (2002). ANSI A14.4-2002 – American National Standard Safety Requirements for Job Made Ladders. Chicago: ALI. Available at: http://www.americanladderinstitute.org/ali/faq.asp.

5. American National Standards Institute/American Ladder Institute. (2000). ANSI A14.5-2000 – American National Standard for Ladders – Portable Reinforced Plastic — Safety Requirements. Chicago: ALI. Available at: http://www.americanladderinstitute.org/ali/faq.asp.

6. American National Standards Institute/American Ladder Institute. (2000). ANSI A14.7-2000 – American National Standard for Mobile Ladder Stands and Mobile Ladder Stand Platforms. Chicago: ALI. Available at: http://www.americanladderinstitute.org/ali/faq.asp.

7. American National Standards Institute/American Ladder Institute. (2000). ANSI A14.10-2000 – American National Standard for Ladders – Portable Special Duty Ladders. Chicago: ALI. Available at: http://www.americanladderinstitute.org/ali/faq.asp.

8. Young, S.L. and Wogalter, M.S. (2000). On improving set-up angle accuracy for extension ladders. Proceedings of the IEA 2000/HFES 2000 Congress. San Diego, CA.

9. Lineberry, G.T., Wiehagen, W.J., Scharf, T., and McCann, M., "Program Toward a Multi-use Educational Intervention for Reducing Injury Risk in the Set-up and Use of Extension Ladders," 6th Int'l Conf., Scientific Committee on Education and Training in Occupational Safety & Health, Baltimore, MD (Oct. 27-30, 2002).

10. The Center to Protect Workers Rights. (2002). *The Construction Chart Book.* Third edition. Silver Spring, MD: CPWR. Chart 36a.

11. The Center to Protect Workers Rights. (2002). *The Construction Chart Book.* Third edition. Silver Spring, MD: CPWR. Chart 37b.

12. Wiehagen W and McCann M., unpublished data

13. McCann M, Hunting KL, Murawski J, Chowdhury R, Welch L. Causes of electrical deaths and injuries among construction workers. Amer. J. Ind. Med. 2003: 43:398-406

14. Cohen HH and Lin L-J. (1991). A scenario analysis of ladder fall accidents. J. Safety Research 22:31-39.

15. Advanced Technologies and Laboratories (ATL) International, Inc. (2003). Current Status and Knowledge Gaps of Fall Protection: Technologies/Research for Falls from Ladders. Germantown, MD: ATL International.

16. New Jersey Department of Health. (1991). New Jersey FACE Report No. 90NJ01201: Worker falls 25 feet to his death from an extension ladder in New Jersey. Morgantown, WV: NIOSH.

17. National Institute for Occupational Safety and Health. (1990). NIOSH FACE 9007: Laborer dies after fall from ladder in South Carolina. Morgantown, WV: NIOSH Massachusetts Department Of Health. (1995). Massachusetts FACE 95-MA-042-01: Massachusetts plumber falls when extension ladder collapses. Morgantown, WV: NIOSH.

18. California Department of Health Services. (1998). California FACE Investigation 98CA006: Sheet metal worker falls off stepladder and dies in California. Morgantown, WV: NIOSH.

19. Massachusetts Department of Health. (1998). Massachusetts FACE 98-MA-030-01: Massachusetts elevator mechanic dies in 20-foot fall from a fixed ladder on the side of the building. Morgantown, WV: NIOSH.

20. National Institute for Occupational Safety and Health. (1992). NIOSH FACE 92-27: Painter electrocuted when metal ladder contacts a powerline — Virginia. Morgantown, WV: NIOSH

21. New Jersey Department Of Health. (1993). New Jersey FACE Investigation #93NJ105: Roofer dies after falling 30 feet from a ladder when an aluminum pole contacts an overhead powerline. Morgantown, WV: NIOSH

Portable Ladder Inspection Checklist

Frequency: Regularly, based on use and exposure

Inspected by: ______________________ Date: ______________________

Responsible Personnel: ______________________ Department: ______________________

Location: ______________________ Area: ______________________

General

Note: Inspection checklists cannot possibly cover all the items and situations that may be encountered in the workplace. Thus, this checklist can only serve as a guide. Refer to the references at the end of this form for more information and details, as well as, other reference sources.

Inspection Item

- Will the ladder support four times the maximum intended load (3.3 times for Type 1A metal or plastic ladders. Check ladder sticker.)
  Yes/No/Comment ______________________
- Are ladder rungs, cleats, and steps parallel, level, secure, and uniformly spaced when in position for use?
  Yes/No/Comment ______________________
- Are rungs spaced 10-14 inches apart? (8-12 inches for step stools; 8-18 inches for base of extension trestle ladders, and 6-12 inches for extension part)
  Yes/No/Comment ______________________
- Is the minimum clear distance between side rails 11.5 inches?
  Yes/No/Comment ______________________
- Are the rungs and steps of portable metal ladders corrugated, knurled, coated with skid-resistant materials, or otherwise treated to minimize slipping?
  Yes/No/Comment ______________________

- Are rungs and steps free of mud, paint, oil, or other slick/sticky substances?
  Yes/No/Comment ____________________
- Are ropes and pulleys on extension ladders in good condition?
  Yes/No/Comment ____________________
- Are slip-resistant feet on portable ladders secure?
  Yes/No/Comment ____________________
- Is the ladder a complete unit (i.e., not 2 or more ladders fastened together to make a longer ladder)?
  Yes/No/Comment ____________________
- Are metal spreaders or locking devices provided on stepladders to lock the front and back sections in open position when used?
  Yes/No/Comment ____________________
- Are platforms or landings provided when two or more ladders are used to access an upper level?
  Yes/No/Comment ____________________
- Are ladder components free of projections or unevenness that could snag clothing?
  Yes/No/Comment ____________________
- Are wood ladders coated with an opaque paint that could hide or fill in cracks?
  Yes/No/Comment ____________________

Submitted by: ____________________ Date: ________ Time: ________

Submitted to: ____________________

Assigned to: ____________________ Date: ________

References

29 CFR 1926.1053, ANSI A14.1-2000, ANSI A14.2-2000, ANSI A14.4-2002, ANSI A15.5-2000

GUIDELINES FOR EXTENSION LADDER SAFETY SET-UP AND REPOSITIONING*

Select the Ladder

- Match ladder duty rating (250, 300, or 375 lbs) to job.
- Choose proper length of ladder for job.
  - Single-section ladder maximum length: up to 30 feet
  - 2 section ladder maximum length: 48 feet
  - 3+ section ladder maximum length: 60 feet
- Choose ladder of proper material (aluminum, fiberglass, wood).
- Inspect ladder for damage and proper operation. Tag damaged ladders: "DO NOT USE"

Scanning the Work Area

- Check for electrical hazards (e.g., overhead power lines).
- Check for other overhead obstructions (e.g., tree limbs).
- Note clutter and traffic patterns in immediate work area; check for tripping/slipping hazards.
- Locate stable surface to secure base of ladder.
- Locate firm and secure surface for top of ladder.
- Note environmental conditions (wind, rain, snow/ice).

On-Site Ladder Inspection

- Check that rungs, safety feet, etc. are secure.
- Check rungs for mud, paint, oil, or other slick/ sticky substances.
- Check for places on ladder that could cause cuts, punctures, or abrasions (e.g., loose screws, bolts, hinges).

- Check for damaged ropes and pulleys.
- Check that rung locks fasten securely.
- Tag damaged ladders: "DO NOT USE".

Ladder Extension/ Set-up

- Use two people to carry and erect ladder whenever possible.
- Place bottom of ladder below point selected for top support (near base of structure).
- Walk ladder up hand-over-hand.
- Extend ladder 3 feet above resting point on roof or rest it against wall.
- Verify minimum overlap of ladder sections and that rung locks are properly engaged.
  - Up to 36', minimum overlap: 3 feet per section
  - Over 36', up to 48', minimum overlap: 4 feet per section
  - Over 48', up to 60', minimum overlap: 5 feet per section
- Pull bottom of ladder away from structure to set proper angle.
  - ¼ rule: 1 foot of horizontal distance from top support for every 4 feet of ladder length to top support
- Clear clutter, level the bottom, and make sure footing is stable.
- Ensure that both rails at the top of ladder are supported.
- Secure ladder at bottom by appropriate means (e.g., staking, digging in).
- Set up traffic barrier or barricade on ground around work area.

Ladder Testing/ Securing

- Recheck setup before or during first climb (e.g., levelness, stability of base, rung locks).

- Recheck shoes for mud, paint, oil, or other slick/sticky substances.
- Use spotter for added security during first climb.
- Check for stability while climbing first couple rungs.
- Tie off top of ladder whenever possible.

General Ladder Handling

- Use two people to carry and erect ladder whenever possible.
- Install a hydraulic or mechanical ladder rack on your vehicle to aid loading/unloading.
- When repositioning ladder more than a few feet, collapse sections before moving.
- If ladder becomes unstable while carrying it, drop it and get out of the way.
- Reinspect dropped ladder; tag damaged ladder: "DO NOT USE".
- Store and lock ladder in covered area, free of contaminants (e.g., mud, water, grease).
- When lowering ladder, keep fingers away from moving or sliding parts.
- Note slipping, tripping, and overhead hazards before moving ladder.

** This checklist is a later version of one presented at the 6th International Conference, Scientific Committee on Education and Training in Occupational Safety & Health, Baltimore, MD (Oct. 27-30, 2002) (9). This checklist is still in the testing stage.*

Guidelines for Extension Ladder Safety – Use

Recheck Setup at Start of Shift and After Breaks

- Recheck setup during climb (e.g., levelness, stability of base, rung locks).
- Use a spotter whenever possible.
- Recheck work area (e.g., ground clutter, traffic barrier, weather).
- For access to upper level, recheck top tie-off.
- For access to upper level, recheck that ladder extends 3 feet above resting point.

Climbing

- Recheck shoes for mud, paint, oil, or other slick/sticky substances.
- Maintain 3-point contact.
- Face ladder when climbing. Note where sections overlap.
- Do not step on top three rungs.

Raising or Lowering Materials and Tools

- Use a rope or lift to raise and lower materials or tools.
- When use of a rope or lift is not possible:
  - Attach tools to belt.
  - Carry a load by hand only if there is no other option and if it will not affect your balance.
  - As you climb, put both feet on each rung before repositioning hand.
  - Recognize that you might have to drop the load.
  - Carefully secure carried items before working or stepping off ladder at height.

Working From a Ladder

- Keep both feet on same rung while working.
- Face the ladder while working.
- Keep body centered between side rails.
- Recheck area below in case tools or materials are dropped.
- Tie off top of ladder, especially when pulling, prying, or reaching.
- Use body harness and separately-anchored lifeline whenever possible.
- Maintain 3-point contact.
- Do not stand on top three rungs.

Stepping On/Off a Ladder at Height

- Free hands of all materials before stepping on/off ladder.
- Assure 3-point contact.
- Check tie-off at top of ladder.
- Check for hazards before stepping on/off ladder

Descending Ladder

- Maintain 3-point contact
- Face ladder as you walk down
- Put both feet on each rung as you descend. Anticipate where sections overlap.
- Check ground for clutter before last step

* *This checklist is a later version of one presented at the 6th International Conference, Scientific Committee on Education and Training in Occupational Safety & Health, Baltimore, MD (Oct. 27-30, 2002) (9). This checklist is still in the testing stage.*

Appendix A

General Requirements for Scaffolding

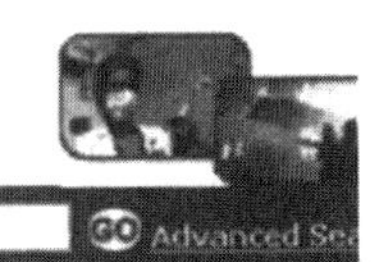

Regulations (Standards - 29 CFR)

General requirements. - 1926.451

Regulations (Standards - 29 CFR) - Table of Contents

- **Part Number:** 1926
- **Part Title:** Safety and Health Regulations for Construction
- **Subpart:** L
- **Subpart Title:** Scaffolds
- **Standard Number:** 1926.451
- **Title:** General requirements.

This section does not apply to aerial lifts, the criteria for which are set out exclusively in 1926.453.

1926.451(a)

"Capacity"

1926.451(a)(1)

Except as provided in paragraphs (a)(2), (a)(3), (a)(4), (a)(5) and (g) of this section, each scaffold and scaffold component shall be capable of supporting, without failure, its own weight and at least 4 times the maximum intended load applied or transmitted to it.

1926.451(a)(2)

Direct connections to roofs and floors, and counterweights used to balance adjustable suspension scaffolds, shall be capable of resisting at least 4 times the tipping moment imposed by the scaffold operating at the rated load of the hoist, or 1.5 (minimum) times the tipping moment imposed by the scaffold operating at the stall load of the hoist, whichever is greater.

1926.451(a)(3)

Each suspension rope, including connecting hardware, used on non-adjustable suspension scaffolds shall be capable of supporting, without failure, at least 6 times the maximum intended load applied or transmitted to that rope.

..1926.451(a)(4)

1926.451(a)(4)

Each suspension rope, including connecting hardware, used on adjustable suspension scaffolds shall be capable of supporting, without failure, at least 6 times the maximum intended load applied or transmitted to that rope with the scaffold operating at either the rated load of the hoist, or 2 (minimum) times the stall load of the hoist, whichever is greater.

1926.451(a)(5)

The stall load of any scaffold hoist shall not exceed 3 times its rated load.

General requirements. - 1926.451

1926.451(a)(6)

Scaffolds shall be designed by a qualified person and shall be constructed and loaded in accordance with that design. Non-mandatory Appendix A to this subpart contains examples of criteria that will enable an employer to comply with paragraph (a) of this section.

1926.451(b)

"Scaffold platform construction."

1926.451(b)(1)

Each platform on all working levels of scaffolds shall be fully planked or decked between the front uprights and the guardrail supports as follows:

1926.451(b)(1)(i)

Each platform unit (e.g., scaffold plank, fabricated plank, fabricated deck, or fabricated platform) shall be installed so that the space between adjacent units and the space between the platform and the uprights is no more than 1 inch (2.5 cm) wide, except where the employer can demonstrate that a wider space is necessary (for example, to fit around uprights when side brackets are used to extend the width of the platform).

..1926.451(b)(1)(ii)

1926.451(b)(1)(ii)

Where the employer makes the demonstration provided for in paragraph (b)(1)(i) of this section, the platform shall be planked or decked as fully as possible and the remaining open space between the platform and the uprights shall not exceed 9 1/2 inches (24.1 cm).

Exception to paragraph (b)(1): The requirement in paragraph (b)(1) to provide full planking or decking does not apply to platforms used solely as walkways or solely by employees performing scaffold erection or dismantling. In these situations, only the planking that the employer establishes is necessary to provide safe working conditions is required.

1926.451(b)(2)

Except as provided in paragraphs (b)(2)(i) and (b)(2)(ii) of this section, each scaffold platform and walkway shall be at least 18 inches (46 cm) wide.

1926.451(b)(2)(i)

Each ladder jack scaffold, top plate bracket scaffold, roof bracket scaffold, and pump jack scaffold shall be at least 12 inches (30 cm) wide. There is no minimum width requirement for boatswains' chairs.

Note to paragraph (b)(2)(i): Pursuant to an administrative stay effective November 29, 1996 and published in the Federal Register on November 25, 1996, the requirement in paragraph (b)(2)(i) that roof bracket scaffolds be at least 12 inches wide is stayed until November 25, 1997 or until rulemaking reguarding the minimum width of roof bracket scaffolds has been completed, whichever is later.

1926.451(b)(2)(ii)

Where scaffolds must be used in areas that the employer can demonstrate are so narrow that platforms and walkways cannot be at least 18 inches (46 cm) wide, such platforms and walkways shall be as wide as feasible, and employees on those platforms and walkways shall be protected from fall hazards by the use of guardrails and/or personal fall arrest systems.

1926.451(b)(3)

Except as provided in paragraphs (b)(3)(i) and (ii) of this section, the front edge of all platforms shall not be more than 14 inches (36 cm) from the face of the work, unless guardrail systems are erected along the front edge and/or personal fall arrest systems are used in accordance with paragraph (g) of this section to protect employees from falling.

..1926.451(b)(3)(i)

1926.451(b)(3)(i)

The maximum distance from the face for outrigger scaffolds shall be 3 inches (8 cm);

1926.451(b)(3)(ii)

The maximum distance from the face for plastering and lathing operations shall be 18 inches (46 cm).

1926.451(b)(4)

Each end of a platform, unless cleated or otherwise restrained by hooks or equivalent means, shall extend over the centerline of its support at least 6 inches (15 cm).

1926.451(b)(5)

1926.451(b)(5)(i)

Each end of a platform 10 feet or less in length shall not extend over its support more than 12 inches (30 cm) unless the platform is designed and installed so that the cantilevered portion of the platform is able to support employees and/or materials without tipping, or has guardrails which block employee access to the cantilevered end.

1926.451(b)(5)(ii)

Each platform greater than 10 feet in length shall not extend over its support more than 18 inches (46 cm), unless it is designed and installed so that the cantilevered portion of the platform is able to support employees without tipping, or has guardrails which block employee access to the cantilevered end.

..1926.451(b)(6)

1926.451(b)(6)

On scaffolds where scaffold planks are abutted to create a long platform, each abutted end shall rest on a separate support surface. This provision does not preclude the use of common support members, such as "T" sections, to support abutting planks, or hook on platforms designed to rest on common supports.

1926.451(b)(7)

On scaffolds where platforms are overlapped to create a long platform, the overlap shall occur only over supports, and shall not be less than 12 inches (30 cm) unless the platforms are nailed together or otherwise restrained to prevent movement.

General requirements. - 1926.451

1926.451(b)(8)

At all points of a scaffold where the platform changes direction, such as turning a corner, any platform that rests on a bearer at an angle other than a right angle shall be laid first, and platforms which rest at right angles over the same bearer shall be laid second, on top of the first platform.

1926.451(b)(9)

Wood platforms shall not be covered with opaque finishes, except that platform edges may be covered or marked for identification. Platforms may be coated periodically with wood preservatives, fire-retardant finishes, and slip-resistant finishes; however, the coating may not obscure the top or bottom wood surfaces.

1926.451(b)(10)

Scaffold components manufactured by different manufacturers shall not be intermixed unless the components fit together without force and the scaffold's structural integrity is maintained by the user. Scaffold components manufactured by different manufacturers shall not be modified in order to intermix them unless a competent person determines the resulting scaffold is structurally sound.

..1926.451(b)(11)

1926.451(b)(11)

Scaffold components made of dissimilar metals shall not be used together unless a competent person has determined that galvanic action will not reduce the strength of any component to a level below that required by paragraph (a)(1) of this section.

1926.451(c)

"Criteria for supported scaffolds."

1926.451(c)(1)

Supported scaffolds with a height to base width (including outrigger supports, if used) ratio of more than four to one (4:1) shall be restrained from tipping by guying, tying, bracing, or equivalent means, as follows:

1926.451(c)(1)(i)

Guys, ties, and braces shall be installed at locations where horizontal members support both inner and outer legs.

1926.451(c)(1)(ii)

Guys, ties, and braces shall be installed according to the scaffold manufacturer's recommendations or at the closest horizontal member to the 4:1 height and be repeated vertically at locations of horizontal members every 20 feet (6.1 m) or less thereafter for scaffolds 3 feet (0.91 m) wide or less, and every 26 feet (7.9 m) or less thereafter for scaffolds greater than 3 feet (0.91 m) wide. The top guy, tie or brace of completed scaffolds shall be placed no further than the 4:1 height from the top. Such guys, ties and braces shall be installed at each end of the scaffold and at horizontal intervals not to exceed 30 feet (9.1 m) (measured from one end [not both] towards the other).

General requirements. - 1926.451

1926.451(c)(1)(iii)

Ties, guys, braces, or outriggers shall be used to prevent the tipping of supported scaffolds in all circumstances where an eccentric load, such as a cantilevered work platform, is applied or is transmitted to the scaffold.

..1926.451(c)(2)

1926.451(c)(2)

Supported scaffold poles, legs, posts, frames, and uprights shall bear on base plates and mud sills or other adequate firm foundation.

1926.451(c)(2)(i)

Footings shall be level, sound, rigid, and capable of supporting the loaded scaffold without settling or displacement.

1926.451(c)(2)(ii)

Unstable objects shall not be used to support scaffolds or platform units.

1926.451(c)(2)(iii)

Unstable objects shall not be used as working platforms.

1926.451(c)(2)(iv)

Front-end loaders and similar pieces of equipment shall not be used to support scaffold platforms unless they have been specifically designed by the manufacturer for such use.

1926.451(c)(2)(v)

Fork-lifts shall not be used to support scaffold platforms unless the entire platform is attached to the fork and the fork-lift is not moved horizontally while the platform is occupied.

1926.451(c)(3)

Supported scaffold poles, legs, posts, frames, and uprights shall be plumb and braced to prevent swaying and displacement.

..1926.451(d)

1926.451(d)

"Criteria for suspension scaffolds."

1926.451(d)(1)

All suspension scaffold support devices, such as outrigger beams, cornice hooks, parapet clamps, and similar devices, shall rest on surfaces capable of supporting at least 4 times the load imposed on them by the scaffold operating at the rated load of the hoist (or at least 1.5 times the load imposed on them by the scaffold at the stall capacity of the hoist, whichever is greater).

1926.451(d)(2)

Suspension scaffold outrigger beams, when used, shall be made of structural metal or equivalent strength material, and shall be restrained to prevent movement.

1926.451(d)(3)

The inboard ends of suspension scaffold outrigger beams shall be stabilized by bolts or other direct connections to the floor or roof deck, or they shall have their inboard ends stabilized by counterweights, except masons' multi-point adjustable suspension scaffold outrigger beams shall not be stabilized by counterweights.

1926.451(d)(3)(i)

Before the scaffold is used, direct connections shall be evaluated by a competent person who shall confirm, based on the evaluation, that the supporting surfaces are capable of supporting the loads to be imposed. In addition, masons' multi-point adjustable suspension scaffold connections shall be designed by an engineer experienced in such scaffold design.

1926.451(d)(3)(ii)

Counterweights shall be made of non-flowable material. Sand, gravel and similar materials that can be easily dislocated shall not be used as counterweights.

..1926.451(d)(3)(iii)

1926.451(d)(3)(iii)

Only those items specifically designed as counterweights shall be used to counterweight scaffold systems. Construction materials such as, but not limited to, masonry units and rolls of roofing felt, shall not be used as counterweights.

1926.451(d)(3)(iv)

Counterweights shall be secured by mechanical means to the outrigger beams to prevent accidental displacement.

1926.451(d)(3)(v)

Counterweights shall not be removed from an outrigger beam until the scaffold is disassembled.

1926.451(d)(3)(vi)

Outrigger beams which are not stabilized by bolts or other direct connections to the floor or roof deck shall be secured by tiebacks.

1926.451(d)(3)(vii)

Tiebacks shall be equivalent in strength to the suspension ropes.

1926.451(d)(3)(viii)

Outrigger beams shall be placed perpendicular to its bearing support (usually the face of the building or structure). However, where the employer can demonstrate that it is not possible to place an outrigger beam perpendicular to the face of the building or structure because of obstructions that cannot be moved, the outrigger beam may be placed at some other angle, provided opposing angle tiebacks are used.

..1926.451(d)(3)(ix)

1926.451(d)(3)(ix)

Tiebacks shall be secured to a structurally sound anchorage on the building or structure. Sound anchorages include structural members, but do not include standpipes, vents, other piping systems, or electrical conduit.

1926.451(d)(3)(x)

Tiebacks shall be installed perpendicular to the face of the building or structure, or opposing angle tiebacks shall be installed. Single tiebacks installed at an angle are prohibited.

1926.451(d)(4)

Suspension scaffold outrigger beams shall be:

1926.451(d)(4)(i)

Provided with stop bolts or shackles at both ends;

1926.451(d)(4)(ii)

Securely fastened together with the flanges turned out when channel iron beams are used in place of I-beams;

1926.451(d)(4)(iii)

Installed with all bearing supports perpendicular to the beam center line;

1926.451(d)(4)(iv)

Set and maintained with the web in a vertical position; and

1926.451(d)(4)(v)

When an outrigger beam is used, the shackle or clevis with which the rope is attached to the outrigger beam shall be placed directly over the center line of the stirrup.

1926.451(d)(5)

Suspension scaffold support devices such as cornice hooks, roof hooks, roof irons, parapet clamps, or similar devices shall be:

..1926.451(d)(5)(i)

1926.451(d)(5)(i)

Made of steel, wrought iron, or materials of equivalent strength;

1926.451(d)(5)(ii)

Supported by bearing blocks; and

1926.451(d)(5)(iii)

Secured against movement by tiebacks installed at right angles to the face of the building or structure, or opposing angle tiebacks shall be installed and secured to a structurally sound point of anchorage on the building or structure. Sound points of anchorage include structural members, but do not include standpipes, vents, other piping systems, or electrical conduit.

1926.451(d)(5)(iv)

Tiebacks shall be equivalent in strength to the hoisting rope.

1926.451(d)(6)

When winding drum hoists are used on a suspension scaffold, they shall contain not less than four wraps of the suspension rope at the lowest point of scaffold travel. When other types of hoists are used, the suspension ropes shall be long enough to allow the scaffold to be lowered to the level below without the rope end passing through the hoist, or the rope end shall be configured or provided with means to prevent the end from passing through the hoist.

1926.451(d)(7)

The use of repaired wire rope as suspension rope is prohibited.

..1926.451(d)(8)

1926.451(d)(8)

Wire suspension ropes shall not be joined together except through the use of eye splice thimbles connected with shackles or coverplates and bolts.

1926.451(d)(9)

The load end of wire suspension ropes shall be equipped with proper size thimbles and secured by eyesplicing or equivalent means.

1926.451(d)(10)

Ropes shall be inspected for defects by a competent person prior to each workshift and after every occurrence which could affect a rope's integrity. Ropes shall be replaced if any of the following conditions exist:

1926.451(d)(10)(i)

Any physical damage which impairs the function and strength of the rope.

1926.451(d)(10)(ii)

Kinks that might impair the tracking or wrapping of rope around the drum(s) or sheave(s).

1926.451(d)(10)(iii)

Six randomly distributed broken wires in one rope lay or three broken wires in one strand in one rope lay.

1926.451(d)(10)(iv)

Abrasion, corrosion, scrubbing, flattening or peening causing loss of more than one-third of the original diameter of the outside wires.

1926.451(d)(10)(v)

Heat damage caused by a torch or any damage caused by contact with electrical wires.

..1926.451(d)(10)(vi)

1926.451(d)(10)(vi)

Evidence that the secondary brake has been activated during an overspeed condition and has engaged the suspension rope.

1926.451(d)(11)

Swaged attachments or spliced eyes on wire suspension ropes shall not be used unless they are made by the wire rope manufacturer or a qualified person.

1926.451(d)(12)

When wire rope clips are used on suspension scaffolds:

1926.451(d)(12)(i)

There shall be a minimum of 3 wire rope clips installed, with the clips a minimum of 6 rope diameters apart;

1926.451(d)(12)(ii)

Clips shall be installed according to the manufacturer's recommendations;

1926.451(d)(12)(iii)

Clips shall be retightened to the manufacturer's recommendations after the initial loading;

1926.451(d)(12)(iv)

Clips shall be inspected and retightened to the manufacturer's recommendations at the start of each workshift thereafter;

1926.451(d)(12)(v)

U-bolt clips shall not be used at the point of suspension for any scaffold hoist;

..1926.451(d)(12)(vi)

1926.451(d)(12)(vi)

When U-bolt clips are used, the U-bolt shall be placed over the dead end of the rope, and the saddle shall be placed over the live end of the rope.

1926.451(d)(13)

Suspension scaffold power-operated hoists and manual hoists shall be tested by a qualified testing laboratory.

1926.451(d)(14)

Gasoline-powered equipment and hoists shall not be used on suspension scaffolds.

1926.451(d)(15)

Gears and brakes of power-operated hoists used on suspension scaffolds shall be enclosed.

1926.451(d)(16)

In addition to the normal operating brake, suspension scaffold power-operated hoists and manually operated hoists shall have a braking device or locking pawl which engages automatically when a hoist makes either of the following uncontrolled movements: an instantaneous change in momentum or an accelerated overspeed.

1926.451(d)(17)

Manually operated hoists shall require a positive crank force to descend.

1926.451(d)(18)

Two-point and multi-point suspension scaffolds shall be tied or otherwise secured to prevent them from swaying, as determined to be necessary based on an evaluation by a competent person. Window cleaners' anchors shall not be used for this purpose.

..1926.451(d)(19)

1926.451(d)(19)

Devices whose sole function is to provide emergency escape and rescue shall not be used as working platforms. This provision does not preclude the use of systems which are designed to function both as suspension scaffolds and emergency systems.

1926.451(e)

"Access." This paragraph applies to scaffold access for all employees. Access requirements for employees erecting or dismantling supported scaffolds are specifically addressed in paragraph (e)(9) of this section.

1926.451(e)(1)

When scaffold platforms are more than 2 feet (0.6 m) above or below a point of access, portable ladders, hook-on ladders, attachable ladders, stair towers (scaffold stairways/towers), stairway-type ladders (such as ladder stands), ramps, walkways, integral prefabricated scaffold access, or direct access from another scaffold, structure, personnel hoist, or similar surface shall be used. Crossbraces shall not be used as a means of access.

General requirements. - 1926.451

1926.451(e)(2)

Portable, hook-on, and attachable ladders (Additional requirements for the proper construction and use of portable ladders are contained in subpart X of this part -- Stairways and Ladders):

1926.451(e)(2)(i)

Portable, hook-on, and attachable ladders shall be positioned so as not to tip the scaffold;

1926.451(e)(2)(ii)

Hook-on and attachable ladders shall be positioned so that their bottom rung is not more than 24 inches (61 cm) above the scaffold supporting level;

..1926.451(e)(2)(iii)

1926.451(e)(2)(iii)

When hook-on and attachable ladders are used on a supported scaffold more than 35 feet (10.7 m) high, they shall have rest platforms at 35-foot (10.7 m) maximum vertical intervals.

1926.451(e)(2)(iv)

Hook-on and attachable ladders shall be specifically designed for use with the type of scaffold used;

1926.451(e)(2)(v)

Hook-on and attachable ladders shall have a minimum rung length of 11 1/2 inches (29 cm); and

1926.451(e)(2)(vi)

Hook-on and attachable ladders shall have uniformly spaced rungs with a maximum spacing between rungs of 16 3/4 inches.

1926.451(e)(3)

Stairway-type ladders shall:

1926.451(e)(3)(i)

Be positioned such that their bottom step is not more than 24 inches (61 cm) above the scaffold supporting level;

1926.451(e)(3)(ii)

Be provided with rest platforms at 12 foot (3.7 m) maximum vertical intervals;

1926.451(e)(3)(iii)

Have a minimum step width of 16 inches (41 cm), except that mobile scaffold stairway-type ladders shall have a minimum step width of 11 1/2 inches (30 cm); and

1926.451(e)(3)(iv)

Have slip-resistant treads on all steps and landings.

..1926.451(e)(4)

1926.451(e)(4)

Stairtowers (scaffold stairway/towers) shall be positioned such that their bottom step is not more than 24 inches (61 cm.) above the scaffold supporting level.

1926.451(e)(4)(i)

A stairrail consisting of a toprail and a midrail shall be provided on each side of each scaffold stairway.

1926.451(e)(4)(ii)

The toprail of each stairrail system shall also be capable of serving as a handrail, unless a separate handrail is provided.

1926.451(e)(4)(iii)

Handrails, and toprails that serve as handrails, shall provide an adequate handhold for employees grasping them to avoid falling.

1926.451(e)(4)(iv)

Stairrail systems and handrails shall be surfaced to prevent injury to employees from punctures or lacerations, and to prevent snagging of clothing.

1926.451(e)(4)(v)

The ends of stairrail systems and handrails shall be constructed so that they do not constitute a projection hazard.

1926.451(e)(4)(vi)

Handrails, and toprails that are used as handrails, shall be at least 3 inches (7.6 cm) from other objects.

..1926.451(e)(4)(vii)

1926.451(e)(4)(vii)

Stairrails shall be not less than 28 inches (71 cm) nor more than 37 inches (94 cm) from the upper surface of the stairrail to the surface of the tread, in line with the face of the riser at the forward edge of the tread.

1926.451(e)(4)(viii)

A landing platform at least 18 inches (45.7 cm) wide by at least 18 inches (45.7 cm) long shall be provided at each level.

1926.451(e)(4)(ix)

Each scaffold stairway shall be at least 18 inches (45.7 cm) wide between stairrails.

1926.451(e)(4)(x)

Treads and landings shall have slip-resistant surfaces.

1926.451(e)(4)(xi)

Stairways shall be installed between 40 degrees and 60 degrees from the horizontal.

1926.451(e)(4)(xii)

Guardrails meeting the requirements of paragraph (g)(4) of this section shall be provided on the open sides and ends of each landing.

1926.451(e)(4)(xiii)

Riser height shall be uniform, within 1/4 inch, (0.6 cm) for each flight of stairs. Greater variations in riser height are allowed for the top and bottom steps of the entire system, not for each flight of stairs.

1926.451(e)(4)(xiv)

Tread depth shall be uniform, within 1/4 inch, for each flight of stairs.

..1926.451(e)(5)

1926.451(e)(5)

Ramps and walkways.

1926.451(e)(5)(i)

Ramps and walkways 6 feet (1.8 m) or more above lower levels shall have guardrail systems which comply with subpart M of this part -- Fall Protection;

1926.451(e)(5)(ii)

No ramp or walkway shall be inclined more than a slope of one (1) vertical to three (3) horizontal (20 degrees above the horizontal).

1926.451(e)(5)(iii)

If the slope of a ramp or a walkway is steeper than one (1) vertical in eight (8) horizontal, the ramp or walkway shall have cleats not more than fourteen (14) inches (35 cm) apart which are securely fastened to the planks to provide footing.

1926.451(e)(6)

Integral prefabricated scaffold access frames shall:

1926.451(e)(6)(i)

Be specifically designed and constructed for use as ladder rungs;

1926.451(e)(6)(ii)

Have a rung length of at least 8 inches (20 cm);

1926.451(e)(6)(iii)

Not be used as work platforms when rungs are less than 11 1/2 inches in length, unless each affected employee uses fall protection, or a positioning device, which complies with 1926.502;

1926.451(e)(6)(iv)

Be uniformly spaced within each frame section;

..1926.451(e)(6)(v)

1926.451(e)(6)(v)

Be provided with rest platforms at 35-foot (10.7 m) maximum vertical intervals on all supported scaffolds more than 35 feet (10.7 m) high; and

1926.451(e)(6)(vi)

Have a maximum spacing between rungs of 16 3/4 inches (43 cm). Non-uniform rung spacing caused by joining end frames together is allowed, provided the resulting spacing does not exceed 16 3/4 inches (43 cm).

1926.451(e)(7)

Steps and rungs of ladder and stairway type access shall line up vertically with each other between rest platforms.

1926.451(e)(8)

Direct access to or from another surface shall be used only when the scaffold is not more than 14 inches (36 cm) horizontally and not more than 24 inches (61 cm) vertically from the other surface.

1926.451(e)(9)

Effective September 2, 1997, access for employees erecting or dismantling supported scaffolds shall be in accordance with the following:

..1926.451(e)(9)(i)

1926.451(e)(9)(i)

The employer shall provide safe means of access for each employee erecting or dismantling a scaffold where the provision of safe access is feasible and does not create a greater hazard. The employer shall have a competent person determine whether it is feasible or would pose a greater hazard to provide, and have employees use a safe means of access. This determination shall be based on site conditions and the type of scaffold being erected or dismantled.

1926.451(e)(9)(ii)

Hook-on or attachable ladders shall be installed as soon as scaffold erection has progressed to a point that permits safe installation and use.

1926.451(e)(9)(iii)

When erecting or dismantling tubular welded frame scaffolds, (end) frames, with horizontal members that are parallel, level and are not more than 22 inches apart vertically may be used as climbing devices for access, provided they are erected in a manner that creates a usable ladder and provides good hand hold and foot space.

1926.451(e)(9)(iv)

Cross braces on tubular welded frame scaffolds shall not be used as a means of access or egress.

1926.451(f)

"Use."

1926.451(f)(1)

Scaffolds and scaffold components shall not be loaded in excess of their maximum intended loads or rated capacities, whichever is less.

1926.451(f)(2)

The use of shore or lean-to scaffolds is prohibited.

..1926.451(f)(3)

1926.451(f)(3)

Scaffolds and scaffold components shall be inspected for visible defects by a competent person before each work shift, and after any occurrence which could affect a scaffold's structural integrity.

1926.451(f)(4)

Any part of a scaffold damaged or weakened such that its strength is less than that required by paragraph (a) of this section shall be immediately repaired or replaced, braced to meet those provisions, or removed from service until repaired.

1926.451(f)(5)

Scaffolds shall not be moved horizontally while employees are on them, unless they have been designed by a registered professional engineer specifically for such movement or, for mobile scaffolds, where the provisions of 1926.452(w) are followed.

1926.451(f)(6)

The clearance between scaffolds and power lines shall be as follows: Scaffolds shall not be erected, used, dismantled, altered, or moved such that they or any conductive material handled on them might come closer to exposed and energized power lines than as follows:

*Insulated Lines

Voltage	Minimum distance	Alternatives

General requirements. - 1926.451

Less than 300 volts.	3 feet (0.9 m)	
300 volts to 50 kv.	10 feet (3.1 m)	
More than 50 kv.....	10 feet (3.1 m) plus 0.4 inches (1.0 cm) for each 1 kv over 50 kv.	2 times the length of the line insulator, but never less than 10 feet (3.1 m).

*Uninsulated lines

Voltage	Minimum distance	Alternatives
Less than 50 kv.....	10 feet (3.1 m).	
More than 50 kv.....	10 feet (3.1 m) plus 0.4 inches (1.0 cm) for each 1 kv over 50 kv.	2 times the length of the line insulator, but never less than 10 feet (3.1 m).

Exception to paragraph (f)(6): Scaffolds and materials may be closer to power lines than specified above where such clearance is necessary for performance of work, and only after the utility company, or electrical system operator, has been notified of the need to work closer and the utility company, or electrical system operator, has deenergized the lines, relocated the lines, or installed protective coverings to prevent accidental contact with the lines.

1926.451(f)(7)

Scaffolds shall be erected, moved, dismantled, or altered only under the supervision and direction of a competent person qualified in scaffold erection, moving, dismantling or alteration. Such activities shall be performed only by experienced and trained employees selected for such work by the competent person.

..1926.451(f)(8)

1926.451(f)(8)

Employees shall be prohibited from working on scaffolds covered with snow, ice, or other slippery material except as necessary for removal of such materials.

1926.451(f)(9)

Where swinging loads are being hoisted onto or near scaffolds such that the loads might contact the scaffold, tag lines or equivalent measures to control the loads shall be used.

1926.451(f)(10)

Suspension ropes supporting adjustable suspension scaffolds shall be of a diameter large enough to provide sufficient surface area for the functioning of brake and hoist mechanisms.

General requirements. - 1926.451

1926.451(f)(11)

Suspension ropes shall be shielded from heat-producing processes. When acids or other corrosive substances are used on a scaffold, the ropes shall be shielded, treated to protect against the corrosive substances, or shall be of a material that will not be damaged by the substance being used.

1926.451(f)(12)

Work on or from scaffolds is prohibited during storms or high winds unless a competent person has determined that it is safe for employees to be on the scaffold and those employees are protected by a personal fall arrest system or wind screens. Wind screens shall not be used unless the scaffold is secured against the anticipated wind forces imposed.

1926.451(f)(13)

Debris shall not be allowed to accumulate on platforms.

..1926.451(f)(14)

1926.451(f)(14)

Makeshift devices, such as but not limited to boxes and barrels, shall not be used on top of scaffold platforms to increase the working level height of employees.

1926.451(f)(15)

Ladders shall not be used on scaffolds to increase the working level height of employees, except on large area scaffolds where employers have satisfied the following criteria:

1926.451(f)(15)(i)

When the ladder is placed against a structure which is not a part of the scaffold, the scaffold shall be secured against the sideways thrust exerted by the ladder;

1926.451(f)(15)(ii)

The platform units shall be secured to the scaffold to prevent their movement;

1926.451(f)(15)(iii)

The ladder legs shall be on the same platform or other means shall be provided to stabilize the ladder against unequal platform deflection, and

1926.451(f)(15)(iv)

The ladder legs shall be secured to prevent them from slipping or being pushed off the platform.

1926.451(f)(16)

Platforms shall not deflect more than 1/60 of the span when loaded.

1926.451(f)(17)

To reduce the possibility of welding current arcing through the suspension wire rope when performing welding from suspended scaffolds, the following precautions shall be taken, as applicable:

..1926.451(f)(17)(i)

1926.451(f)(17)(i)

An insulated thimble shall be used to attach each suspension wire rope to its hanging support (such as cornice hook or outrigger). Excess suspension wire rope and any additional independent lines from grounding shall be insulated;

1926.451(f)(17)(ii)

The suspension wire rope shall be covered with insulating material extending at least 4 feet (1.2 m) above the hoist. If there is a tail line below the hoist, it shall be insulated to prevent contact with the platform. The portion of the tail line that hangs free below the scaffold shall be guided or retained, or both, so that it does not become grounded;

1926.451(f)(17)(iii)

Each hoist shall be covered with insulated protective covers;

1926.451(f)(17)(iv)

In addition to a work lead attachment required by the welding process, a grounding conductor shall be connected from the scaffold to the structure. The size of this conductor shall be at least the size of the welding process work lead, and this conductor shall not be in series with the welding process or the work piece;

1926.451(f)(17)(v)

If the scaffold grounding lead is disconnected at any time, the welding machine shall be shut off; and

1926.451(f)(17)(vi)

An active welding rod or uninsulated welding lead shall not be allowed to contact the scaffold or its suspension system.

..1926.451(g)

1926.451(g)

"Fall protection."

1926.451(g)(1)

Each employee on a scaffold more than 10 feet (3.1 m) above a lower level shall be protected from falling to that lower level. Paragraphs (g)(1)(i) through (vii) of this section establish the types of fall protection to be provided to the employees on each type of scaffold. Paragraph (g)(2) of this section addresses fall protection for scaffold erectors and dismantlers.

Note to paragraph (g)(1): The fall protection requirements for employees installing suspension scaffold support systems on floors, roofs, and other elevated surfaces are set forth in subpart M of this part.

1926.451(g)(1)(i)

Each employee on a boatswains' chair, catenary scaffold, float scaffold, needle beam scaffold, or ladder jack scaffold shall be protected by a personal fall arrest system;

1926.451(g)(1)(ii)

Each employee on a single-point or two-point adjustable suspension scaffold shall be protected by both a personal fall arrest system and guardrail system;

1926.451(g)(1)(iii)

Each employee on a crawling board (chicken ladder) shall be protected by a personal fall arrest system, a guardrail system (with minimum 200 pound toprail capacity), or by a three-fourth inch (1.9 cm) diameter grabline or equivalent handhold securely fastened beside each crawling board;

1926.451(g)(1)(iv)

Each employee on a self-contained adjustable scaffold shall be protected by a guardrail system (with minimum 200 pound toprail capacity) when the platform is supported by the frame structure, and by both a personal fall arrest system and a guardrail system (with minimum 200 pound toprail capacity) when the platform is supported by ropes;

..1926.451(g)(1)(v)

1926.451(g)(1)(v)

Each employee on a walkway located within a scaffold shall be protected by a guardrail system (with minimum 200 pound toprail capacity) installed within 9 1/2 inches (24.1 cm) of and along at least one side of the walkway.

1926.451(g)(1)(vi)

Each employee performing overhand bricklaying operations from a supported scaffold shall be protected from falling from all open sides and ends of the scaffold (except at the side next to the wall being laid) by the use of a personal fall arrest system or guardrail system (with minimum 200 pound toprail capacity).

1926.451(g)(1)(vii)

For all scaffolds not otherwise specified in paragraphs (g)(1)(i) through (g)(1)(vi) of this section, each employee shall be protected by the use of personal fall arrest systems or guardrail systems meeting the requirements of paragraph (g)(4) of this section.

1926.451(g)(2)

Effective September 2, 1997, the employer shall have a competent person determine the feasibility and safety of providing fall protection for employees erecting or dismantling supported scaffolds. Employers are required to provide fall protection for employees erecting or dismantling supported scaffolds where the installation and use of such protection is feasible and does not create a greater hazard.

..1926.451(g)(3)

1926.451(g)(3)

In addition to meeting the requirements of 1926.502(d), personal fall arrest systems used on scaffolds shall be attached by lanyard to a vertical lifeline, horizontal lifeline, or scaffold structural member. Vertical lifelines shall not be used when overhead components, such as overhead protection or additional platform levels, are part of a single-point or two-point adjustable suspension scaffold.

1926.451(g)(3)(i)

When vertical lifelines are used, they shall be fastened to a fixed safe point of anchorage, shall be independent of the scaffold, and shall be protected from sharp edges and abrasion. Safe points of anchorage include structural members of buildings, but do not include standpipes, vents, other piping systems, electrical conduit, outrigger beams, or counterweights.

1926.451(g)(3)(ii)

When horizontal lifelines are used, they shall be secured to two or more structural members of the scaffold, or they may be looped around both suspension and independent suspension lines (on scaffolds so equipped) above the hoist and brake attached to the end of the scaffold. Horizontal lifelines shall not be attached only to the suspension ropes.

1926.451(g)(3)(iii)

When lanyards are connected to horizontal lifelines or structural members on a single-point or two-point adjustable suspension scaffold, the scaffold shall be equipped with additional independent support lines and automatic locking devices capable of stopping the fall of the scaffold in the event one or both of the suspension ropes fail. The independent support lines shall be equal in number and strength to the suspension ropes.

..*1926.451(g)(3)(iv)*

1926.451(g)(3)(iv)

Vertical lifelines, independent support lines, and suspension ropes shall not be attached to each other, nor shall they be attached to or use the same point of anchorage, nor shall they be attached to the same point on the scaffold or personal fall arrest system.

1926.451(g)(4)

Guardrail systems installed to meet the requirements of this section shall comply with the following provisions (guardrail systems built in accordance with Appendix A to this subpart will be deemed to meet the requirements of paragraphs (g)(4)(vii), (viii), and (ix) of this section):

1926.451(g)(4)(i)

Guardrail systems shall be installed along all open sides and ends of platforms. Guardrail systems shall be installed before the scaffold is released for use by employees other than erection/dismantling crews.

1926.451(g)(4)(ii)

The top edge height of toprails or equivalent member on supported scaffolds manufactured or placed in service after January 1, 2000 shall be installed between 38 inches (0.97 m) and 45 inches (1.2 m) above the platform surface. The top edge height on supported scaffolds manufactured and placed in service before January 1, 2000, and on all suspended scaffolds where both a guardrail and a personal fall arrest system are required shall be between 36 inches (0.9 m) and 45 inches (1.2 m). When conditions warrant, the height of the top edge may exceed the 45-inch height, provided the guardrail system meets all other criteria of paragraph (g)(4).

General requirements. - 1926.451

..1926.451(g)(4)(iii)

1926.451(g)(4)(iii)

When midrails, screens, mesh, intermediate vertical members, solid panels, or equivalent structural members are used, they shall be installed between the top edge of the guardrail system and the scaffold platform.

1926.451(g)(4)(iv)

When midrails are used, they shall be installed at a height approximately midway between the top edge of the guardrail system and the platform surface.

1926.451(g)(4)(v)

When screens and mesh are used, they shall extend from the top edge of the guardrail system to the scaffold platform, and along the entire opening between the supports.

1926.451(g)(4)(vi)

When intermediate members (such as balusters or additional rails) are used, they shall not be more than 19 inches (48 cm) apart.

1926.451(g)(4)(vii)

Each toprail or equivalent member of a guardrail system shall be capable of withstanding, without failure, a force applied in any downward or horizontal direction at any point along its top edge of at least 100 pounds (445 n) for guardrail systems installed on single-point adjustable suspension scaffolds or two-point adjustable suspension scaffolds, and at least 200 pounds (890 n) for guardrail systems installed on all other scaffolds.

1926.451(g)(4)(viii)

When the loads specified in paragraph (g)(4)(vii) of this section are applied in a downward direction, the top edge shall not drop below the height above the platform surface that is prescribed in paragraph (g)(4)(ii) of this section.

..1926.451(g)(4)(ix)

1926.451(g)(4)(ix)

Midrails, screens, mesh, intermediate vertical members, solid panels, and equivalent structural members of a guardrail system shall be capable of withstanding, without failure, a force applied in any downward or horizontal direction at any point along the midrail or other member of at least 75 pounds (333 n) for guardrail systems with a minimum 100 pound toprail capacity, and at least 150 pounds (666 n) for guardrail systems with a minimum 200 pound toprail capacity.

1926.451(g)(4)(x)

Suspension scaffold hoists and non-walk-through stirrups may be used as end guardrails, if the space between the hoist or stirrup and the side guardrail or structure does not allow passage of an employee to the end of the scaffold.

1926.451(g)(4)(xi)

General requirements. - 1926.451

Guardrails shall be surfaced to prevent injury to an employee from punctures or lacerations, and to prevent snagging of clothing.

1926.451(g)(4)(xii)

The ends of all rails shall not overhang the terminal posts except when such overhang does not constitute a projection hazard to employees.

1926.451(g)(4)(xiii)

Steel or plastic banding shall not be used as a toprail or midrail.

1926.451(g)(4)(xiv)

Manila or plastic (or other synthetic) rope being used for toprails or midrails shall be inspected by a competent person as frequently as necessary to ensure that it continues to meet the strength requirements of paragraph (g) of this section.

..1926.451(g)(4)(xv)

1926.451(g)(4)(xv)

Crossbracing is acceptable in place of a midrail when the crossing point of two braces is between 20 inches (0.5 m) and 30 inches (0.8 m) above the work platform or as a toprail when the crossing point of two braces is between 38 inches (0.97 m) and 48 inches (1.3 m) above the work platform. The end points at each upright shall be no more than 48 inches (1.3 m) apart.

1926.451(h)

"Falling object protection."

1926.451(h)(1)

In addition to wearing hardhats each employee on a scaffold shall be provided with additional protection from falling hand tools, debris, and other small objects through the installation of toeboards, screens, or guardrail systems, or through the erection of debris nets, catch platforms, or canopy structures that contain or deflect the falling objects. When the falling objects are too large, heavy or massive to be contained or deflected by any of the above-listed measures, the employer shall place such potential falling objects away from the edge of the surface from which they could fall and shall secure those materials as necessary to prevent their falling.

1926.451(h)(2)

Where there is a danger of tools, materials, or equipment falling from a scaffold and striking employees below, the following provisions apply:

1926.451(h)(2)(i)

The area below the scaffold to which objects can fall shall be barricaded, and employees shall not be permitted to enter the hazard area; or

..1926.451(h)(2)(ii)

General requirements. - 1926.451

1926.451(h)(2)(ii)

A toeboard shall be erected along the edge of platforms more than 10 feet (3.1 m) above lower levels for a distance sufficient to protect employees below, except on float (ship) scaffolds where an edging of 3/4 x 1 1/2 inch (2 x 4 cm) wood or equivalent may be used in lieu of toeboards;

1926.451(h)(2)(iii)

Where tools, materials, or equipment are piled to a height higher than the top edge of the toeboard, paneling or screening extending from the toeboard or platform to the top of the guardrail shall be erected for a distance sufficient to protect employees below; or

1926.451(h)(2)(iv)

A guardrail system shall be installed with openings small enough to prevent passage of potential falling objects; or

1926.451(h)(2)(v)

A canopy structure, debris net, or catch platform strong enough to withstand the impact forces of the potential falling objects shall be erected over the employees below.

1926.451(h)(3)

Canopies, when used for falling object protection, shall comply with the following criteria:

1926.451(h)(3)(i)

Canopies shall be installed between the falling object hazard and the employees.

..1926.451(h)(3)(ii)

1926.451(h)(3)(ii)

When canopies are used on suspension scaffolds for falling object protection, the scaffold shall be equipped with additional independent support lines equal in number to the number of points supported, and equivalent in strength to the strength of the suspension ropes.

1926.451(h)(3)(iii)

Independent support lines and suspension ropes shall not be attached to the same points of anchorage.

1926.451(h)(4)

Where used, toeboards shall be:

1926.451(h)(4)(i)

Capable of withstanding, without failure, a force of at least 50 pounds (222 n) applied in any downward or horizontal direction at any point along the toeboard (toeboards built in accordance with Appendix A to this subpart will be deemed to meet this requirement); and

1926.451(h)(4)(ii)

At least three and one-half inches (9 cm) high from the top edge of the toeboard to the level of the walking/working surface. Toeboards shall be securely fastened in place at the outermost edge of the platform and have not more than 1/4 inch (0.7 cm) clearance above the walking/working surface. Toeboards shall be solid or with openings not over one inch (2.5 cm) in the greatest dimension.

[44 FR 8577, Feb. 9, 1979; 44 FR 20940, Apr. 6, 1979, as amended at 58 FR 35182 and 35310, June 30, 1993; 61 FR 46025, Aug. 30 1996; 61 FR 59831, Nov. 25, 1996]

Next Standard (1926.452)

Regulations (Standards - 29 CFR) - Table of Contents

Back to Top http://osha.gov/index.html http://www.dol.gov/

Contact Us | Freedom of Information Act | Customer Survey
Privacy and Security Statement | Disclaimers

Occupational Safety & Health Administration
200 Constitution Avenue, NW
Washington, DC 20210

Appendix B

Safety Standards for Aerial Lifts

U.S. Department of Labor
Occupational Safety & Health Administration

www.osha.gov Search GO Advanced Search

Regulations (Standards - 29 CFR)

Aerial lifts. - 1926.453

Regulations (Standards - 29 CFR) - Table of Contents

- **Part Number:** 1926
- **Part Title:** Safety and Health Regulations for Construction
- **Subpart:** L
- **Subpart Title:** Scaffolds
- **Standard Number:** 1926.453
- **Title:** Aerial lifts.

1926.453(a)

"General requirements."

1926.453(a)(1)

Unless otherwise provided in this section, aerial lifts acquired for use on or after January 22, 1973 shall be designed and constructed in conformance with the applicable requirements of the American National Standards for "Vehicle Mounted Elevating and Rotating Work Platforms," ANSI A92.2-1969, including appendix. Aerial lifts acquired before January 22, 1973 which do not meet the requirements of ANSI A92.2-1969, may not be used after January 1, 1976, unless they shall have been modified so as to conform with the applicable design and construction requirements of ANSI A92.2-1969. Aerial lifts include the following types of vehicle-mounted aerial devices used to elevate personnel to job-sites above ground:

1926.453(a)(1)(i)

Extensible boom platforms;

1926.453(a)(1)(ii)

Aerial ladders;

1926.453(a)(1)(iii)

Articulating boom platforms;

1926.453(a)(1)(iv)

Vertical towers; and

..1926.453(a)(1)(v)

1926.453(a)(1)(v)

A combination of any such devices. Aerial equipment may be made of metal, wood, fiberglass reinforced plastic (FRP), or other material; may be powered or manually operated; and are deemed to be aerial lifts whether or not they are capable of rotating about a substantially vertical axis.

Aerial lifts. - 1926.453

1926.453(a)(2)

Aerial lifts may be "field modified" for uses other than those intended by the manufacturer provided the modification has been certified in writing by the manufacturer or by any other equivalent entity, such as a nationally recognized testing laboratory, to be in conformity with all applicable provisions of ANSI A92.2-1969 and this section and to be at least as safe as the equipment was before modification.

1926.453(b)

"Specific requirements."

1926.453(b)(1)

Ladder trucks and tower trucks. Aerial ladders shall be secured in the lower traveling position by the locking device on top of the truck cab, and the manually operated device at the base of the ladder before the truck is moved for highway travel.

1926.453(b)(2)

Extensible and articulating boom platforms.

1926.453(b)(2)(i)

Lift controls shall be tested each day prior to use to determine that such controls are in safe working condition.

1926.453(b)(2)(ii)

Only authorized persons shall operate an aerial lift.

..1926.453(b)(2)(iii)

1926.453(b)(2)(iii)

Belting off to an adjacent pole, structure, or equipment while working from an aerial lift shall not be permitted.

1926.453(b)(2)(iv)

Employees shall always stand firmly on the floor of the basket, and shall not sit or climb on the edge of the basket or use planks, ladders, or other devices for a work position.

1926.453(b)(2)(v)

A body belt shall be worn and a lanyard attached to the boom or basket when working from an aerial lift.

Note to paragraph (b)(2)(v): As of January 1, 1998, subpart M of this part (1926.502(d)) provides that body belts are not acceptable as part of a personal fall arrest system. The use of a body belt in a tethering system or in a restraint system is acceptable and is regulated under 1926.502(e).

1926.453(b)(2)(vi)

Boom and basket load limits specified by the manufacturer shall not be exceeded.

1926.453(b)(2)(vii)

The brakes shall be set and when outriggers are used, they shall be positioned on pads or a solid surface. Wheel chocks shall be installed before using an aerial lift on an incline, provided they can be safely installed.

1926.453(b)(2)(viii)

An aerial lift truck shall not be moved when the boom is elevated in a working position with men in the basket, except for equipment which is specifically designed for this type of operation in accordance with the provisions of paragraphs (a)(1) and (2) of this section.

..1926.453(b)(2)(ix)

1926.453(b)(2)(ix)

Articulating boom and extensible boom platforms, primarily designed as personnel carriers, shall have both platform (upper) and lower controls. Upper controls shall be in or beside the platform within easy reach of the operator. Lower controls shall provide for overriding the upper controls. Controls shall be plainly marked as to their function. Lower level controls shall not be operated unless permission has been obtained from the employee in the lift, except in case of emergency.

1926.453(b)(2)(x)

Climbers shall not be worn while performing work from an aerial lift.

1926.453(b)(2)(xi)

The insulated portion of an aerial lift shall not be altered in any manner that might reduce its insulating value.

1926.453(b)(2)(xii)

Before moving an aerial lift for travel, the boom(s) shall be inspected to see that it is properly cradled and outriggers are in stowed position except as provided in paragraph (b)(2)(viii) of this section.

1926.453(b)(3)

Electrical tests. All electrical tests shall conform to the requirements of ANSI A92.2-1969 section 5. However equivalent d.c.; voltage tests may be used in lieu of the a.c. voltage specified in A92.2-1969; d.c. voltage tests which are approved by the equipment manufacturer or equivalent entity shall be considered an equivalent test for the purpose of this paragraph (b)(3).

..1926.453(b)(4)

1926.453(b)(4)

Bursting safety factor. The provisions of the American National Standards Institute standard ANSI A92.2-1969, section 4.9 Bursting Safety Factor shall apply to all critical hydraulic and pneumatic components. Critical components are those in which a failure would result in a free fall or free rotation of the boom. All noncritical components shall have a bursting safety factor of at least 2 to 1.

1926.453(b)(5)

Welding standards. All welding shall conform to the following standards as applicable:

1926.453(b)(5)(i)

Standard Qualification Procedure, AWS B3.0-41.

1926.453(b)(5)(ii)

Recommended Practices for Automotive Welding Design, AWS D8.4-61.

1926.453(b)(5)(iii)

Standard Qualification of Welding Procedures and Welders for Piping and Tubing, AWS D10.9-69.

1926.453(b)(5)(iv)

Specifications for Welding Highway and Railway Bridges, AWS D2.0-69.

Note to 1926.453: Non-mandatory Appendix C to this subpart lists examples of national consensus standards that are considered to provide employee protection equivalent to that provided through the application of ANSI A92.2-1969, where appropriate. This incorporation by reference was approved by the Director of the Federal Register in accordance with 5 U.S.C. 552(a) and 1 CFR part 51. Copies may be obtained from the American National Standards Institute. Copies may be inspected at the Docket Office, Occupational Safety and Health Administration, U.S. Department of Labor, 200 Constitution Avenue, NW., room N2634, Washington, DC or at the Office of the Federal Register, 800 North Capitol Street, NW., suite 700, Washington, DC.

[58 FR 35182, June 30, 1993; 61 FR 46025, Aug. 30, 1996; 61 FR 59831, Nov. 25, 1996]

Next Standard (1926.454)

Regulations (Standards - 29 CFR) - Table of Contents

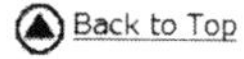
Back to Top http://osha.gov/index.html

Contact Us | Freedom of Information Act | Customer Survey
Privacy and Security Statement | Disclaimers

Occupational Safety & Health Administration
200 Constitution Avenue, NW
Washington, DC 20210

Appendix C

Safety Standards for Specific Scaffolds

Regulations (Standards - 29 CFR)

Additional requirements applicable to specific types of scaffolds. - 1926.452

Regulations (Standards - 29 CFR) - Table of Contents

- **Part Number:** 1926
- **Part Title:** Safety and Health Regulations for Construction
- **Subpart:** L
- **Subpart Title:** Scaffolds
- **Standard Number:** 1926.452
- **Title:** Additional requirements applicable to specific types of scaffolds.

In addition to the applicable requirements of 1926.451, the following requirements apply to the specific types of scaffolds indicated. Scaffolds not specifically addressed by 1926.452, such as but not limited to systems scaffolds, must meet the requirements of 1926.451.

1926.452(a)

"Pole scaffolds."

1926.452(a)(1)

When platforms are being moved to the next level, the existing platform shall be left undisturbed until the new bearers have been set in place and braced, prior to receiving the new platforms.

1926.452(a)(2)

Crossbracing shall be installed between the inner and outer sets of poles on double pole scaffolds.

1926.452(a)(3)

Diagonal bracing in both directions shall be installed across the entire inside face of double-pole scaffolds used to support loads equivalent to a uniformly distributed load of 50 pounds (222 kg) or more per square foot (929 square cm).

1926.452(a)(4)

Diagonal bracing in both directions shall be installed across the entire outside face of all double- and single-pole scaffolds.

1926.452(a)(5)

Runners and bearers shall be installed on edge.

1926.452(a)(6)

Bearers shall extend a minimum of 3 inches (7.6 cm) over the outside edges of runners.

..1926.452(a)(7)

Additional requirements applicable to specific types of scaffolds. - 1926.452

1926.452(a)(7)

Runners shall extend over a minimum of two poles, and shall be supported by bearing blocks securely attached to the poles.

1926.452(a)(8)

Braces, bearers, and runners shall not be spliced between poles.

1926.452(a)(9)

Where wooden poles are spliced, the ends shall be squared and the upper section shall rest squarely on the lower section. Wood splice plates shall be provided on at least two adjacent sides, and shall extend at least 2 feet (0.6 m) on either side of the splice, overlap the abutted ends equally, and have at least the same cross-sectional areas as the pole. Splice plates of other materials of equivalent strength may be used.

1926.452(a)(10)

Pole scaffolds over 60 feet in height shall be designed by a registered professional engineer, and shall be constructed and loaded in accordance with that design. Non-mandatory Appendix A to this subpart contains examples of criteria that will enable an employer to comply with design and loading requirements for pole scaffolds under 60 feet in height.

1926.452(b)

"Tube and coupler scaffolds."

1926.452(b)(1)

When platforms are being moved to the next level, the existing platform shall be left undisturbed until the new bearers have been set in place and braced prior to receiving the new platforms.

..1926.452(b)(2)

1926.452(b)(2)

Transverse bracing forming an "X" across the width of the scaffold shall be installed at the scaffold ends and at least at every third set of posts horizontally (measured from only one end) and every fourth runner vertically. Bracing shall extend diagonally from the inner or outer posts or runners upward to the next outer or inner posts or runners. Building ties shall be installed at the bearer levels between the transverse bracing and shall conform to the requirements of 1926.451(c)(1).

1926.452(b)(3)

On straight run scaffolds, longitudinal bracing across the inner and outer rows of posts shall be installed diagonally in both directions, and shall extend from the base of the end posts upward to the top of the scaffold at approximately a 45 degree angle. On scaffolds whose length is greater than their height, such bracing shall be repeated beginning at least at every fifth post. On scaffolds whose length is less than their height, such bracing shall be installed from the base of the end posts upward to the opposite end posts, and then in alternating directions until reaching the top of the scaffold. Bracing shall be installed as close as possible to the intersection of the bearer and post or runner and post.

1926.452(b)(4)

Where conditions preclude the attachment of bracing to posts, bracing shall be attached to the runners as close to the post as possible.

1926.452(b)(5)

Bearers shall be installed transversely between posts, and when coupled to the posts, shall have the inboard coupler bear directly on the runner coupler. When the bearers are coupled to the runners, the couplers shall be as close to the posts as possible.

1926.452(b)(6)

Bearers shall extend beyond the posts and runners, and shall provide full contact with the coupler.

..1926.452(b)(7)

1926.452(b)(7)

Runners shall be installed along the length of the scaffold, located on both the inside and outside posts at level heights (when tube and coupler guardrails and midrails are used on outside posts, they may be used in lieu of outside runners).

1926.452(b)(8)

Runners shall be interlocked on straight runs to form continuous lengths, and shall be coupled to each post. The bottom runners and bearers shall be located as close to the base as possible.

1926.452(b)(9)

Couplers shall be of a structural metal, such as drop-forged steel, malleable iron, or structural grade aluminum. The use of gray cast iron is prohibited.

1926.452(b)(10)

Tube and coupler scaffolds over 125 feet in height shall be designed by a registered professional engineer, and shall be constructed and loaded in accordance with such design. Non-mandatory Appendix A to this subpart contains examples of criteria that will enable an employer to comply with design and loading requirements for tube and coupler scaffolds under 125 feet in height.

1926.452(c)

"Fabricated frame scaffolds" (tubular welded frame scaffolds).

1926.452(c)(1)

When moving platforms to the next level, the existing platform shall be left undisturbed until the new end frames have been set in place and braced prior to receiving the new platforms.

..1926.452(c)(2)

1926.452(c)(2)

Frames and panels shall be braced by cross, horizontal, or diagonal braces, or combination thereof, which secure vertical members together laterally. The cross braces shall be of such length as will automatically square and align vertical members so that the erected scaffold is always plumb, level, and square. All brace connections shall be secured.

1926.452(c)(3)

Frames and panels shall be joined together vertically by coupling or stacking pins or equivalent means.

1926.452(c)(4)

Where uplift can occur which would displace scaffold end frames or panels, the frames or panels shall be locked together vertically by pins or equivalent means.

1926.452(c)(5)

Brackets used to support cantilevered loads shall:

1926.452(c)(5)(i)

Be seated with side-brackets parallel to the frames and end-brackets at 90 degrees to the frames;

1926.452(c)(5)(ii)

Not be bent or twisted from these positions; and

1926.452(c)(5)(iii)

Be used only to support personnel, unless the scaffold has been designed for other loads by a qualified engineer and built to withstand the tipping forces caused by those other loads being placed on the bracket-supported section of the scaffold.

1926.452(c)(6)

Scaffolds over 125 feet (38.0 m) in height above their base plates shall be designed by a registered professional engineer, and shall be constructed and loaded in accordance with such design.

..1926.452(d)

1926.452(d)

"Plasterers', decorators', and large area scaffolds." Scaffolds shall be constructed in accordance with paragraphs (a), (b), or (c) of this section, as appropriate.

1926.452(e)

"Bricklayers' square scaffolds (squares)."

1926.452(e)(1)

Scaffolds made of wood shall be reinforced with gussets on both sides of each corner.

1926.452(e)(2)

Diagonal braces shall be installed on all sides of each square.

Additional requirements applicable to specific types of scaffolds. - 1926.452

1926.452(e)(3)

Diagonal braces shall be installed between squares on the rear and front sides of the scaffold, and shall extend from the bottom of each square to the top of the next square.

1926.452(e)(4)

Scaffolds shall not exceed three tiers in height, and shall be so constructed and arranged that one square rests directly above the other. The upper tiers shall stand on a continuous row of planks laid across the next lower tier, and shall be nailed down or otherwise secured to prevent displacement.

1926.452(f)

"Horse scaffolds."

1926.452(f)(1)

Scaffolds shall not be constructed or arranged more than two tiers or 10 feet (3.0 m) in height, whichever is less.

1926.452(f)(2)

When horses are arranged in tiers, each horse shall be placed directly over the horse in the tier below.

..1926.452(f)(3)

1926.452(f)(3)

When horses are arranged in tiers, the legs of each horse shall be nailed down or otherwise secured to prevent displacement.

1926.452(f)(4)

When horses are arranged in tiers, each tier shall be crossbraced.

1926.452(g)

"Form scaffolds and carpenters' bracket scaffolds."

1926.452(g)(1)

Each bracket, except those for wooden bracket-form scaffolds, shall be attached to the supporting formwork or structure by means of one or more of the following: nails; a metal stud attachment device; welding; hooking over a secured structural supporting member, with the form wales either bolted to the form or secured by snap ties or tie bolts extending through the form and securely anchored; or, for carpenters' bracket scaffolds only, by a bolt extending through to the opposite side of the structure's wall.

1926.452(g)(2)

Wooden bracket-form scaffolds shall be an integral part of the form panel.

1926.452(g)(3)

Folding type metal brackets, when extended for use, shall be either bolted or secured with a locking-type pin.

..1926.452(h)

1926.452(h)

"Roof bracket scaffolds."

1926.452(h)(1)

Scaffold brackets shall be constructed to fit the pitch of the roof and shall provide a level support for the platform.

1926.452(h)(2)

Brackets (including those provided with pointed metal projections) shall be anchored in place by nails unless it is impractical to use nails. When nails are not used, brackets shall be secured in place with first-grade manila rope of at least three-fourth inch (1.9 cm) diameter, or equivalent.

1926.452(i)

"Outrigger scaffolds."

1926.452(i)(1)

The inboard end of outrigger beams, measured from the fulcrum point to the extreme point of anchorage, shall be not less than one and one-half times the outboard end in length.

1926.452(i)(2)

Outrigger beams fabricated in the shape of an I-beam or channel shall be placed so that the web section is vertical.

1926.452(i)(3)

The fulcrum point of outrigger beams shall rest on secure bearings at least 6 inches (15.2 cm) in each horizontal dimension.

1926.452(i)(4)

Outrigger beams shall be secured in place against movement, and shall be securely braced at the fulcrum point against tipping.

..1926.452(i)(5)

1926.452(i)(5)

The inboard ends of outrigger beams shall be securely anchored either by means of braced struts bearing against sills in contact with the overhead beams or ceiling, or by means of tension members secured to the floor joists underfoot, or by both.

1926.452(i)(6)

The entire supporting structure shall be securely braced to prevent any horizontal movement.

1926.452(i)(7)

To prevent their displacement, platform units shall be nailed, bolted, or otherwise secured to outriggers.

1926.452(i)(8)

Scaffolds and scaffold components shall be designed by a registered professional engineer and shall be constructed and loaded in accordance with such design.

1926.452(j)

"Pump jack scaffolds."

1926.452(j)(1)

Pump jack brackets, braces, and accessories shall be fabricated from metal plates and angles. Each pump jack bracket shall have two positive gripping mechanisms to prevent any failure or slippage.

1926.452(j)(2)

Poles shall be secured to the structure by rigid triangular bracing or equivalent at the bottom, top, and other points as necessary. When the pump jack has to pass bracing already installed, an additional brace shall be installed approximately 4 feet (1.2 m) above the brace to be passed, and shall be left in place until the pump jack has been moved and the original brace reinstalled.

..1926.452(j)(3)

1926.452(j)(3)

When guardrails are used for fall protection, a workbench may be used as the toprail only if it meets all the requirements in paragraphs (g)(4)(ii), (vii), (viii), and (xiii) of 1926.451.

1926.452(j)(4)

Work benches shall not be used as scaffold platforms.

1926.452(j)(5)

When poles are made of wood, the pole lumber shall be straight-grained, free of shakes, large loose or dead knots, and other defects which might impair strength.

1926.452(j)(6)

When wood poles are constructed of two continuous lengths, they shall be joined together with the seam parallel to the bracket.

1926.452(j)(7)

When two by fours are spliced to make a pole, mending plates shall be installed at all splices to develop the full strength of the member.

1926.452(k)

Additional requirements applicable to specific types of scaffolds. - 1926.452

"Ladder jack scaffolds."

1926.452(k)(1)

Platforms shall not exceed a height of 20 feet (6.1 m).

1926.452(k)(2)

All ladders used to support ladder jack scaffolds shall meet the requirements of subpart X of this part -- Stairways and Ladders, except that job-made ladders shall not be used to support ladder jack scaffolds.

..1926.452(k)(3)

1926.452(k)(3)

The ladder jack shall be so designed and constructed that it will bear on the side rails and ladder rungs or on the ladder rungs alone. If bearing on rungs only, the bearing area shall include a length of at least 10 inches (25.4 cm) on each rung.

1926.452(k)(4)

Ladders used to support ladder jacks shall be placed, fastened, or equipped with devices to prevent slipping.

1926.452(k)(5)

Scaffold platforms shall not be bridged one to another.

1926.452(l)

"Window jack scaffolds."

1926.452(l)(1)

Scaffolds shall be securely attached to the window opening.

1926.452(l)(2)

Scaffolds shall be used only for the purpose of working at the window opening through which the jack is placed.

1926.452(l)(3)

Window jacks shall not be used to support planks placed between one window jack and another, or for other elements of scaffolding.

1926.452(m)

"Crawling boards (chicken ladders)."

1926.452(m)(1)

Crawling boards shall extend from the roof peak to the eaves when used in connection with roof construction, repair, or maintenance.

..1926.452(m)(2)

1926.452(m)(2)

Crawling boards shall be secured to the roof by ridge hooks or by means that meet equivalent criteria (e.g., strength and durability).

1926.452(n)

"Step, platform, and trestle ladder scaffolds."

1926.452(n)(1)

Scaffold platforms shall not be placed any higher than the second highest rung or step of the ladder supporting the platform.

1926.452(n)(2)

All ladders used in conjunction with step, platform and trestle ladder scaffolds shall meet the pertinent requirements of subpart X of this part -- Stairways and Ladders, except that job-made ladders shall not be used to support such scaffolds.

1926.452(n)(3)

Ladders used to support step, platform, and trestle ladder scaffolds shall be placed, fastened, or equipped with devices to prevent slipping.

1926.452(n)(4)

Scaffolds shall not be bridged one to another.

1926.452(o)

"Single-point adjustable suspension scaffolds."

1926.452(o)(1)

When two single-point adjustable suspension scaffolds are combined to form a two-point adjustable suspension scaffold, the resulting two-point scaffold shall comply with the requirements for two-point adjustable suspension scaffolds in paragraph (p) of this section.

..1926.452(o)(2)

1926.452(o)(2)

The supporting rope between the scaffold and the suspension device shall be kept vertical unless all of the following conditions are met:

1926.452(o)(2)(i)

The rigging has been designed by a qualified person, and

1926.452(o)(2)(ii)

Additional requirements applicable to specific types of scaffolds. - 1926.452

The scaffold is accessible to rescuers, and

1926.452(o)(2)(iii)

The supporting rope is protected to ensure that it will not chafe at any point where a change in direction occurs, and

1926.452(o)(2)(iv)

The scaffold is positioned so that swinging cannot bring the scaffold into contact with another surface.

1926.452(o)(3)

Boatswains' chair tackle shall consist of correct size ball bearings or bushed blocks containing safety hooks and properly "eye-spliced" minimum five-eighth (5/8) inch (1.6 cm) diameter first-grade manila rope, or other rope which will satisfy the criteria (e.g., strength and durability) of manila rope.

1926.452(o)(4)

Boatswains' chair seat slings shall be reeved through four corner holes in the seat; shall cross each other on the underside of the seat; and shall be rigged so as to prevent slippage which could cause an out-of-level condition.

..1926.452(o)(5)

1926.452(o)(5)

Boatswains' chair seat slings shall be a minimum of five-eight (5/8) inch (1.6 cm) diameter fiber, synthetic, or other rope which will satisfy the criteria (e.g., strength, slip resistance, durability, etc.) of first grade manila rope.

1926.452(o)(6)

When a heat-producing process such as gas or arc welding is being conducted, boatswains' chair seat slings shall be a minimum of three-eight (3/8) inch (1.0 cm) wire rope.

1926.452(o)(7)

Non-cross-laminated wood boatswains' chairs shall be reinforced on their underside by cleats securely fastened to prevent the board from splitting.

1926.452(p)

"Two-point adjustable suspension scaffolds (swing stages)." The following requirements do not apply to two-point adjustable suspension scaffolds used as masons' or stonesetters' scaffolds. Such scaffolds are covered by paragraph (q) of this section.

1926.452(p)(1)

Platforms shall not be more than 36 inches (0.9 m) wide unless designed by a qualified person to prevent unstable conditions.

1926.452(p)(2)

The platform shall be securely fastened to hangers (stirrups) by U-bolts or by other means which satisfy the requirements of 1926.451(a).

1926.452(p)(3)

The blocks for fiber or synthetic ropes shall consist of at least one double and one single block. The sheaves of all blocks shall fit the size of the rope used.

..1926.452(p)(4)

1926.452(p)(4)

Platforms shall be of the ladder-type, plank-type, beam-type, or light-metal type. Light metal-type platforms having a rated capacity of 750 pounds or less and platforms 40 feet (12.2 m) or less in length shall be tested and listed by a nationally recognized testing laboratory.

1926.452(p)(5)

Two-point scaffolds shall not be bridged or otherwise connected one to another during raising and lowering operations unless the bridge connections are articulated (attached), and the hoists properly sized.

1926.452(p)(6)

Passage may be made from one platform to another only when the platforms are at the same height, are abutting, and walk-through stirrups specifically designed for this purpose are used.

1926.452(q)

"Multi-point adjustable suspension scaffolds, stonesetters' multi-point adjustable suspension scaffolds, and masons' multi-point adjustable suspension scaffolds."

1926.452(q)(1)

When two or more scaffolds are used they shall not be bridged one to another unless they are designed to be bridged, the bridge connections are articulated, and the hoists are properly sized.

1926.452(q)(2)

If bridges are not used, passage may be made from one platform to another only when the platforms are at the same height and are abutting.

..1926.452(q)(3)

1926.452(q)(3)

Scaffolds shall be suspended from metal outriggers, brackets, wire rope slings, hooks, or means that meet equivalent criteria (e.g., strength, durability).

1926.452(r)

"Catenary scaffolds."

1926.452(r)(1)

No more than one platform shall be placed between consecutive vertical pickups, and no more than two platforms shall be used on a catenary scaffold.

1926.452(r)(2)

Platforms supported by wire ropes shall have hook-shaped stops on each end of the platforms to prevent them from slipping off the wire ropes. These hooks shall be so placed that they will prevent the platform from falling if one of the horizontal wire ropes breaks.

1926.452(r)(3)

Wire ropes shall not be tightened to the extent that the application of a scaffold load will overstress them.

1926.452(r)(4)

Wire ropes shall be continuous and without splices between anchors.

1926.452(s)

"Float (ship) scaffolds."

1926.452(s)(1)

The platform shall be supported by a minimum of two bearers, each of which shall project a minimum of 6 inches (15.2 cm) beyond the platform on both sides. Each bearer shall be securely fastened to the platform.

1926.452(s)(2)

Rope connections shall be such that the platform cannot shift or slip.

..1926.452(s)(3)

1926.452(s)(3)

When only two ropes are used with each float:

1926.452(s)(3)(i)

They shall be arranged so as to provide four ends which are securely fastened to overhead supports.

1926.452(s)(3)(ii)

Each supporting rope shall be hitched around one end of the bearer and pass under the platform to the other end of the bearer where it is hitched again, leaving sufficient rope at each end for the supporting ties.

1926.452(t)

"Interior hung scaffolds."

1926.452(t)(1)

Scaffolds shall be suspended only from the roof structure or other structural member such as ceiling beams.

1926.452(t)(2)

Overhead supporting members (roof structure, ceiling beams, or other structural members) shall be inspected and checked for strength before the scaffold is erected.

1926.452(t)(3)

Suspension ropes and cables shall be connected to the overhead supporting members by shackles, clips, thimbles, or other means that meet equivalent criteria (e.g., strength, durability).

..1926.452(u)

1926.452(u)

"Needle beam scaffolds."

1926.452(u)(1)

Scaffold support beams shall be installed on edge.

1926.452(u)(2)

Ropes or hangers shall be used for supports, except that one end of a needle beam scaffold may be supported by a permanent structural member.

1926.452(u)(3)

The ropes shall be securely attached to the needle beams.

1926.452(u)(4)

The support connection shall be arranged so as to prevent the needle beam from rolling or becoming displaced.

1926.452(u)(5)

Platform units shall be securely attached to the needle beams by bolts or equivalent means. Cleats and overhang are not considered to be adequate means of attachment.

1926.452(v)

"Multi-level suspended scaffolds."

1926.452(v)(1)

Scaffolds shall be equipped with additional independent support lines, equal in number to the number of points supported, and of equivalent strength to the suspension ropes, and rigged to support the scaffold in the event the suspension rope(s) fail.

1926.452(v)(2)

Independent support lines and suspension ropes shall not be attached to the same points of anchorage.

Additional requirements applicable to specific types of scaffolds. - 1926.452

..1926.452(v)(3)

1926.452(v)(3)

Supports for platforms shall be attached directly to the support stirrup and not to any other platform.

1926.452(w)

"Mobile scaffolds."

1926.452(w)(1)

Scaffolds shall be braced by cross, horizontal, or diagonal braces, or combination thereof, to prevent racking or collapse of the scaffold and to secure vertical members together laterally so as to automatically square and align the vertical members. Scaffolds shall be plumb, level, and squared. All brace connections shall be secured.

1926.452(w)(1)(i)

Scaffolds constructed of tube and coupler components shall also comply with the requirements of paragraph (b) of this section;

1926.452(w)(1)(ii)

Scaffolds constructed of fabricated frame components shall also comply with the requirements of paragraph (c) of this section.

1926.452(w)(2)

Scaffold casters and wheels shall be locked with positive wheel and/or wheel and swivel locks, or equivalent means, to prevent movement of the scaffold while the scaffold is used in a stationary manner.

1926.452(w)(3)

Manual force used to move the scaffold shall be applied as close to the base as practicable, but not more than 5 feet (1.5 m) above the supporting surface.

..1926.452(w)(4)

1926.452(w)(4)

Power systems used to propel mobile scaffolds shall be designed for such use. Forklifts, trucks, similar motor vehicles or add-on motors shall not be used to propel scaffolds unless the scaffold is designed for such propulsion systems.

1926.452(w)(5)

Scaffolds shall be stabilized to prevent tipping during movement.

1926.452(w)(6)

Employees shall not be allowed to ride on scaffolds unless the following conditions exist:

Additional requirements applicable to specific types of scaffolds. - 1926.452

1926.452(w)(6)(i)

The surface on which the scaffold is being moved is within 3 degrees of level, and free of pits, holes, and obstructions;

1926.452(w)(6)(ii)

The height to base width ratio of the scaffold during movement is two to one or less, unless the scaffold is designed and constructed to meet or exceed nationally recognized stability test requirements such as those listed in paragraph (x) of Appendix A to this subpart (ANSI/SIA A92.5 and A92.6);

1926.452(w)(6)(iii)

Outrigger frames, when used, are installed on both sides of the scaffold;

1926.452(w)(6)(iv)

When power systems are used, the propelling force is applied directly to the wheels, and does not produce a speed in excess of 1 foot per second (.3 mps); and

1926.452(w)(6)(v)

No employee is on any part of the scaffold which extends outward beyond the wheels, casters, or other supports.

..1926.452(w)(7)

1926.452(w)(7)

Platforms shall not extend outward beyond the base supports of the scaffold unless outrigger frames or equivalent devices are used to ensure stability.

1926.452(w)(8)

Where leveling of the scaffold is necessary, screw jacks or equivalent means shall be used.

1926.452(w)(9)

Caster stems and wheel stems shall be pinned or otherwise secured in scaffold legs or adjustment screws.

1926.452(w)(10)

Before a scaffold is moved, each employee on the scaffold shall be made aware of the move.

1926.452(x)

"Repair bracket scaffolds."

1926.452(x)(1)

Brackets shall be secured in place by at least one wire rope at least 1/2 inch (1.27 cm) in diameter.

1926.452(x)(2)

Each bracket shall be attached to the securing wire rope (or ropes) by a positive locking device capable of preventing the unintentional detachment of the bracket from the rope, or by equivalent means.

1926.452(x)(3)

Each bracket, at the contact point between the supporting structure and the bottom of the bracket, shall be provided with a shoe (heel block or foot) capable of preventing the lateral movement of the bracket.

..1926.452(x)(4)

1926.452(x)(4)

Platforms shall be secured to the brackets in a manner that will prevent the separation of the platforms from the brackets and the movement of the platforms or the brackets on a completed scaffold.

1926.452(x)(5)

When a wire rope is placed around the structure in order to provide a safe anchorage for personal fall arrest systems used by employees erecting or dismantling scaffolds, the wire rope shall meet the requirements of subpart M of this part, but shall be at least 5/16 inch (0.8 cm) in diameter.

1926.452(x)(6)

Each wire rope used for securing brackets in place or as an anchorage for personal fall arrest systems shall be protected from damage due to contact with edges, corners, protrusions, or other discontinuities of the supporting structure or scaffold components.

1926.452(x)(7)

Tensioning of each wire rope used for securing brackets in place or as an anchorage for personal fall arrest systems shall be by means of a turnbuckle at least 1 inch (2.54 cm) in diameter, or by equivalent means.

1926.452(x)(8)

Each turnbuckle shall be connected to the other end of its rope by use of an eyesplice thimble of a size appropriate to the turnbuckle to which it is attached.

1926.452(x)(9)

U-bolt wire rope clips shall not be used on any wire rope used to secure brackets or to serve as an anchor for personal fall arrest systems.

..1926.452(x)(10)

1926.452(x)(10)

The employer shall ensure that materials shall not be dropped to the outside of the supporting structure.

1926.452(x)(11)

Scaffold erection shall progress in only one direction around any structure.

1926.452(y)

"Stilts." Stilts, when used, shall be used in accordance with the following requirements:

1926.452(y)(1)

An employee may wear stilts on a scaffold only if it is a large area scaffold.

1926.452(y)(2)

When an employee is using stilts on a large area scaffold where a guardrail system is used to provide fall protection, the guardrail system shall be increased in height by an amount equal to the height of the stilts being used by the employee.

1926.452(y)(3)

Surfaces on which stilts are used shall be flat and free of pits, holes and obstructions, such as debris, as well as other tripping and falling hazards.

1926.452(y)(4)

Stilts shall be properly maintained. Any alteration of the original equipment shall be approved by the manufacturer.

[44 FR 8577, Feb. 9, 1979; 44 FR 20940. Apr. 6, 1979, as amended at 55 FR 47687, Nov. 14, 1990; 61 FR 46025, Aug. 30, 1996]

Next Standard (1926.453)

Regulations (Standards - 29 CFR) - Table of Contents

Back to Top http://osha.gov/index.html

Contact Us | Freedom of Information Act | Customer Survey
Privacy and Security Statement | Disclaimers

Occupational Safety & Health Administration
200 Constitution Avenue, NW
Washington, DC 20210

Appendix D

Scaffold Construction

BRACING - TUBE & COUPLER SCAFFOLDS

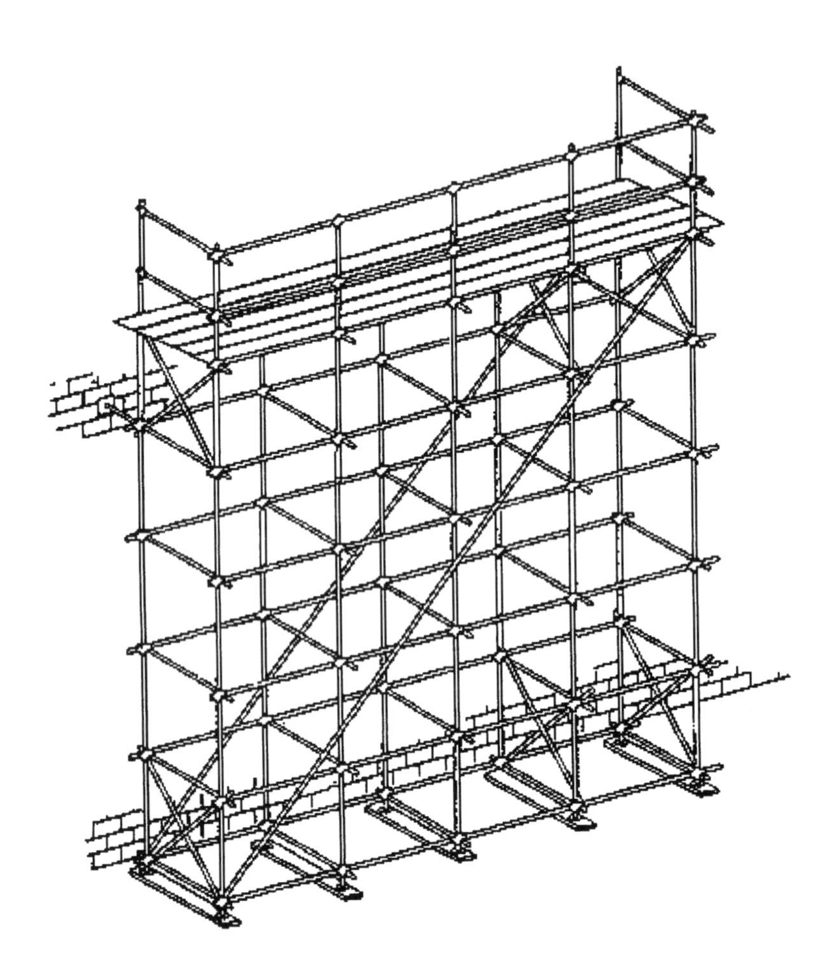

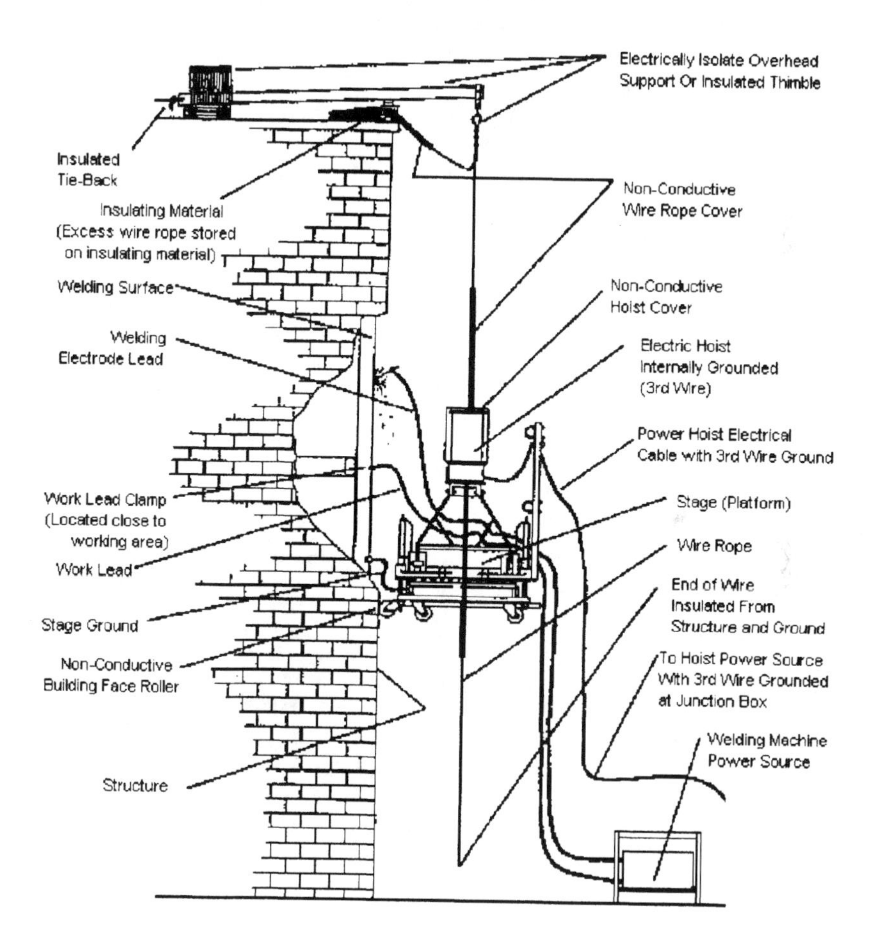
SUSPENDED SCAFFOLD PLATFORM WELDING PRECAUTIONS
Electrically Isolate Overhead Support Or Insulated Thimble
Insulated Tie-Back
Insulating Material (Excess wire rope stored on insulating material)
Non-Conductive Wire Rope Cover
Welding Surface
Non-Conductive Hoist Cover
Welding Electrode Lead
Electric Hoist Internally Grounded (3rd Wire)
Power Hoist Electrical Cable with 3rd Wire Ground
Work Lead Clamp (Located close to working area)
Stage (Platform)
Work Lead
Wire Rope
End of Wire Insulated From Structure and Ground
Stage Ground
Non-Conductive Building Face Roller
To Hoist Power Source With 3rd Wire Grounded at Junction Box
Welding Machine Power Source
Structure

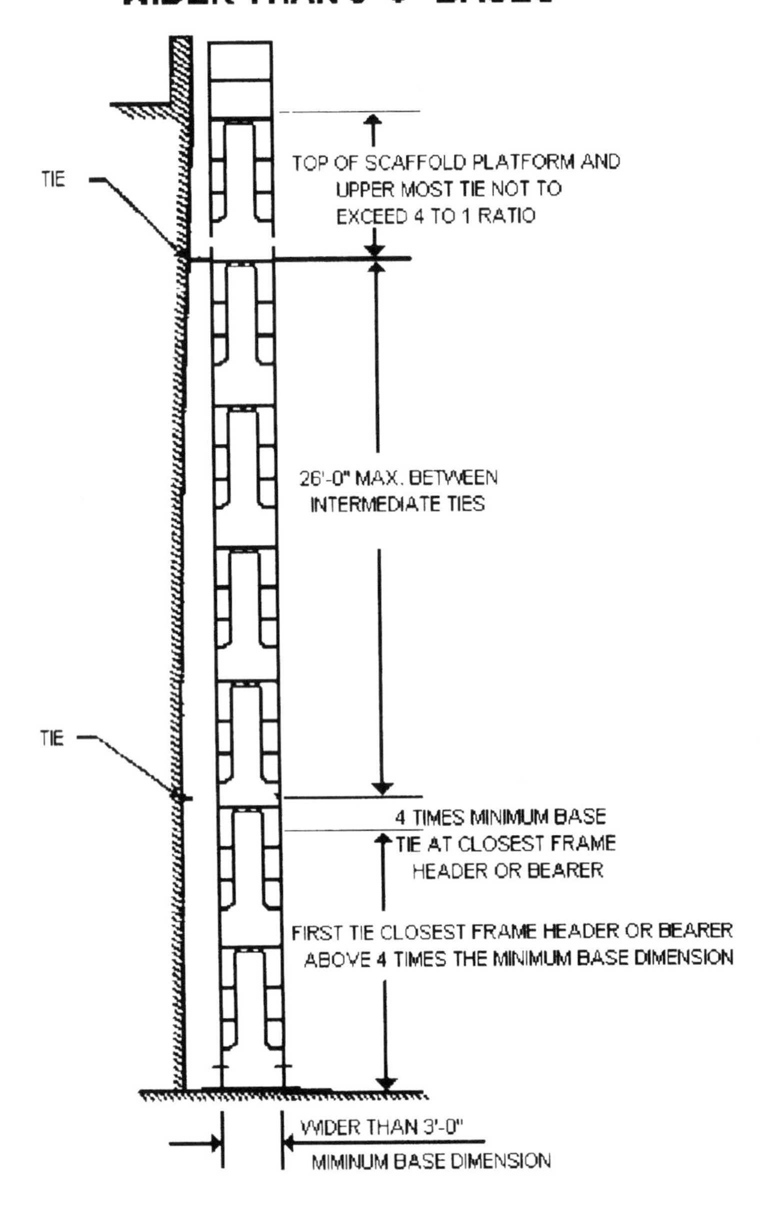
MAXIMUM VERTICAL TIE SPACING
WIDER THAN 3'-0" BASES
TIE
TOP OF SCAFFOLD PLATFORM AND
UPPER MOST TIE NOT TO
EXCEED 4 TO 1 RATIO
26'-0" MAX. BETWEEN
INTERMEDIATE TIES
TIE
4 TIMES MINIMUM BASE
TIE AT CLOSEST FRAME
HEADER OR BEARER
FIRST TIE CLOSEST FRAME HEADER OR BEARER
ABOVE 4 TIMES THE MINIMUM BASE DIMENSION
WIDER THAN 3'-0"
MIMINUM BASE DIMENSION

MAXIMUM VERTICAL TIE SPACING
3'-0" AND NARROWER BASES

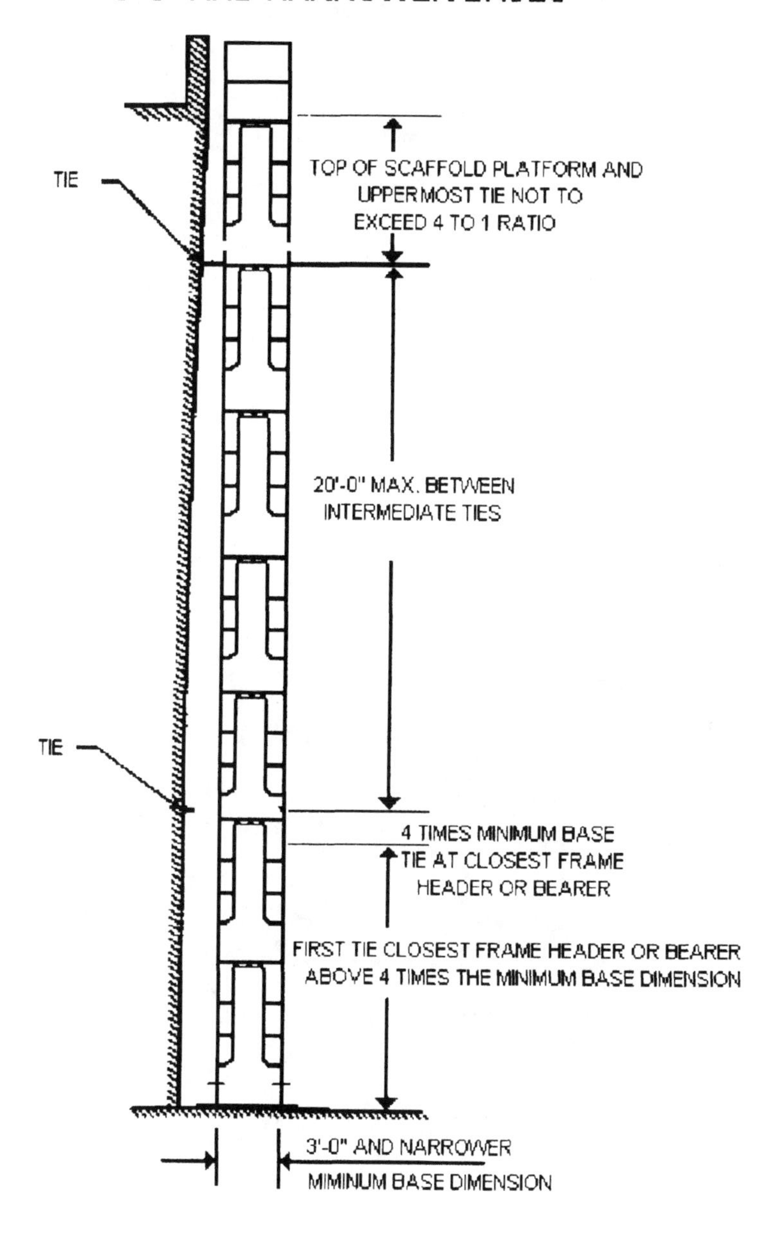

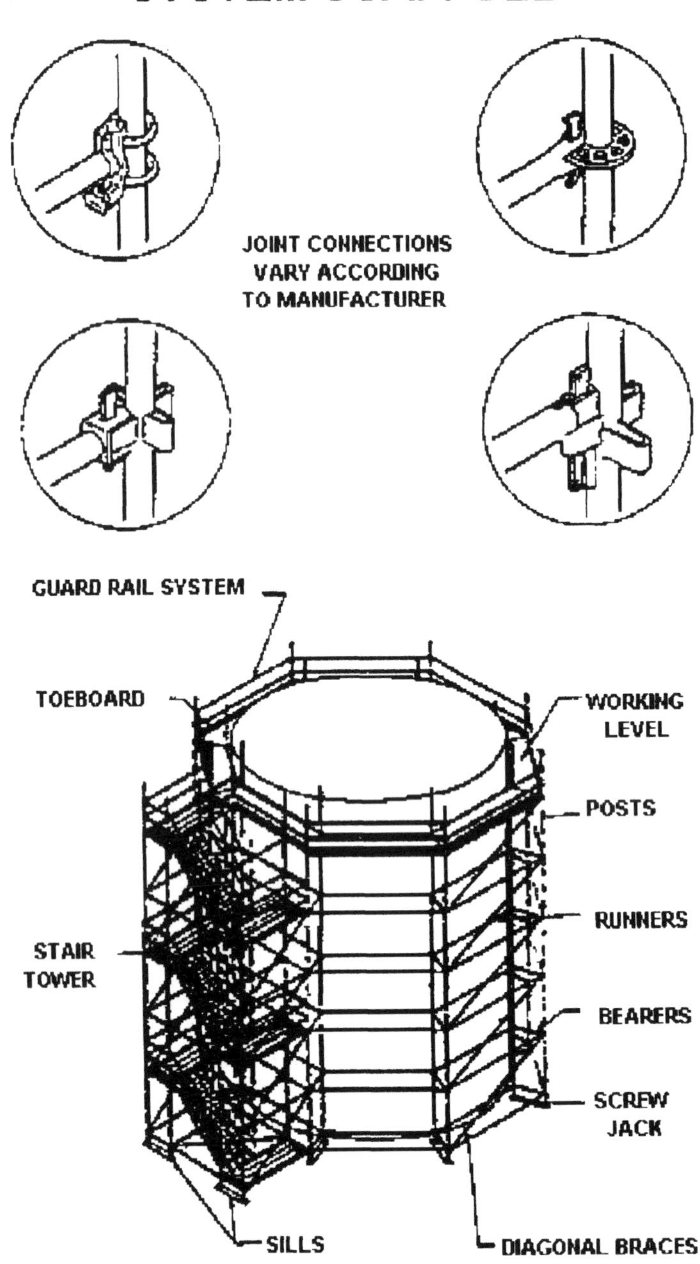
SYSTEM SCAFFOLD
JOINT CONNECTIONS
VARY ACCORDING
TO MANUFACTURER
GUARD RAIL SYSTEM
TOEBOARD
WORKING
LEVEL
POSTS
RUNNERS
STAIR
TOWER
BEARERS
SCREW
JACK
SILLS
DIAGONAL BRACES

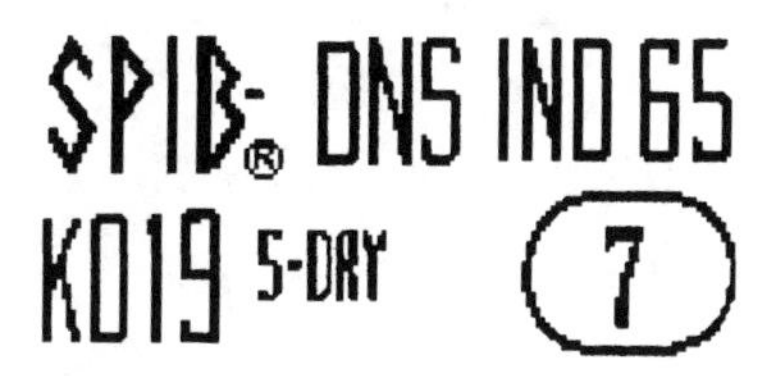

Grade stamp courtesy of Southern Pine Inspection Bureau

Grade stamp courtsey of West Coast Lumber Inspection Bureau

TUBE and COUPLER SCAFFOLD

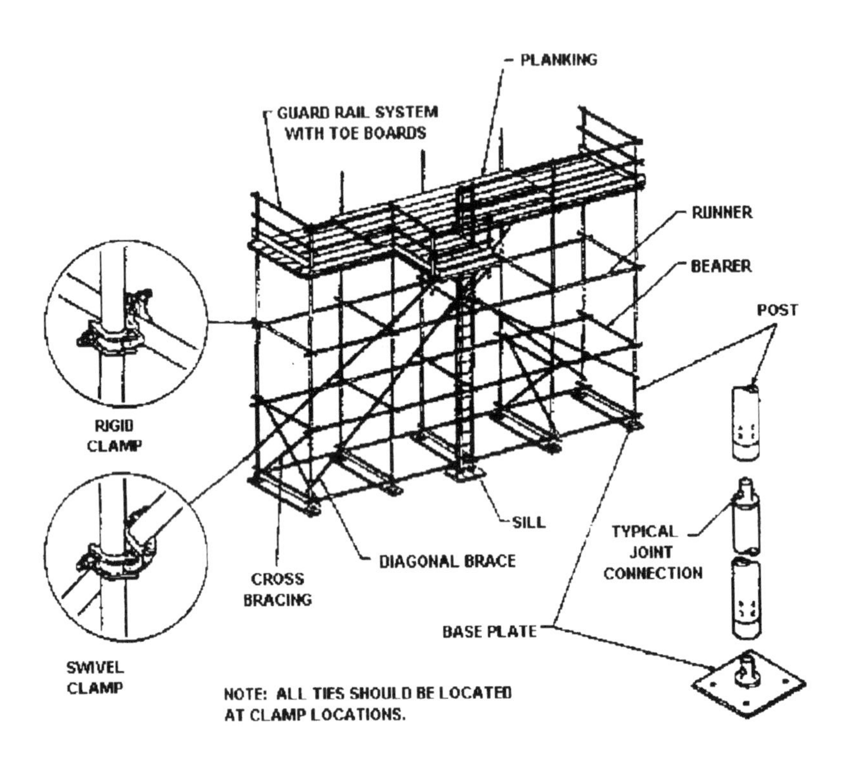

SCAFFOLDING WORK SURFACES

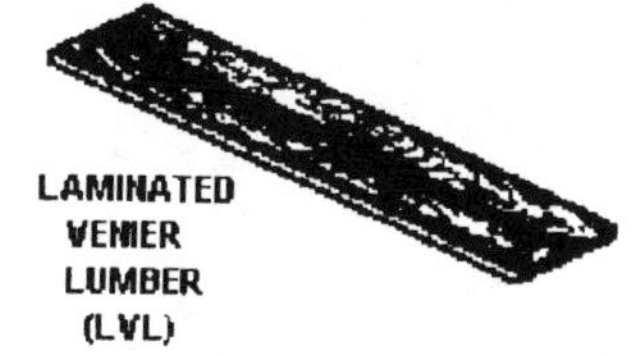

SCAFFOLD PLANKS

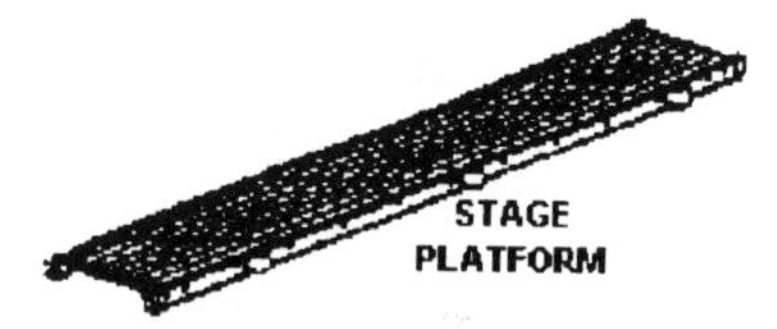

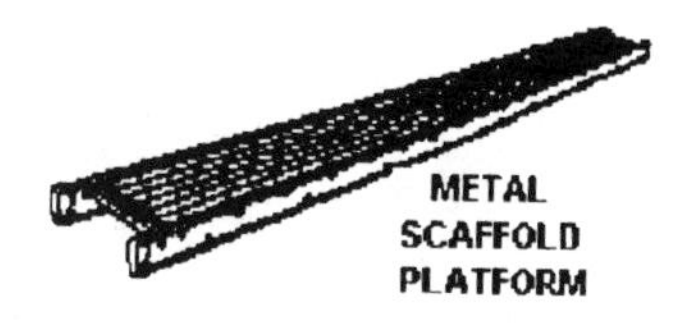

Appendix E

Scaffold Hazards

U.S. Department of Labor
Occupational Safety & Health Administration

Regulations (Standards - 29 CFR)

Training requirements. - 1926.454

Regulations (Standards - 29 CFR) - Table of Contents

- **Part Number:** 1926
- **Part Title:** Safety and Health Regulations for Construction
- **Subpart:** L
- **Subpart Title:** Scaffolds
- **Standard Number:** 1926.454
- **Title:** Training requirements.

This section supplements and clarifies the requirements of 1926.21(b)(2) as these relate to the hazards of work on scaffolds.

1926.454(a)

The employer shall have each employee who performs work while on a scaffold trained by a person qualified in the subject matter to recognize the hazards associated with the type of scaffold being used and to understand the procedures to control or minimize those hazards. The training shall include the following areas, as applicable:

1926.454(a)(1)

The nature of any electrical hazards, fall hazards and falling object hazards in the work area;

1926.454(a)(2)

The correct procedures for dealing with electrical hazards and for erecting, maintaining, and disassembling the fall protection systems and falling object protection systems being used;

1926.454(a)(3)

The proper use of the scaffold, and the proper handling of materials on the scaffold;

1926.454(a)(4)

The maximum intended load and the load-carrying capacities of the scaffolds used; and

1926.454(a)(5)

Any other pertinent requirements of this subpart.

..1926.454(b)

1926.454(b)

The employer shall have each employee who is involved in erecting, disassembling, moving, operating, repairing, maintaining, or inspecting a scaffold trained by a competent person to recognize any hazards associated with the work in question. The training shall include the following topics, as applicable:

Training requirements. - 1926.454

1926.454(b)(1)

The nature of scaffold hazards;

1926.454(b)(2)

The correct procedures for erecting, disassembling, moving, operating, repairing, inspecting, and maintaining the type of scaffold in question;

1926.454(b)(3)

The design criteria, maximum intended load-carrying capacity and intended use of the scaffold;

1926.454(b)(4)

Any other pertinent requirements of this subpart.

1926.454(c)

When the employer has reason to believe that an employee lacks the skill or understanding needed for safe work involving the erection, use or dismantling of scaffolds, the employer shall retrain each such employee so that the requisite proficiency is regained. Retraining is required in at least the following situations:

1926.454(c)(1)

Where changes at the worksite present a hazard about which an employee has not been previously trained; or

..1926.454(c)(2)

1926.454(c)(2)

Where changes in the types of scaffolds, fall protection, falling object protection, or other equipment present a hazard about which an employee has not been previously trained; or

1926.454(c)(3)

Where inadequacies in an affected employee's work involving scaffolds indicate that the employee has not retained the requisite proficiency.

[61 FR 46025, Aug. 30, 1996]

Next Standard (1926 Subpart L App A)

Regulations (Standards - 29 CFR) - Table of Contents

Back to Top

Appendix F

Scaffold Specifications

U.S. Department of Labor
Occupational Safety & Health Administration

www.osha.gov Search GO Advanced Searc

Regulations (Standards - 29 CFR)

Scaffold Specifications - 1926 Subpart L App A

Regulations (Standards - 29 CFR) - Table of Contents

- **Part Number:** 1926
- **Part Title:** Safety and Health Regulations for Construction
- **Subpart:** L
- **Subpart Title:** Scaffolds
- **Standard Number:** 1926 Subpart L App A
- **Title:** Scaffold Specifications

This Appendix provides non-mandatory guidelines to assist employers in complying with the requirements of subpart L of this part. An employer may use these guidelines and tables as a starting point for designing scaffold systems. However, the guidelines do not provide all the information necessary to build a complete system, and the employer is still responsible for designing and assembling these components in such a way that the completed system will meet the requirements of 1926.451(a). Scaffold components which are not selected and loaded in accordance with this Appendix, and components for which no specific guidelines or tables are given in this Appendix (e.g., joints, ties, components for wood pole scaffolds more than 60 feet in height, components for heavy-duty horse scaffolds, components made with other materials, and components with other dimensions, etc.) must be designed and constructed in accordance with the capacity requirements of 1926.451(a), and loaded in accordance with 1926.451(d)(1).

Index to Appendix A for Subpart L

1. General guidelines and tables.
2. Specific guidelines and tables.

(a) Pole scaffolds:
Single-pole wood pole scaffolds.
Independent wood pole scaffolds.
(b) Tube and coupler scaffolds.
(c) Fabricated frame scaffolds.
(d) Plasterers', decorators' and large area scaffolds.
(e) Bricklayers' square scaffolds.
(f) Horse scaffolds.
(g) Form scaffolds and carpenters' bracket scaffolds.
(h) Roof bracket scaffolds.
(i) Outrigger scaffolds (one level).
(j) Pump jack scaffolds.
(k) Ladder jack scaffolds.
(l) Window jack scaffolds.
(m) Crawling boards (chicken ladders).
(n) Step, platform and trestle ladder scaffolds.
(o) Single-point adjustable suspension scaffolds.
(p) Two-point adjustable suspension scaffolds.
(q)(1) Stonesetters' multi-point adjustable suspension scaffolds.
(q)(2) Masons' multi-point adjustable suspension scaffolds.
(r) Catenary scaffolds.
(s) Float (ship) scaffolds.

(t) Interior hung scaffolds.
(u) Needle beam scaffolds.
(v) Multi-level suspension scaffolds.
(w) Mobile scaffolds.
(x) Repair bracket scaffolds.
(y) Stilts.
(z) Tank builders' scaffolds.

1. General Guidelines and Tables

(a) The following tables, and the tables in Part 2 -- Specific guidelines and tables, assume that all load-carrying timber members (except planks) of the scaffold are a minimum of 1,500 lb-f/in(2) (stress grade) construction grade lumber. All dimensions are nominal sizes as provided in the American Softwood Lumber Standards, dated January 1970, except that, where rough sizes are noted, only rough or undressed lumber of the size specified will satisfy minimum requirements.

(b) Solid sawn wood used as scaffold planks shall be selected for such use following the grading rules established by a recognized lumber grading association or by an independent lumber grading inspection agency. Such planks shall be identified by the grade stamp of such association or agency. The association or agency and the grading rules under which the wood is graded shall be certified by the Board of Review, American Lumber Standard Committee, as set forth in the American Softwood Lumber Standard of the U. S. Department of Commerce.

(i) Allowable spans shall be determined in compliance with the National Design Specification for Wood Construction published by the National Forest Products Association; paragraph 5 of ANSI A10.8-1988 Scaffolding-Safety Requirements published by the American National Standards Institute; or for 2 x 10 inch (nominal) or 2 x 9 inch (rough) solid sawn wood planks, as shown in the following table:

Maximum intended nominal load (lb/ft(2))	Maximum permissible span using full thickness undressed lumber (ft)	Maximum permissible span using nominal thickness lumber (ft)
25............	10	8
50............	8	6
75............	6	

(ii) The maximum permissible span for 1 1/4 x 9-inch or wider wood plank of full thickness with a maximum intended load of 50 lb/ft.(2) shall be 4 feet.

(c) Fabricated planks and platforms may be used in lieu of solid sawn wood planks. Maximum spans for such units shall be as recommended by the manufacturer based on the maximum intended load being calculated as follows:

Rated load capacity	Intended load

Scaffold Specifications - 1926 Subpart L App A

Light-duty.............	* 25 pounds per square foot applied uniformly over the entire span area.
Medium-duty...	* 50 pounds per square foot applied uniformly over the entire span area.
Heavy-duty.............	* 75 pounds per square foot applied uniformly over the entire span area.
One-person.............	* 250 pounds placed at the center of the span (total 250 pounds).
Two-person.............	* 250 pounds placed 18 inches to the left and right of the center of the span (total 500 pounds).
Three-person............	* 250 pounds placed at the center of the span and 250 pounds placed 18 inches to the left and right of the center of the span (total 750 pounds).

Note: Platform units used to make scaffold platforms intended for light-duty use shall be capable of supporting at least 25 pounds per square foot applied uniformly over the entire unit-span area, or a 250-pound point load placed on the unit at the center of the span, whichever load produces the greater shear force.

(d) Guardrails shall be as follows:

(i) Toprails shall be equivalent in strength to 2 inch by 4 inch lumber; or

1 1/4 inch x 1/8 inch structural angle iron; or
1 inch x .070 inch wall steel tubing; or 1.990 inch x .058 inch wall aluminum tubing.

(ii) Midrails shall be equivalent in strength to 1 inch by 6 inch lumber; or

1 1/4 inch x 1 1/4 inch x 1/8 inch structural angle iron; or
1 inch x .070 inch wall steel tubing; or
1.990 inch x .058 inch wall aluminum tubing.

(iii) Toeboards shall be equivalent in strength to 1 inch by 4 inch lumber; or

1 1/4 inch x 1 1/4 inch structural angle iron; or
1 inch x .070 inch wall steel tubing; or
1.990 inch x .058 inch wall aluminum tubing.

(iv) Posts shall be equivalent in strength to 2 inch by 4 inch lumber; or

1 1/4 inch x 1 1/4 inch x 1/8 structural angle iron; or
1 inch x .070 inch wall steel tubing; or
1.990 inch x .058 inch wall aluminum tubing.

(v) Distance between posts shall not exceed 8 feet.

(e) Overhead protection shall consist of 2 inch nominal planking laid tight, or 3/4-inch plywood.

(f) Screen installed between toeboards and midrails or toprails shall consist of No. 18 gauge U.S. Standard wire one inch mesh.

2. Specific guidelines and tables.

(a) Pole Scaffolds.

Single Pole Wood Pole Scaffolds

	Light duty up to 20 feet high	Light duty up to 60 feet high	Medium duty up to 60 feet high	Heavy duty up to 60 feet high
Maximum intended load (lbs/ft(2))......	25.........	25.........	50..........	75
Poles or uprights	2 x 4 in...	4 x 4 in...	4 x 4 in....	4 x 6 in.
Maximum pole spacing (longitudinal)...	6 feet.....	10 feet....	8 feet......	6 feet
Maximum pole spacing (transverse).....	5 feet.....	5 feet.....	5 feet......	5 feet
Runners...........	1 x 4 in...	1 1/4 x 9 in	2 x 10 in...	2 x 10 in.
Bearers and maximum spacing of bearers:				
3 feet......	2 x 4 in...	2 x 4 in...	2 x 10 in.. or 3 x 4 in.	2 x 10 in. or 3 x 5 in.
5 feet......	2 x 6 in. or 3 x 4 in...	2 x 6 in. or 3 x 4 in.. (rough).	2 x 10 in. or 3 x 4 in....	2 x 10 in. or 3 x 5 in.
6 feet......	....	....	2 x 10 in. or 3 x 4 in....	2 x 10 in. or 3 x 5 in.
8 feet......	....	....	2 x 10 in. or 3 x 4 in....	....
Planking..........	1 1/4 x 9 in	2 x 10 in..	2 x 10 in....	2 x 10 in.
Maximum vertical spacing of horizontal members.	7 feet......	9 feet......	7 feet.......	6 ft. 6 in.

Scaffold Specifications - 1926 Subpart L App A

Bracing horizontal....	1 x 4 in....	1 x 4 in....	1 x 6 in. or 1 1/4 x 4 in	2 x 4 in.
Bracing diagonal..	1 x 4 in....	1 x 4 in....	1 x 4 in.....	2 x 4 in.
Tie-ins..........	1 x 4 in....	1 x 4 in....	1 x 4 in.....	1 x 4 in.

Note: All members except planking are used on edge. All wood bearers shall be reinforced with 3/16 x 2 inch steel strip, or the equivalent, secured to the lower edges for the entire length of the bearer.

Independent Wood Pole Scaffolds

	Light duty up to 20 feet high	Light duty up to 60 feet high	Medium duty up to 60 feet high	Heavy duty up to 60 feet high
Maximum intended load..........	25 lbs/ft(2)	25 lbs/ft(2)	50 lbs/ft(2)	75 lbs/ft(2).
Poles or uprights	2 x 4 in....	4 x 4 in....	4 x 4 in.....	4 x 4 in.
Maximum pole spacing (longitudinal).	6 feet......	10 feet.....	8 feet.......	6 feet.
Maximum (transverse)..	6 feet......	10 feet.....	8 feet.......	8 feet.
Runners.........	1 1/4 x 4 in	1 1/4 x 9 in	2 x 10 in....	2 x 10 in.
Bearers and maximum spacing of bearers:				
3 feet......	2 x 4 in....	2 x 4 in....	2 x 10 in....	2 x 10 in. (rough).
6 feet......	2 x 6 in. or 3 x 4 in...	2 x 10 in.. (rough) or 3 x 8 in.	2 x 10 in....	2 x 10 in. (rough).
8 feet......	2 x 6 in. or 3 x 4 in...	2 x 10 in.. (rough) or 3 x 8 in.	2 x 10 in....	....
10 feet.....	2 x 6 in. or 3 x 4 in...	2 x 10 in.. (rough) or 3 x 3 in..	....	....
Planking........	1 1/4 x 9 in	2 x 10 in...	2 x 10 in....	2 x 10 in.
Maximum vertical spacing of				

Scaffold Specifications - 1926 Subpart L App A

horizontal members.	7 feet......	7 feet......	6 feet.......	6 feet.
Bracing horizontal....	1 x 4 in....	1 x 4 in....	1 x 6 in. or 1 1/4 x 4 in.	2 x 4 in.
Bracing diagonal.	1 x 4 in....	1 x 4 in....	1 x 4 in.....	2 x 4 in.
Tie-ins.........	1 x 4 in....	1 x 4 in....	1 x 4 in.....	1 x 4 in.

Note: All members except planking are used on edge. All wood bearers shall be reinforced with 3/16 x 2 inch steel strip, or the equivalent, secured to the lower edges for the entire length of the bearer.

(b) Tube and coupler scaffolds.

Minimum Size of Members

	Light duty	Medium duty	Heavy duty
Maximum intended load............	25 lbs/ft(2)	50 lbs/ft(2)...	75 lbs/ft(2).
Posts, runners and braces...........	Nominal 2 in. (1.90 inches) OD steel tube or pipe.	Nominal 2 in. (1.90 inches) OD steel tube or pipe.	Nominal 2 in. (1.90 inches) OD steel tube or pipe.
Bearers............	Nominal 2 in. (1.90 inches) OD steel tube or pipe and a maximum post spacing of 4 ft. x 10 ft.	Nominal 2 in. (1.90 inches).. OD steel tube or pipe and a maximum post spacing of 4 ft. x 7 ft. or Nominal 2 1/2 in. (2.375 in.). OD steel tube or pipe and a maximum post spacing of 6 ft. x8 ft.(*)	Nominal 2 1/2 in. (2.375 in.). OD steel tube or pipe and a maximum post spacing of 6 ft. x 6 ft.
Maximum runner spacing vertically.......	6 ft. 6 in..	6 ft. 6 in.....	6 ft. 6 in.

Footnote(*) Bearers shall be installed in the direction of the shorter dimension.

Scaffold Specifications - 1926 Subpart L App A

Note: Longitudinal diagonal bracing shall be installed at an angle of 45 deg. (+/- 5 deg.).

Maximum Number of Planked Levels

	Maximum number of additional planked levels			Maximum height of scaffold (in feet)
	Light duty	Medium duty	Heavy duty	
Number of Working Levels:				
1....................	16	11	6	125
2....................	11	1	0	125
3....................	6	0	0	125
4....................	1	0	0	125

(c) "Fabricated frame scaffolds." Because of their prefabricated nature, no additional guidelines or tables for these scaffolds are being adopted in this Appendix.

(d) "Plasterers', decorators', and large area scaffolds." The guidelines for pole scaffolds or tube and coupler scaffolds (Appendix A (a) and (b)) may be applied.

(e) "Bricklayers' square scaffolds."

Maximum intended load: 50 lb/ft.(2)(*)

Footnote(*) The squares shall be set not more than 8 feet apart for light duty scaffolds and not more than 5 feet apart for medium duty scaffolds.

Maximum width: 5 ft.
Maximum height: 5 ft.
Gussets: 1 x 6 in.
Braces: 1 x 8 in.
Legs: 2 x 6 in.
Bearers (horizontal members): 2 x 6 in.

(f) Horse scaffolds.

Maximum intended load (light duty): 25 lb/ft.(2)(**)

Footnote(**) Horses shall be spaced not more than 8 feet apart for light duty loads, and not more than 5 feet apart for medium duty loads.

Maximum intended load (medium duty): 50 lb/ft.(2)(**)

Footnote(**) Horses shall be spaced not more than 8 feet apart for light duty loads, and not more than 5 feet apart for medium duty loads.

Horizontal members or bearers:

Light duty: 2 x 4 in.
Medium duty: 3 x 4 in.

Legs: 2 x 4 in.
Longitudinal brace between legs: 1 x 6 in.
Gusset brace at top of legs: 1 x 8 in.
Half diagonal braces: 2 x 4 in.

(g) "Form scaffolds and carpenters' bracket scaffolds."

(1) Brackets shall consist of a triangular-shaped frame made of wood with a cross-section not less than 2 inches by 3 inches, or of 1 1/4 inch x 1 1/4 inch x 1/8 inch structural angle iron.

(2) Bolts used to attach brackets to structures shall not be less than 5/8 inches in diameter.

(3) Maximum bracket spacing shall be 8 feet on centers.

(4) No more than two employees shall occupy any given 8 feet of a bracket or form scaffold at any one time. Tools and materials shall not exceed 75 pounds in addition to the occupancy.

(5) Wooden figure-four scaffolds:

Maximum intended load: 25 lb/ft.(2)
Uprights: 2 x 4 in. or 2 x 6 in.
Bearers (two): 1 x 6 in.
Braces: 1 x 6 in.
Maximum length of bearers (unsupported): 3 ft. 6 in.

(i) Outrigger bearers shall consist of two pieces of 1 x 6 inch lumber nailed on opposite sides of the vertical support.

(ii) Bearers for wood figure-four brackets shall project not more than 3 feet 6 inches from the outside of the form support, and shall be braced and secured to prevent tipping or turning. The knee or angle brace shall intersect the bearer at least 3 feet from the form at an angle of approximately 45 degrees, and the lower end shall be nailed to a vertical support.

(6) Metal bracket scaffolds:

Maximum intended load: 25 lb/ft.(2)
Uprights: 2 x 4 inch
Bearers: As designed.
Braces: As designed.

```
 (7) Wood bracket scaffolds:

Maximum intended load: 25 lb/ft.(2)
Uprights: 2 x 4 in or 2 x 6 in
Bearers: 2 x 6 in
Maximum scaffold width: 3 ft 6 in
Braces: 1 x 6 in
```

(h) "Roof bracket scaffolds." No specific guidelines or tables are given.

(i) "Outrigger scaffolds (single level)." No specific guidelines tables are given.

(j) "Pump jack scaffolds." Wood poles shall not exceed 30 feet in height. Maximum intended load -- 500 lbs between poles; applied at the center of the span. Not more than two employees shall be on a pump jack scaffold at one time between any two supports. When 2 x 4's are spliced together to make a 4 x 4 inch wood pole, they shall be spliced with "10 penny" common nails no more than 12 inches center to center, staggered uniformly from the opposite outside edges.

(k) "Ladder jack scaffolds." Maximum intended load -- 25 lb/ft(2). However, not more than two employees shall occupy any platform at any one time. Maximum span between supports shall be 8 feet.

(l) "Window jack scaffolds." Not more than one employee shall occupy a window jack scaffold at any one time.

(m) "Crawling boards (chicken ladders)." Crawling boards shall be not less than 10 inches wide and 1 inch thick, with cleats having a minimum 1 x 1 1/2 inch cross-sectional area. The cleats shall be equal in length to the width of the board and spaced at equal intervals not to exceed 24 inches.

(n) "Step, platform, and trestle ladder scaffolds." No additional guidelines or tables are given.

(o) "Single-point adjustable suspension scaffolds." Maximum intended load -- 250 lbs. Wood seats for boatswains' chairs shall be not less than 1 inch thick if made of non-laminated wood, or 5/8 inches thick if made of marine quality plywood.

(p) "Two-point adjustable suspension scaffolds." (1) In addition to direct connections to buildings (except window cleaners' anchors) acceptable ways to prevent scaffold sway include angulated roping and static lines. Angulated roping is a system of platform suspension in which the upper wire rope sheaves or suspension points are closer to the plane of the building face than the corresponding attachment points on the platform, thus causing the platform to press against the face of the building. Static lines are separate ropes secured at their top and bottom ends closer to the plane of the building face than the outermost edge of the platform. By drawing the static line taut, the platform is drawn against the face of the building.

(2) On suspension scaffolds designed for a working load of 500 pounds, no more than two employees shall be permitted on the scaffold at one time. On suspension scaffolds with a working load of 750 pounds, no more than three employees shall be permitted on the scaffold at one time.

(3) Ladder-type platforms. The side stringer shall be of clear straight-grained spruce. The rungs shall be of straight-grained oak, ash, or hickory, at least 1 1/8 inches in diameter, with 7/8 inch tenons mortised into the side stringers at least 7/8 inch. The stringers shall be tied together with tie rods not less than 1/4 inch in diameter, passing through the stringers and riveted up tight against washers on both ends. The flooring strips shall be spaced not more than 5/8 inch apart, except at the side rails where the space may be 1 inch. Ladder-type platforms shall be constructed in accordance with the following table:

```
              Schedule for Ladder-Type Platforms
```

Scaffold Specifications - 1926 Subpart L App A

Length of Platform.	12 feet..........	14 & 16 feet.....	18 & 20 feet.
Side stringers, minimum cross section (finished sizes):			
At ends........	1 3/4 x 2 3/4 in.	1 3/4 x 2 3/4 in.	1 3/4 x 3 in.
At middle......	1 3/4 x 3 3/4 in.	1 3/4 x 3 3/4 in.	1 3/4 x 4 in.
Reinforcing strip (minimum)........	A 1/8 x 7/8 inch steel reinforcing strip shall be attached to the side or underside, full length.		
Rungs..............	Rungs shall be 1 1/8 inch minimum diameter with at least 7/8 inch in diameter tenons, and the maximum spacing shall be 12 inches to center.		
Tie rods:			
Number (minimum).	3................	4................	4
Diameter (minimum)......	1/4 inch.........	1/4 inch.........	1/4 inch
Flooring, minimum finished size....	1/2 x 2 3/4 in...	1/2 x 2 3/4 in...	1/2 x 2 3/4 in.

Schedule for Ladder-Type Platforms

Length of Platform.........	22 & 24 ft..........	28 & 30 ft.
Side stringers, minimum cross section (finished sizes):		
At ends................	1 3/4 x 3 in........	1 3/4 x 3 1/2 in.
At middle..............	1 3/4 x 4 1/4 in....	1 3/4 x 5 in.
Reinforcing strip (minimum)	A 1/8 x 7/8-inch steel reinforcing strip shall be attached to the side or underside, full length.	
Rungs......................	Rungs shall be 1 1/8 inch minimum diameter with at least 7/8 inch in diameter tenons, and the maximum spacing shall be 12 inches to center. Tie rods.	
Number (minimum).......	5..................	6.
Diameter (minimum).....	1/4 in..............	1/4 in.
Flooring, minimum finished size....................	1/2 x 2 3/4 in......	1/2 x 2 3/4 in.

(4) Plank-Type Platforms. Plank-type platforms shall be composed of not less than nominal 2 x 8 inch unspliced planks, connected together on the underside with cleats at intervals not exceeding 4 feet, starting 6 inches from each end. A bar or other effective means shall be securely fastened to the platform at each end to prevent the platform from slipping off the hanger. The span between hangers for plank-type platforms shall not exceed 10 feet.

(5) Beam-Type Platforms. Beam platforms shall have side stringers of lumber not less than 2 x 6 inches set

on edge. The span between hangers shall not exceed 12 feet when beam platforms are used. The flooring shall be supported on 2 x 6 inch cross beams, laid flat and set into the upper edge of the stringers with a snug fit, at intervals of not more than 4 feet, securely nailed to the cross beams. Floor-boards shall not be spaced more than 1/2 inch apart.

(q)(1) "Multi-point adjustable suspension scaffolds and stonesetters' multi-point adjustable suspension scaffolds." No specific guidelines or tables are given for these scaffolds.

(q)(2) "Masons' multi-point adjustable suspension scaffolds." Maximum intended load -- 50 lb/ft(2). Each outrigger beam shall be at least a standard 7 inch, 15.3 pound steel I-beam, at least 15 feet long. Such beams shall not project more than 6 feet 6 inches beyond the bearing point. Where the overhang exceeds 6 feet 6 inches, outrigger beams shall be composed of stronger beams or multiple beams.

(r) "Catenary scaffolds." (1) Maximum intended load -- 500 lbs.

(2) Not more than two employees shall be permitted on the scaffold at one time.

(3) Maximum capacity of come-along shall be 2,000 lbs.

(4) Vertical pickups shall be spaced not more than 50 feet apart.

(5) Ropes shall be equivalent in strength to at least 1/2 inch (1.3 cm) diameter improved plow steel wire rope.

(s) "Float (ship) scaffolds." (1) Maximum intended load -- 750 lbs.

(2) Platforms shall be made of 3/4 inch plywood, equivalent in rating to American Plywood Association Grade B-B, Group I, Exterior.

(3) Bearers shall be made from 2 x 4 inch, or 1 x 10 inch rough lumber. They shall be free of knots and other flaws.

(4) Ropes shall be equivalent in strength to at least 1 inch (2.5 cm) diameter first grade manila rope.

(t) Interior hung scaffolds.

```
Bearers (use on edge): 2 x 10 in.
Maximum intended load: Maximum span
25 lb/ft.(2): 10 ft.
50 lb/ft.(2): 10 ft.
75 lb/ft.(2): 7 ft.
```

```
  (u) "Needle beam scaffolds."

Maximum intended load: 25 lb/ft.(2)
Beams: 4 x 6 in.
Maximum platform span: 8 ft.
Maximum beam span: 10 ft.
```

(1) Ropes shall be attached to the needle beams by a scaffold hitch or an eye splice. The loose end of the rope shall be tied by a bowline knot or by a round turn and a half hitch.

(2) Ropes shall be equivalent in strength to at least 1 inch (2.5 cm) diameter first grade manila rope.

(v) "Multi-level suspension scaffolds." No additional guidelines or tables are being given for these scaffolds.

(w) "Mobile Scaffolds." Stability test as described in the ANSI A92 series documents, as appropriate for the type of scaffold, can be used to establish stability for the purpose of 1926.452(w)(6).

(x) "Repair bracket scaffolds." No additional guidelines or tables are being given for these scaffolds.

(y) "Stilts." No specific guidelines or tables are given.

(z) "Tank builder's scaffold."

(1) The maximum distance between brackets to which scaffolding and guardrail supports are attached shall be no more than 10 feet 6 inches.

(2) Not more than three employees shall occupy a 10 feet 6 inch span of scaffold planking at any time.

(3) A taut wire or synthetic rope supported on the scaffold brackets shall be installed at the scaffold plank level between the innermost edge of the scaffold platform and the curved plate structure of the tank shell to serve as a safety line in lieu of an inner guardrail assembly where the space between the scaffold platform and the tank exceeds 12 inches (30.48 cm). In the event the open space on either side of the rope exceeds 12 inches (30.48 cm), a second wire or synthetic rope appropriately placed, or guardrails in accordance with 1926.451(e)(4), shall be installed in order to reduce that open space to less than 12 inches (30.48 cm).

(4) Scaffold planks of rough full-dimensioned 2-inch (5.1 cm) x 12-inch (30.5 cm) Douglas Fir or Southern Yellow Pine of Select Structural Grade shall be used. Douglas Fir planks shall have a fiber stress of at least 1900 lb/in(2) (130,929 n/cm(2)) and a modulus of elasticity of at least 1,900,000 lb/in(2) (130,929,000 n/cm(2)), while Yellow Pine planks shall have a fiber stress of at least 2500 lb/in(2) (172,275 n/cm(2)) and a modulus of elasticity of at least 2,000,000 lb/in(2) (137,820,000 n/cm(2)).

(5) Guardrails shall be constructed of a taut wire or synthetic rope, and shall be supported by angle irons attached to brackets welded to the steel plates. These guardrails shall comply with 1926.451(e)(4). Guardrail supports shall be located at no greater than 10 feet 6 inch intervals.

[61 FR 46025, Aug. 30, 1996]

Next Standard (1926 Subpart L App B)

Regulations (Standards - 29 CFR) - Table of Contents

Back to Top http://osha.gov/index.html

Contact Us | Freedom of Information Act | Customer Survey
Privacy and Security Statement | Disclaimers

Occupational Safety & Health Administration
200 Constitution Avenue, NW
Washington, DC 20210

Appendix G

Scaffold Shoring and Forming Institute

We would like to give our thanks to the Scaffolding, Shoring & Forming Institute for giving us permission to use the material in this Appendix. They do so with the following comments, however:

The publication, "Compression Testing of Welded Frame Scaffolds and Shoring Equipment" is obsolete. It is being superseded by new standards for testing and rating scaffolds and shoring.

These new standards are in the process of writing and approval and should be approved by the end of the year. Since this standard applies until approval of the new standards, we are including it with the note that you can go to the Scaffolding, Shoring & Forming Institute website (www.ssfi.org) and get the new standard after its ANSI procedure is completed. Once the new standard is approved, it will supersede the standard included with this appendix.

Once again, we thank the SSFI for giving us permission to use their material in this publication.

Approved American National Standard
ANSI

ANSI/SSFI SPS1.1-1/03

American National Standard

STANDARD REQUIREMENTS AND TEST METHODS FOR TESTING AND RATING PORTABLE RIGGING DEVICES FOR SUSPENDED SCAFFOLDS

Scaffolding, Shoring & Forming Institute, Inc.

Sponsor:

Scaffolding, Shoring
& Forming Institute, Inc.
1300 Sumner Ave
Cleveland, Ohio 44115-2851

ANSI/SSFI SPS1.1-1/03

SCAFFOLDING, SHORING & FORMING INSTITUTE
AMERICAN NATIONAL STANDARD
Standard Requirements and Test Methods for Testing and Rating Portable Rigging Devices for Suspended Scaffolds

Sponsor

Scaffolding, Shoring & Forming Institute, Inc.

ANSI/SSFI SPS1.1-1/03

Scaffolding, Shoring & Forming Institute

A Scaffolding, Shoring & Forming Institute standard is intended as a guide to aid the manufacturer, the consumer, and the general public. The existence of a Scaffolding, Shoring & Forming Institute standard does not in any respect preclude anyone, whether he has approved the standard or not, from manufacturing, marketing, purchasing or using products, processes, or procedures not conforming to the standard. Scaffolding, Shoring & Forming Institute standards are subject to periodic review and users are cautioned to obtain the latest editions.

CAUTION NOTICE:

This Scaffolding, Shoring & Forming Institute standard may be revised or withdrawn at any time. The procedures of the Scaffolding, Shoring & Forming Institute require that action be taken to reaffirm, revise, or withdraw this standard no later than five years from the date of publication. Purchasers of Scaffolding, Shoring & Forming Institute standards may receive current information on all standards by calling or writing the Scaffolding, Shoring & Forming Institute.

Sponsored and published by:
SCAFFOLDING, SHORING & FORMING INSTITUTE
1300 Sumner Avenue
Cleveland, OH 44115-2851
Phn: 216/241-7333
Fax: 216/241-0105
E-Mail: ssfi@ssfi.org
URL: www.ssfi.org

Suggestions for improvement of this standard will be welcome. They should be sent to the Scaffolding, Shoring & Forming Institute, Inc.

Printed in the United States of America

ANSI/SSFI SPS1.1-1/03

CONTENTS

PAGE

Foreword v
1. Scope and Application 1
2. Definitions 1
3. Design & Use 1
4. Material Requirements 2
5. Manufacturing Requirements 2
6. Marking & Labeling 2
7. Test Procedures 3

TABLES

1. Cornice Hook Test Data 11
2. Parapet Hook Test Data 12
3. Parapet Clamp Test Data 13
4. Outrigger Beam Test Data 14
5. Beam and Column Clamp Test Data 15
6. Trolley Test Data 15
7. Tank Top Roller Test Data 15
8. Outrigger Carriage Test Data 16

FIGURES

1. Cornice Hook Test Setup 17
2. Parapet Hook Test Setup 18
3. Parapet Clamp Test Setup 19
4. Outrigger Beam Test Setup 20
5. Trolley or Clamp Test Setup 21
6. Tank Top Roller Test Setup 22
7. Outrigger Carriage Test Setup 23

ANSI/SSFI SPS1.1-1/03

Foreword (This foreword is included for information only and is not part of ANSI/SSFI SPS1.1-, *Standard Requirements and Test Methods for Testing and Rating Portable Rigging Devices for Suspended Scaffolds.*)

The following standard has been formulated by the Suspended Powered Scaffolding Engineering Committee of the Scaffolding, Shoring & Forming Institute, Inc. as an assistance and guide to the manufacturers, purchasers, and users of rigging devices.

SSFI recognizes the need to periodically review and update this standard. Suggestions for improvement should be forwarded to the Scaffolding, Shoring & Forming Institute, Inc., 1300 Sumner Avenue, Cleveland, Ohio, 44115-2851. All constructive suggestions for expansion and revision of this standard are welcome.

The existence of a Scaffolding, Shoring & Forming Institute (SSFI) standard does not in any respect preclude any member or non-member from manufacturing or selling products not conforming to this standard nor is the SSFI responsible for the use of this standard.

At the time this standard was approved, the following were members of the SSFI Suspended Powered Scaffolding Section:

Hi-Lo Climbers
930 North Shore Drive
Lake Bluff, IL 60044

Nihon Bisoh
3788 Hinamigo, Togitsu-cho
Nagasaki, 851-21 Japan

Patent Construction Systems/SGB U.S.A.
One Mack Centre Drive
Paramus, NJ 07652

SafeWorks/Power Climber/Spider
365 Upland Drive
Tukwila, WA 98188

Safway Steel Products
N14 W23833 Stone Ridge Dr., Suite 400
Waukesha, WI 53188

Sky Climber
1501 Rock Mountain Blvd.
Stone Mountain, GA 30083

Tractel, Inc., Griphoist Division
110 Shawmut Road
P. O. Box 188
Canton, MA 02021

ANSI/SSFI SPS1.1-1/03

Standard Requirements and Test Methods for Testing and Rating Portable Rigging Devices for Suspended Scaffolds

1. SCOPE AND APPLICATION

1.1 This standard establishes methods for testing and rating portable rigging devices used to support transportable suspended scaffolds for construction, alteration, demolition, and maintenance of buildings or structures.

1.2 This standard does not cover permanently installed suspended scaffold systems (davits or roofcars).

1.3 Specially engineered temporary rigging devices shall comply with sections 3, 4, 5 and 6 and shall be evaluated by a professional engineer.

2. DEFINITIONS

2.1 Beam Clamp – A support device for suspended scaffold equipment that may be stationary or rolling (see trolley) and attaches to a flange, angle or channel, or other shape. The clamp transfers the suspended load to that member.

2.2 Column Clamp – A support device for suspended scaffold equipment that attaches to a vertical column and transfers the suspended load to the structure.

2.3 Cornice Hook – A curved support device for suspended scaffold equipment that is normally point loaded and transfers the suspended load to the building or structure.

2.4 Failure – Load refusal, breakage, or separation of component parts.

2.5 Outrigger Beam – A support device for suspended scaffold equipment that consists of a beam that extends out from the edge of the building or structure. Counterweights or engineered equivalent prevent the beam from overturning.

2.6 Outrigger Carriage – A stationary or rolling outrigger beam support system usually consisting of a front fulcrum carriage and a rear counterweight carriage. The carriage may be engineered scaffold frames, or part of an outrigger beam system. They may incorporate jacks on the fulcrum carriage to apply the load to the structure.

2.7 Parapet Clamp – A support device for suspended scaffold equipment that clamps to a building parapet and transfers the suspended load to the parapet.

2.8 Parapet Hook – A curved support device that fits over a parapet and supports suspended scaffold equipment. It may be provided with a standoff device to keep the suspension rope away from the building.

2.9 Rated Load – The manufacturer's specified maximum load to be applied to the rigging device. Typically the same as rated hoist capacity on suspended scaffolding (e.g., 1000, 1500 lb. capacity).

2.10 Tank top Roller (pin type) – A support device for suspended scaffold equipment that normally rests on the edge of a storage tank and is restrained by a wire rope. It keeps the suspension rope away from the tank wall and transfers the suspended load to the tank.

2.11 Trolley – A support device for suspended scaffold equipment that rolls on a flange, angle or channel, or other shape and transfers the suspended load to that member. It may be moved by pushing, or by means of geared wheels, or motorization.

3. DESIGN & USE

3.1 General Design and Use Requirements

3.1.1 All rigging devices shall be designed to satisfy the test requirements specified in section 7.

3.1.2 All suspension scaffold support devices such as outrigger beams, cornice hooks, parapet clamps and hooks shall rest on surfaces capable of supporting at least 4 times the load imposed on them.

3.2 Stability

3.2.1 Carriages shall have a maximum height to base ratio of 2 to 1.

3.3 Strength

3.3.1 All rigging devices shall support without failure loads equal to or greater than four times the rated capacity of the device.

3.3.2 All rigging devices shall be designed so no plastic deformation (no yield) occurs when subjected to four (4) times the rated load of the device as specified in section 7.

4. MATERIAL REQUIREMENTS

4.1 Structural and mechanical components shall be fabricated from suitable materials. Structural and mechanical components shall be protected from corrosion.

5. MANUFACTURING REQUIREMENTS

5.1 All mechanical connections (bolts, pins, etc.) shall be of a secure type to prevent loosening during use. When welding is employed for structural connections, the welding shall be done by an American Welding Society certified welder for that particular process. For foreign manufactured equipment, when welding is employed for structural connections, a registered professional engineer shall ascertain that weld designs employ weld configurations, materials, sizes, and processes that are listed by the American Welding Society. A United States registered professional engineer shall review foreign manufactured equipment to ensure that it complies with this standard.

6. MARKING AND LABELING

6.1 Parapet Clamps and Hooks

6.1.1 Parapet clamps and hooks shall be marked permanently or shall have a corrosion-resistant data tag or label securely attached to the clamp or hook, readily visible to interested persons. The marking(s), tag(s) or label(s) shall bear the following information:

a. Rated load and maximum reach.

b. Parapet use and size restrictions (min./max. opening of clamp/hook).

c. Name of manufacturer.

d. Notice that a tieback or tiedown is required.

e. "Caution - Before use comply with manufacturer's instructions."

f. "Design tested to ANSI/SSFI Standard SPS1.1."

6.2 Cornice Hooks

6.2.1 Cornice hooks shall be marked permanently or shall have a corrosion-resistant data tag or label securely attached to the hook, readily visible to interested persons. The marking(s), tag(s) or label(s) shall bear the following information:

a. Rated load and size restrictions.

b. Name of manufacturer.

c. Notice that tieback is required.

d. "Caution - Before use comply with manufacturer's instructions."

e. "Design tested to ANSI/SSFI Standard SPS1.1."

6.3 Outrigger Beams

6.3.1 Outrigger beams shall be marked permanently or shall have a corrosion-resistant data tagb or label securely attached to the beam, readily visible to interested persons. The marking(s), tag(s) or label(s) shall bear the following information:

a. Rated load and size restrictions.

b. Name of manufacturer.

c. Notice that tieback is required.

d. Counterweight chart or formula.

e. "Caution - Before use comply with manufacturer's instructions."

f. "Design tested to ANSI/SSFI Standard SPS1.1."

6.4 Beam and Column Clamps

6.4.1 Beam and column clamps shall be marked permanently or shall have a corrosion-resistant data tag or label securely attached to the clamp, readily visible to interested persons. The marking(s), tag(s) or label(s) shall bear the following information:

a. Rated load and size restrictions.
b. Name of manufacturer.

c. "Caution - Before use comply with manufacturer's instructions."

d. "Design tested to ANSI/SSFI Standard SPS1.1."

6.5 Trolleys

6.5.1 Trolleys shall be marked permanently or shall have a corrosion-resistant data tag or label securely attached to the trolley, readily visible to interested persons. The marking(s), tag(s) or label(s) shall bear the following information:

a. Rated load and size restrictions.

b. Name of manufacturer.

c. "Caution - Before use comply with manufacturer's instructions."

d. "Design tested to ANSI/SSFI Standard SPS1.1."

6.6 Tank Top Rollers

6.6.1 Tank top rollers shall be marked permanently or shall have a corrosion-resistant data tag or label securely attached, readily visible to interested persons. The marking(s), tag(s) or label(s) shall bear the following information:

a. Rated load and size restrictions.

b. Name of manufacturer.

c. "Caution - Before use comply with manufacturer's instructions."

d. "Design tested to ANSI/SSFI Standard SPS1.1."

6.7 Outrigger Carriages

6.7.1 Outrigger Carriages shall be marked permanently or shall have a corrosion-resistant data tag or label securely attached, readily visible to interested persons. The marking(s), tag(s) or label(s) shall bear the following information:

a. Rated load and size restrictions.

b. Name of manufacturer.

c. "Caution - Before use comply with manufacturer's instructions."

d. "Design tested to ANSI/SSFI Standard SPS1.1."

7. TEST PROCEDURES

In order to validate the design and construction of a particular rigging device, at least one copy of the device must be destructively tested. The copy selected must be of "typical" materials and fabrication, with no special attention paid during its manufacture.

CAUTION: The testing process described herein may result in catastrophic failure of the device. Care must be taken to protect personnel and property during the testing.

7.1 Cornice Hook Test

This test shall be used to test rigging hooks which are intended to be supported at only one point (point loading).

7.1.1 Test Preparation and Setup

7.1.1.1 Determine the test load to be applied to the cornice hook. As a minimum, this will be four (4) times rated load of the hook.

7.1.1.2 Determine the maximum load that the point load fixture will be subjected to. Design and fabricate a point load fixture which will safely withstand this load.

7.1.1.3 Obtain the remainder of the equipment needed to perform the test: load cell, loading device, measuring device, and weights for the anchor. Ensure that the load cell and measuring device used are calibrated for the range of the testing.

7.1.1.4 Set up the test in a manner similar to Figure 1.

7.1.2 Test Process

7.1.2.1 Using the loading device, apply 100 pounds (0.44 kN) of load (as measured by the load cell) to the cornice hook.

7.1.2.2 As indicated in Figure 1, take measurement "D" to the nearest 0.030" (0.80mm). Record this value as the Nominal Hook Opening (NHO) on Table 1.

7.1.2.3 Using the loading device, increase the load on the cornice hook to 500 pounds (2.22 kN). Hold for at least one (1) minute, then measure and record "D" (Loaded Hook Opening - LHO).

7.1.2.4 Increase the load on the cornice hook in 500 pound (2.22 kN) increments until the load applied is two (2) times the rated load. Measure and record "D" at each increment. NOTE: During the application of load, elastic or plastic deformation is occurring. Apply the load gradually, at approximately 500 pounds/minute (2.22 kN/min), to allow the deformation to keep pace with the load.

7.1.2.5 At two (2) times rated load, hold for five (5) minutes after the first measurement, then re-measure and record "D." Verify that there has been no reduction in the load as indicated by the load cell. If necessary, increase the load back to two (2) times rated load, then recommence the five (5) minute wait.

7.1.2.6 Reduce the load on the cornice hook to 100 pounds (0.44 kN), then measure and record "D."

7.1.2.7 Increase the load on the hook gradually until two (2) times rated load is re-achieved. There is no need to pause and take measurements during this step. Once the two (2) times rated load reading is reached, measure and record "D".

7.1.2.8 Increase the load on the cornice hook in 500 pound (2.22 kN) increments until the required test load is achieved. Measure and record "D" at each increment.

CAUTION: Depending on the actual capacity of the hook, you may be approaching catastrophic failure. Pay close attention to how the hook is reacting to prevent sudden failure and the possibility of personal injury or equipment damage.

7.1.2.9 At the test load, hold for five (5) minutes after the first measurement. Verify that there has been no reduction in the load as indicated by the load cell. If necessary, increase the load back to the test load, then recommence the five (5) minute wait. Failure to hold the load for at least five (5) minutes indicates that the hook is yielding, resulting in a test failure.

7.1.2.10 Reduce the load to zero and de-rig the test apparatus. Note that the test hook should never be placed in service, regardless of its condition following this test. The application of the test load has, by definition, "failed" the hook.

7.1.3 Test Interpretation

7.1.3.1 By dividing the maximum load the hook can withstand for five (5) minutes by its rated load, the factor of safety is determined. As required by 29 CFR 1926, this value must be at least four (4). If it is, the hook has passed this test.

7.2 Parapet Hook Test

This test shall be used to test any rigging hook which is designed and constructed to be used on parapets (not point loaded). If the hook is also designed to be used in a point loaded condition, then the testing of 7.1 must also be performed.

If the hook being tested is designed for use with a stand off bracket, then the testing shall include the bracket, with the bracket set at its maximum designed stand off condition.

7.2.1 Test Preparation and Setup

7.2.1.1 Determine the test load to be applied to the parapet hook. As a minimum, this will be four (4) times the rated load of the hook.

7.2.1.2 Determine the maximum load that the parapet fixture will be subjected to. Design and fabricate a parapet fixture which will safely withstand this load.

7.2.1.3 Obtain the remainder of the equipment needed to perform the test: load cell, loading device, measuring device, and weights for the anchor. Ensure that the load cell and measuring device used are calibrated for the range of the testing.

7.2.1.4 Set up the test in a manner similar to Figure 2.

7.2.2 Test Process

7.2.2.1 Using the loading device, apply 100 pounds (0.44 kN) of load (as measured by the load cell) to the parapet hook.

7.2.2.2 As indicated on Figure 2, take measurement "D" to the nearest 0.030" (0.80mm). Record this value as the Nominal Hook Opening (NHO) on Table 2.

7.2.2.3 Using the loading device, increase the load on the parapet hook to 500 pounds (2.22 kN). Hold for at least one (1) minute, then measure and record "D" (Loaded Hook Opening - LHO).

7.2.2.4 Increase the load on the parapet hook in 500 pound (2.22 kN) increments until the load applied is two (2) times the rated load. Measure and record "D" at each increment. NOTE: During the application of load, elastic or plastic deformation is occurring. Apply the load gradually, at approximately 500 pounds/minute (2.22 kN/min), to allow the deformation to keep pace with the load.

7.2.2.5 At two (2) times rated load, hold for five (5) minutes after the first measurement, then re-measure and record "D." Verify that there has been no reduction in the load as indicated by the load cell. If necessary, increase the load back to two (2) times rated load, then recommence the five (5) minute wait.

7.2.2.6 Reduce the load on the parapet hook to 100 pounds (0.44 kN), then measure and record "D."

7.2.2.7 Increase the load on the hook gradually until two (2) times rated load is re-achieved. There is no need to pause and take measurements during this step. Once the two (2) times rated load reading is reached, measure and record "D".

7.2.2.8 Increase the load on the parapet hook in 500 pound (2.22 kN) increments until the required test load is achieved. Measure and record "D" at each increment.

CAUTION: Depending on the actual capacity of the hook, you may be approaching catastrophic failure. Pay close attention to how the hook is reacting to prevent sudden failure and the possibility of personal injury or equipment damage.

7.2.2.9 At the test load, hold for five (5) minutes after the first measurement. Verify that there has been no reduction in the load as indicated by the load cell. If necessary, increase the load back to the test load, then recommence the five (5) minute wait. Failure to hold the load for at least five (5) minutes indicates that the hook is yielding, resulting in a test failure.

7.2.2.10 Reduce the load to zero and de-rig the test apparatus. Note that the test hook should never be placed in service, regardless of its condition following this test. The application of the test load has, by definition, "failed" the hook.

7.2.3 Test Interpretation

7.2.3.1 By dividing the maximum load the hook can withstand for five (5) minutes by its rated load, the factor of safety is determined. As required by 29 CFR 1926, this value must be at least four (4). If it is, the hook has passed this test.

7.3 Parapet Clamp Test

This test shall be used on any parapet clamp. Test procedure shall include the parapet clamp set to the minimum throat opening and maximum reach as prescribed by the manufacturer.

This test does not apply to applications in which the load is applied in a direction other than vertical. For applications in which the load is applied to the clamp in a direction other than vertical, refer to Section 1.3 of the standard.

7.3.1 Test Preparation and Setup

7.3.1.1 Determine the test load to be applied to the parapet clamp. As a minimum, this will be four (4) times rated load of the clamp.

7.3.1.2 Determine the maximum load that the parapet fixture will be subjected to. Design and fabricate a parapet fixture which will safely withstand this load and will not deflect during testing.

7.3.1.3 Obtain the remainder of the equipment needed to perform the test: load cell, loading device, measuring device, and weights for the anchor. Ensure that the load cell and measuring device used are calibrated for the range of the testing.

7.3.1.4 Set up the test in a manner similar to Figure 3.

7.3.2 Test Process

7.3.2.1 Using the loading device, apply 100 pounds (0.44 kN) of load (as measured by the load cell) to the parapet clamp.

7.3.2.2 As indicated on Figure 3, take measurements "D" and "E" to the nearest 0.030" (0.80mm). Record these values as the Nominal Deflection (ND) and Stand Deflection (SD), respectively, on Table 3.

7.3.2.3 Using the loading device, increase the load on the parapet clamp to 500 pounds (2.22 kN). Hold for at least one (1) minute, then measure and record "D" (Loaded Deflection - LD) and "E" (Stand Deflection - SD). During the conduct of this test, the Stand Deflection should not change, with the exception of normal measurement variations.

7.3.2.4 Increase the load on the parapet clamp in 500 pound (2.22 kN) increments until the load applied is two (2) times the rated load. Measure and record at each increment. NOTE: During the application of load, elastic or plastic deformation is occurring. Apply the load gradually, at approximately 500 pounds/minute (2.22 kN/min), to allow the deformation to keep pace with the load.

7.3.2.5 At two (2) times rated load, hold for five (5) minutes after the first measurement, then re-measure and record "D" and "E." Verify that there has been no reduction in the load as indicated by the load cell. If necessary, increase the load back to two (2) times rated load, then recommence the five (5) minute wait.

7.3.2.6 Reduce the load on the parapet clamp to 100 pounds (0.44 kN), then measure and record "D" and "E."

7.3.2.7 Increase the load on the clamp gradually until two (2) times rated load is re-achieved. There is no need to pause and take measurements during this step. Once the two (2) times rated load reading is reached, measure and record.

7.3.2.8 Increase the load on the parapet clamp in 500 pound increments (2.22 kN) until the required test load is achieved. Measure and record at each increment.

CAUTION: Depending on the actual capacity of the clamp, you may be approaching catastrophic failure. Pay close attention to how the clamp is reacting to prevent sudden failure and the possibility of personal injury or equipment damage.

7.3.2.9 At the test load, hold for five (5) minutes after the first measurement. Verify that there has been no reduction in the load as indicated by the load cell. If necessary, increase the load back to the test load, then recommence the five (5) minute wait. Failure to hold the load for at least five (5) minutes indicates that the clamp is yielding, resulting in a test failure.

7.3.2.10 Reduce the load to zero and de-rig the test apparatus. Note that the test clamp should never be placed in service, regardless of its condition following this test. The application of the test load has, by definition, "failed" the clamp.

7.3.3 Test Interpretation

7.3.3.1 By dividing the maximum load, the clamp can withstand for five (5) minutes by its rated load, the factor of safety is determined. As required by 29 CFR 1926, this value must be at least four (4). If it is, the clamp has passed this test.

7.4 Outrigger Beam Test

This test shall be used to test outrigger beams, either stationary or rolling. Test procedure shall include the outrigger beam reach and load set to the worst case load condition (such as, but not limited to, maximum overturning moment, shear, torsion, back span bending moment, buckling) allowed by the manufacturer.

7.4.1 Test Preparation and Setup

7.4.1.1 Determine the test load to be applied to the outrigger beam. As a minimum, this will be four (4) time the rated load of the beam.

7.4.1.2 Determine the maximum load that the Outrigger Test Fixture will be subjected to. Design and fabricate an Outrigger Test Fixture, which will safely withstand this load and not deflect during testing.

7.4.1.3 Determine the maximum moment that the outrigger beam will be subjected to. Multiply this figure by 1.5 and then divide by the outrigger backspan length. This amount, in pounds (kg), is how much counterweight should be applied to the back of the beam to prevent it from overturning.

7.4.1.4 Obtain the remainder of the equipment needed to perform the test: load cell, loading device, measuring device, and weights for the anchor. Ensure that the load cell and measuring device used are calibrated for the range of the testing.

7.4.1.5 Set up the test in a manner similar to Figure 4.

7.4.2 Test Process

7.4.2.1 Using the loading device, apply 100 pounds (0.44 kN) of load (as measured by the load cell) to the outrigger beam.

7.4.2.2 As indicated on Figure 4, take measurements "D" and "E" to the nearest 0.030" (0.80mm). Record these values as the Nominal Deflection (ND) and Stand Deflection (SD), respectively, on Table 4.

7.4.2.3 Using the load device, increase the load on the outrigger beam to 500 pounds (2.22 kN). Hold for at least one (1) minute, then measure and record "D" (Loaded Beam Deflection - LBD) and "E" (Stand Deflection - D). During conduct of the test, the Stand Deflection should not change, with the exception of normal measurement variations.

7.4.2.4 Increase the load on the outrigger beam in 500 pound (2.22 kN) increments until the load applied is two (2) times the rated load. Measure and record at each increment. NOTE: During the application of load, elastic or plastic deformation is occurring. Apply the load gradually, at approximately 500 pounds/minute (2.22 kN/min), to allow the deformation to keep pace with the load.

7.4.2.5 At two (2) times rated load, hold for five (5) minutes after the first measurement, then remeasure and record "D" and "E". Verify that there has been no reduction in the load as indicated by the load cell. If necessary, increase the load back to two (2) times rated load, then recommence the five (5) minute wait.

7.4.2.6 Reduce the load on the outrigger beam to 100 pounds (0.44 kN), then measure and record "D" and "E".

7.4.2.7 Increase the load on the beam gradually until two (2) times rated load is reachieved. There is no need to pause and take measurements during this step. Once the two (2) times rated load reading is reached, measure and record.

7.4.2.8 Increase the load on the outrigger beam in 500 pound increments (2.22 kN) until the required test load is achieved. Measure and record at each increment.

CAUTION: Depending on the actual capacity of the beam, you may be approaching catastrophic failure. Pay close attention to how the beam is reacting to prevent sudden failure and the possibility of personal injury or equipment damage.

7.4.2.9 At the test load, hold for five (5) minutes after the first measurement. Verify that there has been no reduction in the load as indicated by the load cell. If necessary, increase the load back to the test load, and then recommence the five (5) minute wait. Failure to hold the load for at least five (5) minutes indicates that the beam is yielding, resulting in test failure.

7.4.2.10 Reduce the load to zero and de-rig the test apparatus. Note that the test beam should never be placed in service, regardless of its condition following this test. The application of the test load has, by definition, "failed" the beam.

7.4.3 Test Interpretation

7.4.3.1 By dividing the maximum load the beam can withstand for five (5) minutes by its rated load, the factor of safety is determined. As required by 29 CFR 1926, this value must be at least four (4). If it is, the beam has passed this test.

7.5 Beam and Column Clamp Test

This test shall be used to test beam and column clamps used for support of suspended scaffolds. Test procedure shall include the clamp and load in the most unfavorable position (shear or bending) as allowed by the manufacturer.

7.5.1 Test Preparation and Setup

7.5.1.1 Determine the test load to be applied to the clamp. As a minimum, this will be four (4) times the rated load of the clamp.

7.5.1.2 Determine the maximum load to which the test fixture will be subjected. Design and fabricate the test fixture, which will safely support this load and will resist deflection.

7.5.1.3 Obtain the remainder of the equipment needed to perform the test: load cell, weights for the anchor, tensioning device. Ensure that the load cell used is calibrated and rated for the range of testing loads. Ensure that the tensioning device used is rated for the testing loads.

7.5.1.4 Set-up test similar to Figure 5.

7.5.2 Test Process

7.5.2.1 Using the tensioning device, apply 100 lbs. (0.44kN) of load (as measured by the load cell) to the clamp. The load shall be applied to the most unfavorable position to achieve maximum stress to the component. The clamp shall be tested at its widest opening and its smallest as allowed by the manufacturer.

7.5.2.2 Using the tensioning device, increase the load on the clamp to 500 pounds (2.22 kN). Hold for at least one (1) minute.

7.5.2.3 Using the tensioning device, increase the load on the clamp in increments not to exceed 50% of rated load until the load applied is two (2) times the rated load. NOTE: During the application of load, elastic or plastic deformation is occurring. Apply the load gradually to allow the deformation to keep pace with the load.

7.5.2.4 Increase the load gradually, in increments not to exceed 50% of rated load, until the test load is achieved (minimum four times rated load).

CAUTION: Depending on the capacity of the clamp you may be approaching catastrophic failure of the clamp. Pay close attention to how the clamp is reacting to avoid sudden failure and the possibility of personal injury or equipment damage.

7.5.2.5 At the test load, hold for five (5) minutes. Verify that there has been no reduction in load as measured by the load cell. If necessary, increase the load back to the test load, and restart the five (5) minute timing. Failure to hold the test load for five (5) minutes indicates that the clamp is yielding resulting in test failure.

7.5.2.6 Reduce the load to zero and de-rig the apparatus. Note the test clamp should never be placed in service, regardless of its condition after the test. The application of the test load has, by definition, "failed" the clamp.

7.5.3 Test Interpretation

7.5.3.1 By dividing the maximum load the clamp can withstand for five (5) minutes by its rated load, the factor of safety is established. As required by 29 CFR 1926, this value must be at least four (4). If it is, the clamp has passed this test.

7.6 Trolley Test

This test shall be used to test trolleys used to support suspended platforms or other devices used to support people. If the trolley has a range of flange sizes, it shall be tested in its most unfavorable position, generally the widest width. If the trolley is specific to a shape or profile, it shall be tested with that profile used as support. For applications in which the trolley is intended to be loaded at any angle other than vertical, it shall be tested at the maximum angle allowed by the manufacturer.

7.6.1 Test Preparation and Setup

7.6.1.1 Determine the test load to be applied to the trolley. As a minimum, this will be four (4) times the rated load of the trolley.

7.6.1.2 Determine the maximum load to which the trolley test fixture will be subjected. Design and fabricate a trolley test fixture which will safely withstand this load.

7.6.1.3 Obtain the remainder of the equipment needed to perform the test: load cell, loading device, and weights for the anchor. Ensure that the load cell is calibrated for the range of testing.

7.6.1.4 Set-up the test in a manner similar to figure 5.

7.6.2 Test Process

7.6.2.1 Affix the trolley to the beam or support profile as instructed by the manufacturer, paying attention to any clearances required between the wheels and the flange.

7.6.2.2 Using the loading device, apply 100 pounds (0.44kN) of load (as measured by the load cell) to the trolley.

7.6.2.3 Using the loading device, increase the load on the trolley to one half (½) of its rated load. Hold for at least one (1) minute. Increase to rated load.

7.6.2.4 Using the loading device, increase the load, in increments not to exceed 50% of rated load, to two (2) times its rated load. Hold for five (5) minutes.

7.6.2.5 Increase the load gradually, in increments not to exceed 50% of rated load, until the test load is achieved (minimum four times rated load).

CAUTION: Depending on the actual capacity of the trolley, beam, or profile, you may be approaching catastrophic failure. Pay close attention to how the trolley is reacting to prevent sudden failure and the possibility of personal injury or equipment damage.

7.6.2.6 At the test load, hold for five (5) minutes. Verify that there has been no reduction in the load as indicated by the load measuring device. Failure to hold the load for at least five (5) minutes indicates that the trolley is yielding, resulting in test failure.

7.6.2.7 Reduce the load to zero and de-rig the test apparatus. Note that the test trolley should never be placed in service, regardless of its condition after the test. The application of the test load by definition "failed" the trolley.

7.6.3 Test Interpretation

7.6.3.1 By dividing the maximum load the trolley can withstand for five (5) minutes by its rated load, the factor of safety is determined. As required by 29 CFR 1926, this value must be at least four (4). If it is, the trolley has passed the test.

7.7 Tank Top Roller (Pin Style) Test

This test shall be used to test the pin style tank top roller used to support suspended scaffolding.

7.7.1 Test Preparation and Setup

7.7.1.1 Determine the rated load and maximum reach for which the equipment is rated. The minimum test load will be four (4) times the rated load.

7.7.1.2 Determine the maximum load to which the test fixture will be subjected. Design and fabricate the test fixture so that it will safely support test, tieback and rim loads.

7.7.1.3 Obtain the remainder of the equipment needed to perform the test: Load cell, tensioning device and weights for the anchor. Ensure that the load cell used is calibrated and is rated for the range of testing loads. Ensure that the tensioning device used is rated for the testing loads

7.7.1.4 Set-up test similar to Figure 6.

7.7.2 Test Process

7.7.2.1 Set-up the tank top roller per manufacturer's instructions.

7.7.2.2 Using the tensioning device, apply a 100 lb. (0.44kN) load (as measured by the load cell) to the tank top roller. Measure the reach. The tank top roller shall be tested at the maximum reach allowed by the manufacturer.

7.7.2.3 Using the tensioning device, increase the load on the tank top roller to one half (1/2) its rated load. Hold for at least one (1) minute. Increase the load to rated load.

7.7.2.4 Using the tensioning device, increase the load on the tank top roller in increments not to exceed 50% of rated load until the load applied is two (2) times the rated load.

7.7.2.5 At two (2) times rated load, hold for five (5) minutes after the first measurement. Verify that there has been no reduction in the load as indicated by the load cell. If necessary, increase the load back to two (2) times rated load, then recommence the five (5) minute wait. Check the reach of the tank top roller.

7.7.2.6 Increase the load gradually, in increments not to exceed 50% of rated load, until the test load is achieved (minimum four times rated load).

CAUTION: Depending on the actual capacity of the tank top roller, you may be approaching catastrophic failure. Pay close attention to how the tank top roller is reacting to prevent sudden failure and the possibility of personal injury or equipment damage.

7.7.2.7 At the test load, hold for five (5) minutes. Verify that there has been no reduction in load as indicated by the load cell. If necessary, increase the load to maintain the test load and then restart the five (5) minute timing. Failure to hold the test load for five (5) minutes indicates that the tank top roller is yielding, resulting in test failure.

7.7.2.8 Reduce the load to zero and de-rig the apparatus. Note that the test tank top roller should never be placed in service, regardless of its condition after the test. The application of the test load by definition "failed" the tank top roller.

7.7.3 Test Interpretation

7.7.3.1 By dividing the maximum load the tank top roller can withstand for five (5) minutes by its rated load, the factor of safety is established. As required by 29 CFR 1926, this value must be at least four (4). If it is, the tank top roller has passed this test.

7.8 Outrigger Carriage (Rolling And Stationary) Test

This test shall be used to test the outrigger support carriages (stationary or rolling) used for the support of outrigger beams.

7.8.1 Test Procedure and Setup

7.8.1.1 Determine the maximum rated fulcrum load of the outrigger support carriage. The minimum test load shall be four (4) times the rated fulcrum load.

7.8.1.2 Determine the maximum load to which the test fixture will be subjected. Design and fabricate the test fixture so that it will safely support the maximum load.

7.8.1.3 Obtain the remainder of the equipment needed to perform the test: load cell, measuring device and weights or ram. Ensure that the load cell used is calibrated and is rated for the range of testing loads. Ensure that the ram used is adequately rated for the test load.

7.8.1.4 Set up the test similar to Figure 7. The bearing area shall simulate the outrigger beam bearing area.

7.8.2 Test Process

7.8.2.1 Assemble the carriage per manufacturer's instructions. Separate tests shall be conducted if the carriage is used in both stationary and rolling applications.

7.8.2.2 Using the loading device, apply 100 pounds (0.44kN) of load (as measured by the load cell) to the carriage bearing member.

7.8.2.3 Using the measuring device, take an initial measurement reading, "D", to the nearest 0.030" (0.80mm).

7.8.2.4 Using the loading device, increase the load on the carriage to one half (1/2) its rated load. Hold the load for at least one minute and take a measurement reading. Increase the load to rated load, hold for one minute and take measurement reading, "D".

7.8.2.5 Increase the load gradually, in increments not to exceed 50% of rated load, to two (2) times the rated load. Take measurement reading, "D".

7.8.2.6 At two (2) times the rated load, hold for five (5) minutes after the measurement reading and take a second measurement reading, "D". Verify that there has been no reduction in the load as indicated by the load cell. If necessary, increase the load back to two (2) times the rated load and recommence the five (5) minute wait.

7.8.2.7 Increase the load gradually, in increments not to exceed 50% of rated load, until the test load is achieved (minimum four times rated load). Take a measurement reading, "D", at each increment.

CAUTION: Depending on the actual capacity of the carriage, you may be approaching catastrophic failure. Pay close attention to how the carriage is reacting to prevent sudden failure and the possibility of personal injury or equipment damage.

7.8.2.8 At the test load, hold for five (5) minutes. Verify that there has been no reduction in load as indicated by the load cell. If necessary, increase the load to maintain the test load then recommence the five (5) minute wait. Failure of the carriage to hold the test load for five (5) minutes indicates that the carriage is yielding, resulting in test failure.

7.8.2.9 Reduce the load to zero and de-rig the apparatus. Note that the carriage should never be placed in service, regardless of its condition after testing. The application of the test load has, by definition, failed the carriage.

7.8.3 Test Interpretation

7.8.3.1 By dividing the maximum load the carriage can withstand for five (5) minutes by its rated load, the factor of safety can be determined. As required by 29 CFR 1926, this value must be at least four (4). If it is, the carriage has passed this test.

ANSI/SSFI SPS1.1-1/03

Table 1. Cornice Hook Test Data

Load Rating of the Cornice Hook ______ pounds

2 times Load Rating ______ pounds (Multiply the figure above by 2)

Test Load ______ pounds (Multiply the load rating by at least 4)

Load	1st meas. "D"		2nd meas. "D"	3rd meas. "D"
100 pounds	________	NHO	________	________
pounds	________		________	________
pounds	________		________	________
pounds	________		________	________
pounds	________		________	________
pounds	________		________	________
pounds	________		________	________
pounds	________		________	________
pounds	________		________	________
pounds	________		________	________
pounds	________		________	________
pounds	________		________	________
pounds	________		________	________
pounds	________		________	________
pounds	________		________	________
pounds	________		________	________
pounds	________		________	________
pounds	________		________	________
pounds	________		________	________
pounds	________		________	________
pounds	________		________	________

A Nominal Hook Opening (NHO) ______ inches

B LHO at 2 times Rated Load (first time) ______ inches

C NHO following application of 2 times Rated Load ______ inches

D LHO at 2 times Rated Load (second time) ______ inches

Test Date:____________________

Test Part Number:____________________ **Test Number:**____________________

Tested By:____________________ **Witnessed By:**____________________

Table 2. Parapet Hook Test Data

Load Rating of the Parapet Hook _______ pounds

2 times Load Rating _______ pounds (Multiply the figure above by 2)

Test Load _______ pounds (Multiply the load rating by at least 4)

Load	1st meas. "D"		2nd meas. "D"	3rd meas. "D"
100 pounds	________	NHO	________	________
pounds	________		________	________
pounds	________		________	________
pounds	________		________	________
pounds	________		________	________
pounds	________		________	________
pounds	________		________	________
pounds	________		________	________
pounds	________		________	________
pounds	________		________	________
pounds	________		________	________
pounds	________		________	________
pounds	________		________	________
pounds	________		________	________
pounds	________		________	________
pounds	________		________	________
pounds	________		________	________
pounds	________		________	________
pounds	________		________	________
pounds	________		________	________
pounds	________		________	________

A	Nominal Hook Opening (NHO)	______ inches
B	LHO at 2 times Rated Load (first time)	______ inches
C	NHO following application of 2 times Rated Load	______ inches
D	LHO at 2 times Rated Load (second time)	______ inches

Test Date:____________________________

Test Part Number:____________________ **Test Number:**____________________

Tested By:____________________________ **Witnessed By:**____________________

Table 3. Parapet Clamp Test Data

Load Rating of the Parapet Clamp _______ pounds

2 times Load Rating _______ pounds (Multiply the figure above by 2)

Test Load _______ pounds (Multiply the load rating by at least 4)

Load		1st meas. "D"		2nd meas. "D"	3rd meas. "D"	Stand defl. "E"
100	pounds	________	ND	________	________	________
	pounds	________		________	________	________
	pounds	________		________	________	________
	pounds	________		________	________	________
	pounds	________		________	________	________
	pounds	________		________	________	________
	pounds	________		________	________	________
	pounds	________		________	________	________
	pounds	________		________	________	________
	pounds	________		________	________	________
	pounds	________		________	________	________
	pounds	________		________	________	________
	pounds	________		________	________	________
	pounds	________		________	________	________
	pounds	________		________	________	________
	pounds	________		________	________	________
	pounds	________		________	________	________
	pounds	________		________	________	________
	pounds	________		________	________	________
	pounds	________		________	________	________
	pounds	________		________	________	________

A Nominal Deflection (ND) _______ inches

B LD at 2 times Rated Load (first time) _______ inches

C ND following application of 2 times Rated Load _______ inches

D LD at 2 times Rated Load (second time) _______ inches

Test Date:____________________________

Test Part Number:____________________ **Test Number:**________________________

Tested By:____________________________ **Witnessed By:**________________________

Table 4. Outrigger Beam Test Data

Load Rating of the Outrigger Beam ______ pounds (The load at the maximum allowed moment)

2 times Load Rating ______ pounds (Multiply the figure above by 2)

Test Load ______ pounds (Multiply the load rating by at least 4)

Load	1st meas. "D"		2nd meas. "D"	3rd meas. "D"	Stand defl.(SD) "E"
100 pounds	______	ND	______	______	______
pounds	______		______	______	______
pounds	______		______	______	______
pounds	______		______	______	______
pounds	______		______	______	______
pounds	______		______	______	______
pounds	______		______	______	______
pounds	______		______	______	______
pounds	______		______	______	______
pounds	______		______	______	______
pounds	______		______	______	______
pounds	______		______	______	______
pounds	______		______	______	______
pounds	______		______	______	______
pounds	______		______	______	______
pounds	______		______	______	______
pounds	______		______	______	______
pounds	______		______	______	______
pounds	______		______	______	______
pounds	______		______	______	______
pounds	______		______	______	______

A	Nominal Deflection (ND)	______ inches
B	LBD at 2 times Rated Load (first time)	______ inches
C	ND following application of 2 times Rated Load	______ inches
D	LBD at 2 times Rated Load (second time)	______ inches

Test Date:______________________

Test Part Number:______________________ **Test Number:**______________________

Tested By:______________________ **Witnessed By:**______________________

Table 5. Beam and Column Clamp Test Data

Load Rating of the Clamp ______ pounds

2 times Load Rating ______ pounds (Multiply the figure above by 2)

Test Load ______ pounds (Multiply the load rating by at least 4)

Test Date:______________________

Test Part Number:______________________ **Test Number:**______________________

Tested By:______________________ **Witnessed By:**______________________

Table 6. Trolley Test Data

Load Rating of the Trolley ______ pounds

2 times Load Rating ______ pounds (Multiply the figure above by 2)

Test Load ______ pounds (Multiply the load rating by at least 4)

Test Date:______________________

Test Part Number:______________________ **Test Number:**______________________

Tested By:______________________ **Witnessed By:**______________________

Table 7. Tank Top Roller Test Data

Load Rating of the Tank Top Roller ______ pounds

2 times Load Rating ______ pounds (Multiply the figure above by 2)

Test Load ______ pounds (Multiply the load rating by at least 4)

Initial Reach (at 100 lbs.) ______ inches

Test Date:______________________

Test Part Number:______________________ **Test Number:**______________________

Tested By:______________________ **Witnessed By:**______________________

Table 8. Outrigger Carriage Test Data

Load Rating of the Outrigger Carriage ______ pounds

2 times Load Rating ______ pounds (Multiply the figure above by 2)

Test Load ______ pounds (Multiply the load rating by at least 4)

	Load		Measure "D"
	100	pounds	________
½ Rated Load	________	pounds	________
	________	pounds	________
	________	pounds	________
	________	pounds	________
	________	pounds	________
	________	pounds	________
	________	pounds	________
	________	pounds	________
	________	pounds	________
	________	pounds	________
	________	pounds	________
	________	pounds	________
	________	pounds	________
	________	pounds	________
	________	pounds	________
	________	pounds	________
	________	pounds	________
	________	pounds	________
	________	pounds	________
	________	pounds	________

A Initial Reading ______ inches

B Reading at 2 times Rated Load ______ inches

C Reading at 4 times Rated Load ______ inches

Test Date:______________________

Test Part Number:______________________ **Test Number:**______________________

Tested By:______________________ **Witnessed By:**______________________

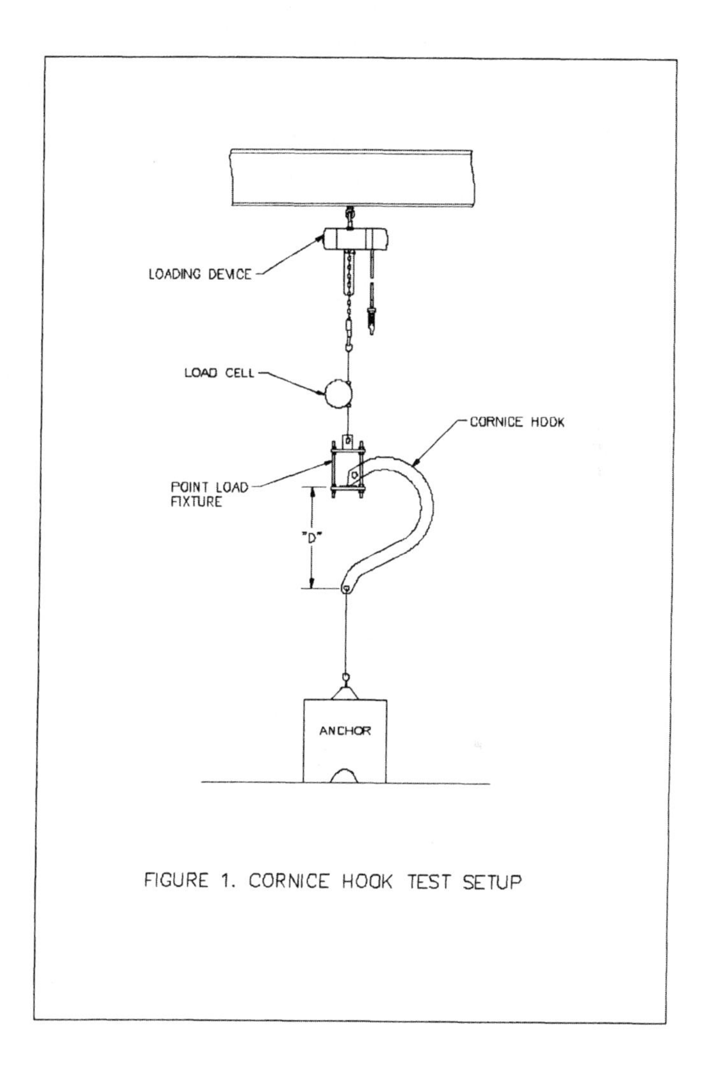

FIGURE 1. CORNICE HOOK TEST SETUP

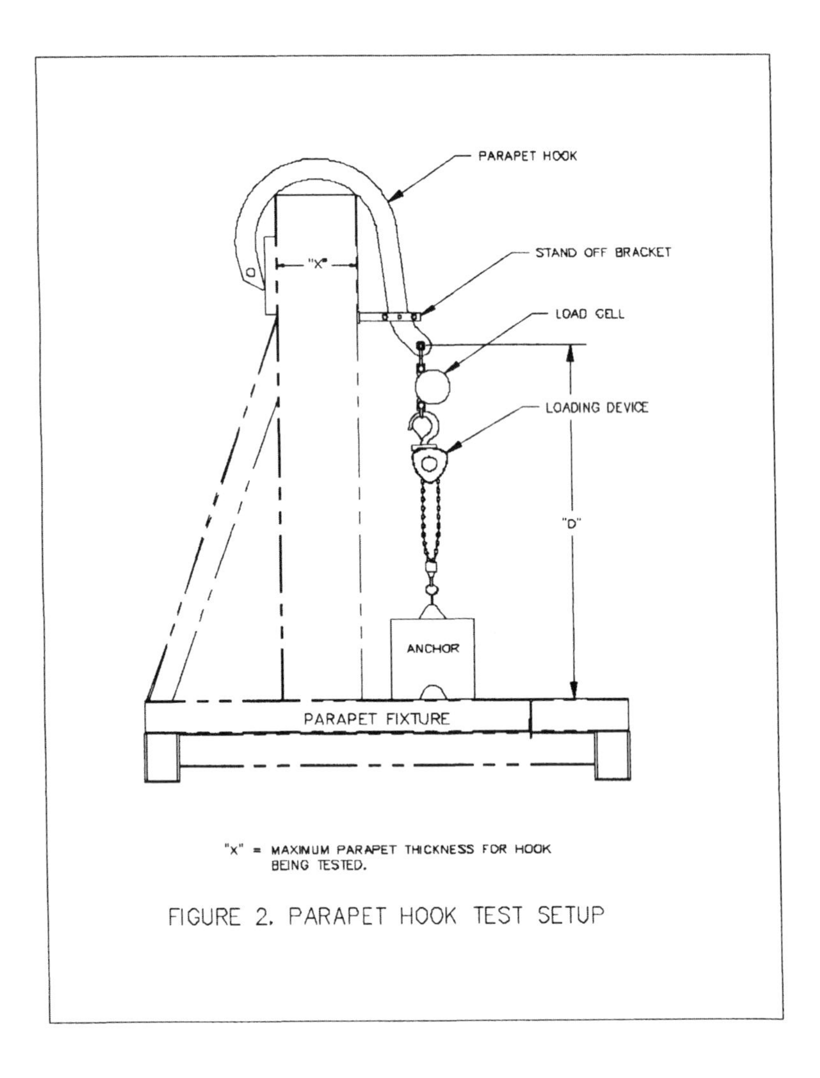

FIGURE 2. PARAPET HOOK TEST SETUP

ANSI/SSFI SPS1.1-1/03

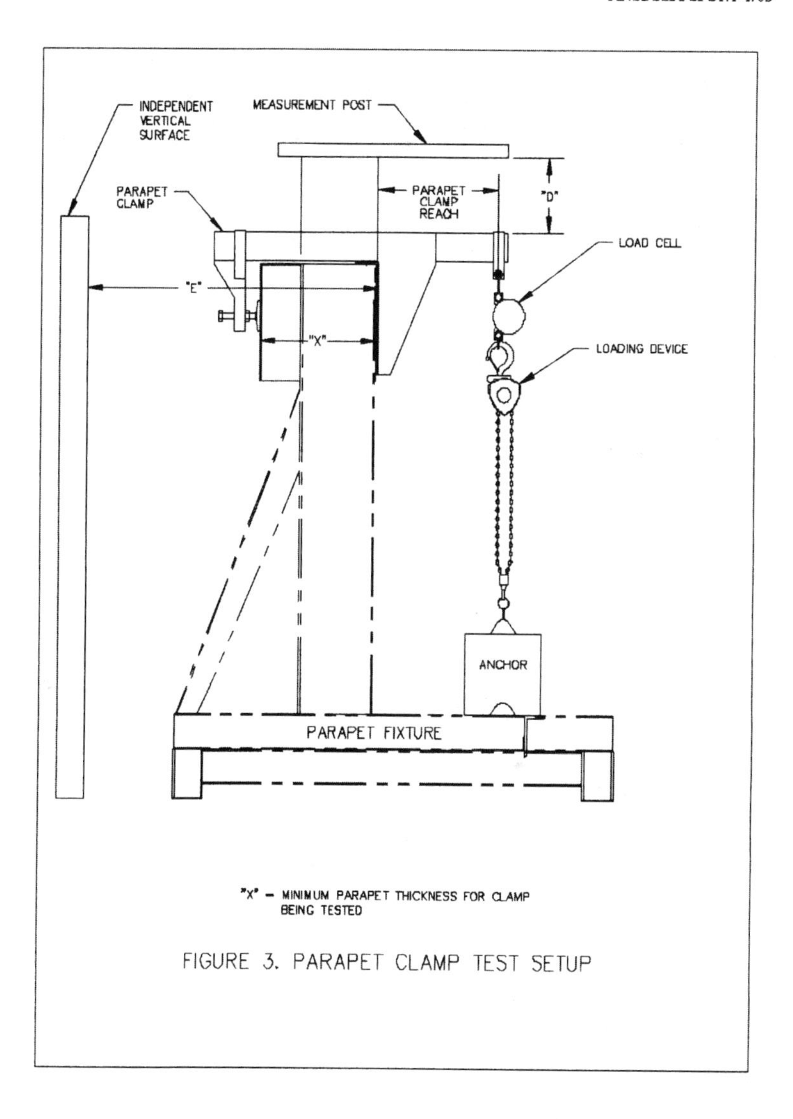

FIGURE 3. PARAPET CLAMP TEST SETUP

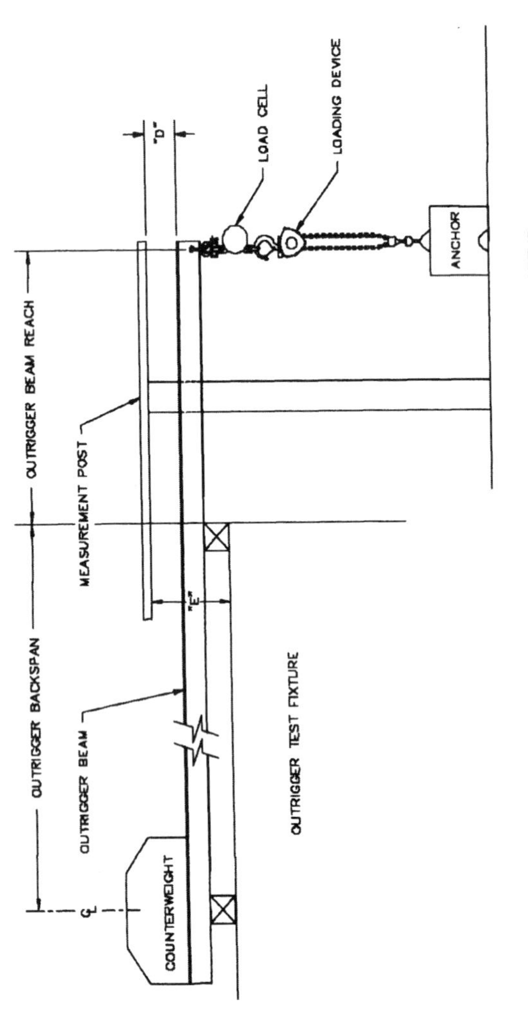

FIGURE 4. OUTRIGGER BEAM TEST SETUP

ANSI/SSFI SPS1.1-1/03

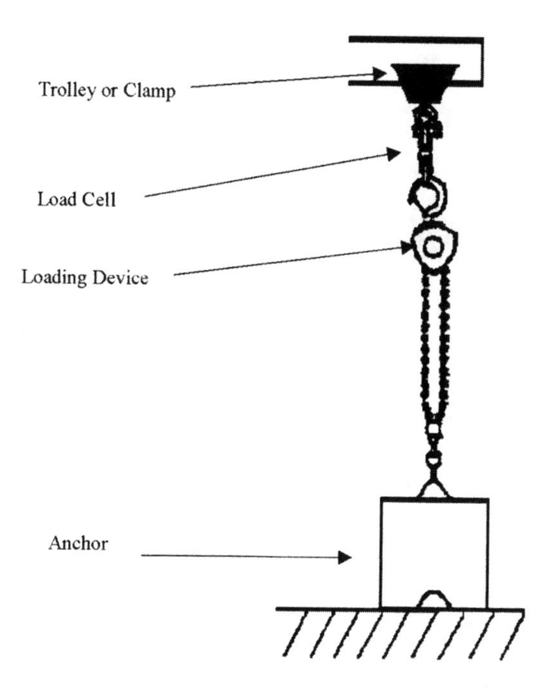

FIGURE 5 TROLLEY OR CLAMP TEST SETUP

ANSI/SSFI SPS1.1-1/03

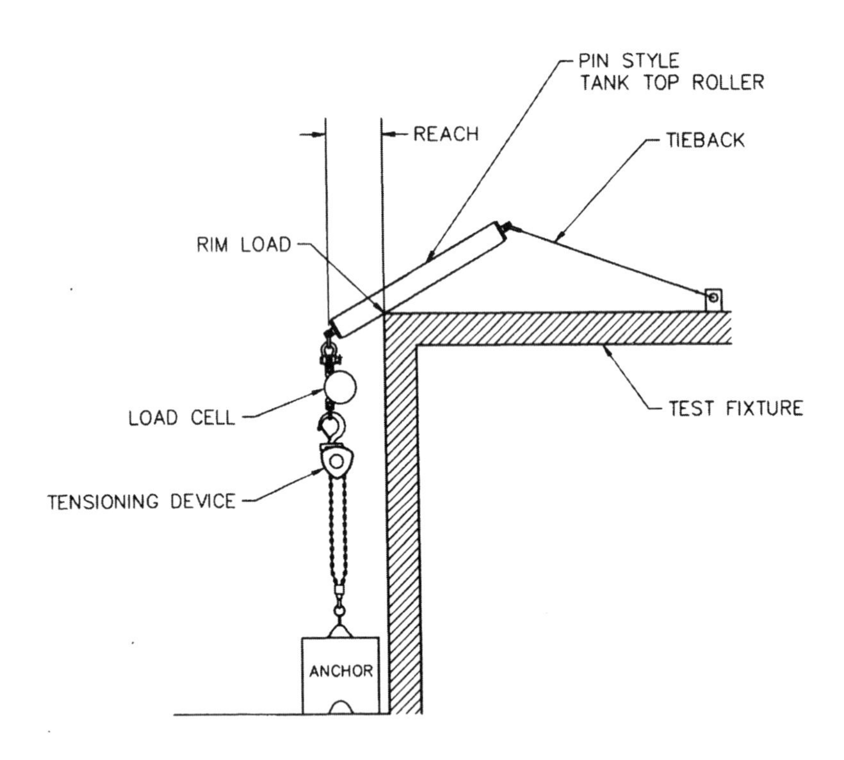

FIGURE 6 TANK TOP ROLLER TEST SETUP

ANSI/SSFI SPS1.1-1/03

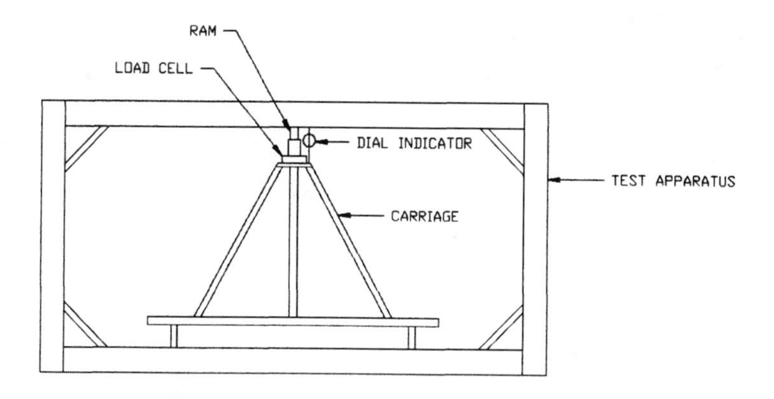

FIGURE 7 OUTRIGGER CARRIAGE TEST SETUP

CODE OF SAFE PRACTICES FOR FRAME SCAFFOLDS, SYSTEM SCAFFOLDS, TUBE AND CLAMP SCAFFOLDS & ROLLING SCAFFOLDS

DEVELOPED FOR INDUSTRY BY SCAFFOLDING, SHORING & FORMING INSTITUTE (SSFI)
and SCAFFOLD INDUSTRY ASSOCIATION, INC. (SIA)

It shall be the responsibility of all users to read and comply with the following common sense guidelines which are designed to promote safety in the erecting, dismantling and use of Scaffolds. These guidelines do not purport to be all inclusive nor to supplant or replace other additional safety and precautionary measures to cover usual or unusual conditions. If these guidelines in any way conflict with any state, local, provincial, federal or other government statute or regulation, said statute or regulation shall supersede these guidelines and it shall be the responsibility of each user to comply wherewith.

I. GENERAL GUIDELINES

A. **POST THESE SCAFFOLDING SAFETY GUIDELINES** in a conspicuous place and be sure that all persons who erect, dismantle, or use scaffolding are aware of them, and also use them in tool box safety meetings.

B. **FOLLOW ALL STATE, LOCAL AND FEDERAL CODES, ORDINANCES AND REGULATIONS** pertaining to scaffolding.

C. **SURVEY THE JOB SITE.** A survey shall be made of the job site by a competent person for hazards, such as untamped earth fills, ditches, debris, high tension wires, unguarded openings, and other hazardous conditions created by other trades. These conditions should be corrected or avoided as noted in the following sections.

D. **INSPECT ALL EQUIPMENT BEFORE USING.** Never use any equipment that is damaged or defective in any way. Mark it or tag it as defective. Remove it from the job site.

E. **SCAFFOLDS MUST BE ERECTED IN ACCORDANCE WITH DESIGN AND/OR MANUFACTURERS' RECOMMENDATIONS.**

F. **DO NOT ERECT, DISMANTLE OR ALTER A SCAFFOLD** unless under the supervision of a competent person.

G. **DO NOT ABUSE OR MISUSE THE SCAFFOLD EQUIPMENT.**

H. **ERECTED SCAFFOLDS SHOULD BE CONTINUALLY INSPECTED** by users to be sure that they are maintained in safe condition. Report any unsafe condition to your supervisor.

I. **NEVER TAKE CHANCES! IF IN DOUBT REGARDING THE SAFETY OR USE OF THE SCAFFOLD, CONSULT YOUR SCAFFOLD SUPPLIER.**

J. **NEVER USE EQUIPMENT FOR PURPOSES OR IN WAYS FOR WHICH IT WAS NOT INTENDED.**

K. **DO NOT WORK ON SCAFFOLDS** if your physical condition is such that you feel dizzy or unsteady in any way.

L. **DO NOT WORK UNDER THE INFLUENCE** of alcohol or illegal drugs.

II. GUIDELINES FOR ERECTION AND USE OF SCAFFOLDS

A. **SCAFFOLD BASE MUST BE SET ON BASE PLATES AND AN ADEQUATE SILL OR PAD** to prevent slipping or sinking and fixed thereto where required. Any part of a building or structure used to support the scaffold shall be capable of supporting the maximum intended load to be applied.

B. **USE ADJUSTING SCREWS** or other approved methods to adjust to uneven grade conditions.

C. **BRACING, LEVELING & PLUMBING OF FRAME SCAFFOLDS-**
   1. Plumb and level all scaffolds as erection proceeds. Do not force frames or braces to fit. Level the scaffold until proper fit can be easily made.
   2. Each frame or panel shall be braced by horizontal bracing, cross bracing, diagonal bracing or any combination thereof for securing vertical members together laterally. All brace connections shall be made secure, in accordance with the manufacturer's recommendations.

D. **BRACING, LEVELING & PLUMBING OF TUBE & CLAMP AND SYSTEM SCAFFOLDS-**
   1. Posts shall be erected plumb in all directions, with the first level of runners and bearers positioned as close to the base as feasible. The distance between bearers and runners shall not exceed manufacturer's recommendations.
   2. Plumb and level all scaffolds as erection proceeds.
   3. Fasten all couplers and/or connections securely before assembly of next level.
   4. Vertical and/or horizontal diagonal bracing must be installed according to manufacturer's recommendations.

E. **WHEN FREE STANDING SCAFFOLD TOWERS** exceed a height of four (4) times their minimum base dimension, they must be restrained from tipping. (CAL/OSHA and some government agencies require stricter ratio of 3 to 1.)

F. **TIE CONTINUOUS (RUNNING) SCAFFOLDS TO THE WALL OR STRUCTURE** at each end and at least every 30 feet of length in between when scaffold height exceeds the maximum allowable free standing dimension. Install additional ties on taller scaffolds as follows: On scaffolds 3 feet or narrower in width, subsequent vertical ties shall be repeated at intervals no greater than every 20 feet. On scaffolds wider than 3 feet, subsequent vertical ties shall be repeated at intervals not greater than 26 feet. The top tie shall be installed as close to the top of the platform as possible; however, no lower from the top than 4 times the scaffold's minimum base dimension. Ties must prevent the scaffold from tipping either into or away from the structure. Stabilize circular or irregular scaffolds in such a manner that the completed scaffold is secure from tipping. Place ties near horizontal members. When scaffolds are fully or partially enclosed, or when scaffolds are subjected to overturning loads, additional ties may be required. Consult a qualified person.

G. **DO NOT ERECT SCAFFOLDS NEAR ELECTRICAL POWER LINES.** Consult a qualified person for advice.

H. **ACCESS SHALL BE PROVIDED TO ALL PLATFORMS.** Do not climb crossbraces or diagonal braces.

I. **PROVIDE A GUARDRAIL SYSTEM, FALL PROTECTION AND TOEBOARDS WHERE REQUIRED BY THE PREVAILING CODE.**

J. **BRACKETS AND CANTILEVERED PLATFORMS-**
   1. Brackets for system scaffolds shall be installed and used in accordance with manufacturer's recommendations.
   2. Brackets for frame scaffolds shall be seated correctly with side bracket parallel to the frames and end brackets at 90 degrees to the frames. Brackets shall not be bent or twisted from normal position. Brackets (except mobile brackets designed to carry materials) are to be used as work platforms only and shall not be used for storage of material or equipment.
   3. Cantilevered platforms shall be designed, installed and used in accordance with manufacturers' recommendations.

K. **ALL SCAFFOLDING COMPONENTS** shall be installed and used in accordance with the manufacturers' recommended procedure. Components shall not be altered. Scaffold frames and their components manufactured by different companies shall not be intermixed, unless the component parts readily fit together and the resulting scaffold's structural integrity is maintained by the user.

L. **PLANKING-**
   1. Working platforms shall cover scaffold bearer as completely as possible. Only scaffold grade wood planking, or fabricated planking and decking meeting scaffold use requirements shall be used. **Planks and platforms should rest on bearers only.**
   2. Check each plank prior to use to be sure plank is not warped, damaged, or otherwise unsafe.
   3. Planking shall have at least 12" overlap and extend 6" beyond center of support, or be cleated or restrained at both ends to prevent sliding off supports.
   4. Solid sawn lumber, LVL (laminated veneer lumber) or fabricated scaffold planks and platforms (unless cleated or restrained) shall extend over their end supports not less than 6" nor more than 18". This overhang should be guardrailed to prevent access.

M. **FOR "PUTLOGS" AND "TRUSSES" THE FOLLOWING ADDITIONAL GUIDELINES APPLY:**
   1. Do not cantilever or extend putlogs/trusses as side brackets without thorough consideration of loads to be applied.
   2. Install and brace putlogs and trusses in accordance with manufacturer's instructions.

N. **FOR ROLLING SCAFFOLDS THE FOLLOWING ADDITIONAL GUIDELINES APPLY:**
   1. **RIDING A ROLLING SCAFFOLD IS VERY HAZARDOUS.** The SSFI and the SIA do not recommend nor encourage this practice.
   2. Casters with plain stems shall be attached to the frames or adjustment screws by pins or other suitable means.
   3. No more than 12 inches of the screw jack shall extend between the bottom of the adjusting nut and the top of the caster.
   4. Wheels or casters shall be locked to prevent caster rotation and scaffold movement when scaffold is in use.
   5. Joints shall be restrained from separation.
   6. Use horizontal diagonal bracing near the bottom and at 20 foot intervals measured from the rolling surface.
   7. Do not use brackets or other platform extensions without compensating for the overturning effect.
   8. The top platform height **as measured from the rolling surface** of a rolling scaffold must not exceed four (4) times the smallest base dimension (CAL/OSHA and some government agencies require a stricter ratio of 3:1).
   9. Cleat or secure all plank.
   10. Secure or remove all materials and equipment from platform before moving.

11. Do not attempt to move a rolling scaffold without sufficient help - watch out for holes in floor and overhead obstructions - stabilize against tipping.

O. **SAFE USE OF SCAFFOLD.**

1. Prior to use, inspect scaffold to insure it has not been altered and is in safe working condition.
2. Erected scaffolds and platforms should be inspected continuously by those using them.
3. Exercise caution when entering or leaving a work platform.
4. Do not overload scaffold. Follow manufacturer's safe working load recommendations.
5. Do not jump onto planks or platforms.
6. **DO NOT USE** ladders or makeshift devices to increase the working height of a scaffold. Do not plank guardrails to increase the height of a scaffold.
7. Climb in access areas only and use both hands.

III. WHEN DISMANTLING SCAFFOLDING THE FOLLOWING ADDITIONAL GUIDELINES APPLY:

A. Check to assure scaffolding has not been structurally altered in a way which would make it unsafe and, if it has, reconstruct and/or stabilize where necessary before commencing with dismantling procedures. This includes all scaffold ties.

B. Visually inspect planks prior to dismantling to be sure they are safe.

C. Do not remove a scaffold component without considering the effect of that removal.

D. Do not accumulate excess components or equipment on the level being dismantled.

E. Do not remove ties until scaffold above has been dismantled to that level.

F. Lower dismantled components in an orderly manner. Do not throw off of scaffold.

G. Dismantled equipment should be stockpiled in an orderly manner.

Since field conditions vary and are beyond the control of the SSFI and the SIA, safe and proper use of scaffolding is the sole responsibility of the user.

Publication S100 Printed in U.S.A 4/00
(2707)

CODE OF SAFE PRACTICES FOR SUSPENDED SCAFFOLDS

DEVELOPED FOR INDUSTRY BY **SCAFFOLDING, SHORING & FORMING INSTITUTE (SSFI)**
and SCAFFOLD INDUSTRY ASSOCIATION, INC. (SIA)

It shall be the responsibility of all users to read and comply with the following common sense guidelines which are designed to promote safety in the erecting, dismantling and use of suspended scaffolds. These guidelines do not purport to be all-inclusive nor to supplant or replace other additional safety and precautionary measures to cover usual or unusual conditions. If these guidelines in any way conflict with any state, local, provincial, federal or other government statute or regulation, said statute or regulation shall supersede these guidelines and it shall be the responsibility of each user to comply therewith.

I. GENERAL GUIDELINES

A. **POST THESE SAFETY GUIDELINES** in a conspicuous place and be sure that all persons who erect, use, locate, or dismantle suspended scaffold systems are fully aware of them and also use them in tool box safety meetings.

B. **FOLLOW ALL EQUIPMENT MANUFACTURERS' RECOMMENDATIONS** as well as all state, local and federal codes, ordinances and regulations relating to suspended scaffolding.

C. **SURVEY THE JOB SITE.** A survey shall be made of the job site by a competent person for hazards such as exposed electrical wires, obstructions that could overload or tip the suspended scaffold when it is raised or lowered, unguarded roof edges or openings, inadequate or missing tiebacks. Those conditions should be corrected before installing or using suspended scaffold systems.

D. **INSPECT ALL EQUIPMENT BEFORE EACH USE.** Never use any equipment that is damaged or defective in any way. Mark it or tag it as damaged or defective equipment and remove it from the jobsite.

E. **ERECT AND DISMANTLE SUSPENDED SCAFFOLD EQUIPMENT** in accordance with design and / or manufacturer's recommendations.

F. **DO NOT ERECT, DISMANTLE, OR ALTER SUSPENDED SCAFFOLD SYSTEMS** unless under the supervision of a competent person.

G. **DO NOT ABUSE OR MISUSE SUSPENDED SCAFFOLD EQUIPMENT.** Never overload platforms or hoists.

H. **ERECTED SUSPENDED SCAFFOLDS SHOULD BE CONTINUOUSLY INSPECTED** by the user to be sure that they are maintained in a safe condition. Report any unsafe condition to your supervisor.

I. **NEVER TAKE CHANCES! IF IN DOUBT REGARDING THE SAFETY OR USE OF SUSPENDED SCAFFOLDS, CONSULT YOUR SCAFFOLD SUPPLIER.**

J. **NEVER USE SUSPENDED SCAFFOLD EQUIPMENT FOR PURPOSES OR IN OTHER WAYS FOR WHICH IT WAS NOT INTENDED.**

K. **CARE SHOULD BE TAKEN WHEN OPERATING AND STORING EQUIPMENT DURING WINDY CONDITIONS.**

L. **SUSPENDED SCAFFOLD SYSTEMS** should be installed and used in accordance with the manufacturer's recommended procedures. Do not alter components in the field.

M. **SUSPENDED PLATFORMS MUST NEVER BE OPERATED NEAR LIVE POWER LINES** unless proper precautions are taken. Consult the power service company for advice.

N. **ALWAYS ATTACH FALL ARREST EQUIPMENT** when working on suspended scaffolds.

O. **DO NOT WORK ON OR INSTALL SUSPENDED SCAFFOLDS** if your physical condition is such that you feel dizzy or unsteady in any way.

P. **DO NOT WORK ON SUSPENDED SCAFFOLDS** when under the influence of alcohol or illegal drugs.

II. GUIDELINES FOR ERECTION AND USE OF SUSPENDED SCAFFOLD SYSTEMS

A. RIGGING:

1. **WEAR FALL PREVENTION EQUIPMENT** when rigging on exposed roofs or floors.
2. **ROOF HOOKS, PARAPET CLAMPS, OUTRIGGER BEAMS, OR OTHER SUPPORTING DEVICES** must be capable of supporting the hoist machine rated load with a factor of safety of 4.
3. **VERIFY THAT THE BUILDING OR STRUCTURE WILL SUPPORT** the suspended loads with a factor of safety of 4.
4. **ALL OVERHEAD RIGGING** must be secured from movement in any direction.
5. **COUNTERWEIGHTS USED WITH OUTRIGGER BEAMS** must be of a non-flowable material and must be secured to the beam to prevent accidental displacement.
6. **OUTRIGGER BEAMS THAT DO NOT USE COUNTERWEIGHTS** must be installed and secured on the roof structure with devices specifically designed for that purpose. Direct connections shall be evaluated by a competent person.
7. **TIE BACK ALL TRANSPORTABLE RIGGING DEVICES.** Tiebacks shall be equivalent in strength to suspension ropes.
8. **INSTALL TIEBACKS AT RIGHT ANGLES TO THE FACE OF THE BUILDING** and secure, without slack, to a structurally sound portion of the structure, capable of supporting the hoisting machine rated load with a safety factor of 4. **IN THE EVENT THAT TIEBACKS CANNOT BE INSTALLED AT RIGHT ANGLES,** two tiebacks at opposing angles must be used to prevent movement.
9. **RIG AND USE HOISTING MACHINES DIRECTLY UNDER THEIR SUSPENSION POINTS.**

B. WIRE ROPE AND HARDWARE:

1. **USE ONLY WIRE ROPE AND ATTACHMENTS** as specified by the hoisting machine manufacturer.
2. **ASSURE THAT WIRE ROPE IS LONG ENOUGH** to reach to the lowest possible landing.
3. **CLEAN AND LUBRICATE WIRE ROPE** in accordance with the wire rope manufacturer's instructions.
4. **HANDLE WIRE ROPE WITH CARE.**
5. **COIL AND UNCOIL WIRE ROPE** in accordance with manufacturer's instructions in order to avoid kinks or damage.
6. **TIGHTEN WIRE ROPE CLAMPS** in accordance with the clamp manufacturer's instructions.
7. **INSPECT WIRE ROPE IN ACCORDANCE WITH MANUFACTURER'S INSTRUCTIONS. DO NOT USE WIRE ROPE THAT IS KINKED, BIRDCAGED, CORRODED, UNDERSIZED, OR DAMAGED IN ANY WAY.** Do not expose wire rope to fire, undue heat, corrosive atmosphere, electricity, chemicals or damage by tool handling.
8. **USE THIMBLES AND SHACKLES AT ALL WIRE ROPE SUSPENSION TERMINATIONS.**
9. **USE J-TYPE CLAMPS OR SWEDGE FITTINGS.** Do not use U-bolts. Retighten J Clamps under load and retighten daily.
10. **WIRE ROPES USED WITH TRACTION HOISTS MUST HAVE PREPARED ENDS.** Follow manufacturer's recommendations.

C. POWER SUPPLY FOR MOTORIZED EQUIPMENT:

1. **GROUND ALL ELECTRICAL POWER SOURCES AND POWER CORD CONNECTIONS** and protect them with circuit breakers.
2. **USE POWER CORDS OR AIR HOSES OF THE PROPER SIZE THAT ARE LONG ENOUGH** for the job.
3. **POWER CORD OR AIR HOSE CONNECTIONS MUST BE RESTRAINED** to prevent their separation.
4. **USE STRAIN RELIEF DEVICES TO ATTACH POWER CORDS OR AIR SUPPLY HOSES TO THE SUSPENDED SCAFFOLD** to prevent them from falling.
5. **PROTECT POWER CORDS OR AIR HOSES AT SHARP EDGES.**
6. **USE GFI WITH POWER TOOLS.**

D. FALL ARREST EQUIPMENT:

1. **EACH PERSON ON A SUSPENDED SCAFFOLD** must be attached to a separate fall arrest system unless the installation was specifically designed not to require one.
2. **EACH LIFELINE MUST BE FASTENED IN ACCORDANCE WITH MANUFACTURER'S INSTRUCTIONS** to a separate anchorage capable of holding a minimum of 5000 pounds.
3. **DO NOT WRAP LIFELINES AROUND STRUCTURAL MEMBERS** unless lifelines are protected and a suitable anchorage connection is used.
4. **PROTECT LIFELINES AT SHARP CORNERS** to prevent chafing.
5. **RIG FALL ARREST SYSTEMS** to prevent free fall in excess of six feet.
6. **SUSPEND LIFELINES FREELY** without contact with structural members or building façade.
7. **USE LIFELINES OF SIZE AND CONSTRUCTION** that are compatible with the rope grab use.
8. **ASSURE A PROPERLY ATTACHED ROPE GRAB IS INSTALLED ON EACH LIFELINE IN THE PROPER DIRECTION.** Install in accordance with the manufacturer's recommendations.

9. **KEEP ROPE GRAB POSITIONED ABOVE YOUR HEAD LEVEL.**
10. **USE ONLY FULL BODY HARNESSES** of the proper size and that are tightly fastened.
11. **ASSURE FULL BODY HARNESS HAS LANYARD** attachment with D-ring at the center of your back.
12. **CONSULT FALL PROTECTION SUPPLIER FOR INSPECTION PROCEDURE. INSPECT FALL PROTECTION ANCHORAGE / EQUIPMENT BEFORE EACH USE.**
13. **WHEN A SECONDARY WIRE ROPE SYSTEM IS USED,** a horizontal lifeline secured to two or more structural members of the scaffold in lieu of vertical lifelines.

E. **DURING USE:**

1. **USE ALL EQUIPMENT AND ALL DEVICES** in accordance with the manufacturer's instructions.
2. **DO NOT OVERLOAD, MODIFY, OR SUBSTITUTE EQUIPMENT.**
3. **BEFORE COMMENCING WORK OPERATIONS** preload wire rope and equipment with the maximum working load, then retighten wire rope rigging clamps and recheck rigging to manufacturer's recommendations.
4. **INSPECT ALL RIGGING EQUIPMENT AND SUSPENDED SCAFFOLD SYSTEMS DAILY.**
5. **INSPECT WIRE ROPE DURING EACH ASCENT OR DESCENT FOR DAMAGE.**
6. **USE CARE TO PREVENT DAMAGE TO EQUIPMENT** by corrosive or other damaging substances.
7. **CLEAN AND SERVICE EQUIPMENT REGULARLY.**
8. **ALWAYS MAINTAIN AT LEAST (4) FOUR WRAPS OF WIRE ROPE ON DRUM TYPE HOISTS.**
9. **DO NOT JOIN PLATFORMS** unless the installation was designed for that purpose.
10. **ONLY MOVE SUSPENDED SCAFFOLDS HORIZONTALLY WHEN NOT OCCUPIED.**
11. **WHEN RIGGING FOR ANOTHER DROP** assure sufficient wire rope is available before moving the suspended scaffold system horizontally.
12. **WHEN WELDING FROM SUSPENDED SCAFFOLDS:**
    a. Assure platform is grounded to structure.
    b. Insulate wire rope above and below the platform.
    c. Insulate wire rope at suspension point and assure wire does not contact structure along its entire length.
    d. Prevent the bitter end from touching the welding ground.

Since field conditions vary and are beyond the control of the SSFI and the SIA, safe and proper use of suspended scaffolding is the sole responsibility of the user.

Publication SP201

Printed in U.S.A 2/99
(2700)

SSFI RP102

SSFI Recommended Procedure

Compression Testing of Welded Frame Scaffolds and Shoring Equipment

SSFI RP102

Scaffolding, Shoring & Forming Institute, Inc.

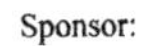
Sponsor:

Scaffolding, Shoring
& Forming Institute, Inc.
1300 Sumner Ave
Cleveland, Ohio 44115-2851

SSFI RP102

SCAFFOLDING, SHORING & FORMING INSTITUTE

Recommended Procedure for Compression Testing of Welded Frame Scaffolds and Shoring Equipment

Sponsor

Scaffolding, Shoring & Forming Institute, Inc.

A Scaffolding, Shoring & Forming Institute (SSFI) recommended procedure is intended as a guide to aid the manufacturer, the consumer, and the general public. The existence of an SSFI recommended procedure does not in any respect preclude anyone, whether he has approved the procedure or not, from manufacturing, marketing, purchasing or using products, processes, or procedures not conforming to the recommended procedure. Scaffolding, Shoring & Forming Institute recommended procedures are subject to periodic review and users are cautioned to obtain the latest editions.

CAUTION NOTICE:

This Scaffolding, Shoring & Forming Institute recommended procedure may be revised or withdrawn at any time. The procedures of the SSFI require that action be taken to reaffirm, revise, or withdraw this procedure no later than five years from the date of publication. Purchasers of Scaffolding, Shoring & Forming Institute standards and procedures may receive current information on all documents by calling or writing the Scaffolding, Shoring & Forming Institute.

Sponsored and published by:
SCAFFOLDING, SHORING & FORMING INSTITUTE
1300 Sumner Avenue
Cleveland, OH 44115-2851
Phn: 216/241-7333
Fax: 216/241-0105
E-Mail: ssfi@ssfi.org
URL: www.ssfi.org

Suggestions for improvement of this standard will be welcome.
They should be sent to the Scaffolding, Shoring & Forming Institute, Inc.

Printed in the United States of America

CONTENTS **PAGE**

Foreword iv

1. Scope 1

2. Definitions 1

3. Calibration of Testing Devices 3

4. Test Specimens 3

5. Procedure of Test 3

6. Duration of Test 4

7. Speed of Testing 4

8. Types of Tests 4

9. Witness of Testing 5

10. Report of Test Results 5

Appendix

A. Compression Testing Test Summary Form 6

Foreword (This foreword is included for information only and is not part of SSFI RP102 *Recommended Procedure for Compression Testing of Welded Frame Scaffolds and Shoring Equipment.*)

This SSFI Recommended Procedure was developed by the Scaffolding and Shoring Sections and Engineering Committees of the Scaffolding, Shoring & Forming Institute, Inc. as an assistance and guide to the manufacturers, purchasers, and users of rigging devices.

SSFI recognizes the need to periodically review and update this procedure. Suggestions for improvement should be forwarded to the Scaffolding, Shoring & Forming Institute, Inc., 1300 Sumner Avenue, Cleveland, Ohio, 44115-2851. All constructive suggestions for expansion and revision of this procedure are welcome.

The existence of a Scaffolding, Shoring & Forming Institute (SSFI) standard or recommended procedure does not in any respect preclude any member or non-member from manufacturing or selling products not conforming to the standard or recommended procedure and the SSFI is not responsible for the use of the standard or recommended procedure.

SSFI RP 102

Recommended Procedure
Compression Testing Of Welded Frame Scaffolds And Shoring Equipment

1 Scope

This procedure is intended to cover the compression testing of equipment used for scaffolding and vertical shoring.

2 Definition of Terms

The definition of terms below, relating to the compression testing of equipment used for scaffolding and shoring should be considered as applying to the terms used in these methods of compression testing.

2.1 ACCESSORIES – Those items other than frames, braces, or post shores used to facilitate the construction of scaffold and shoring.

2.2 ADJUSTMENT SCREW – A leveling device or jack composed of a threaded screw and an adjusting handle used for the vertical adjustment of the scaffolding, shoring and formwork.

2.3 ALLOWABLE LOAD – The ultimate load divided by factor of safety.

2.4 BASE PLATE – A device used to distribute the vertical load.

2.5 COUPLER OR CLAMP* – A device for locking together the component parts of a tubular metal scaffold. (*These terms may be used synonymously.)

2.6 COUPLING PIN – An insert device used to connect lifts or tiers vertically.

2.7 CROSS BRACING – A system of members which connect frames or panels of scaffolding or shoring laterally to make a tower or continuous structure.

2.8 DEAD LOAD SHORING – The load of forms, stringers, joists, reinforcing rods, and the actual concrete to be placed.

2.9 EXTENSION DEVICE – Any device, other than an adjustment screw, used to obtain vertical adjustment of shoring or scaffolding equipment.

2.10 FACTOR OF SAFETY – The ratio of ultimate load to the allowable load.

2.11 FORMWORK – The material used to give the required shape and support of poured concrete, consisting primarily of the following:
Sheathing – Material which is in direct contact with the concrete such as wood, plywood, metal sheet or plastic sheet.
Joists – Members which directly support sheathing.
Stringers or ledgers – Members which directly support the joists, usually wood or steel load-bearing members.

2.12 FRAME OR PANEL* – The principal prefabricated, welded structural unit in a tower. (*These terms may be used synonymously.)

2.13 HORIZONTAL SHORING – Metal or wood load-carrying beams of fabricated trussed section used to carry a shoring load from one bearing point, column, frame, post, or wall to another.

2.14 JOISTS – See Formwork

2.15 LIFTS OR TIERS* – The number of frames erected one above each other in a vertical direction. (*These terms may be used synonymously.)

2.16 LIVE LOAD – The total weight of workmen, equipment, buggies, vibrators, and other loads that will exist and move about on the scaffold or shoring equipment.

2.17 LOAD BEARING MEMBER – Any component of a scaffold or shoring structure which is directly subjected to load.

2.18 LOCKING DEVICE – A device used to secure the cross brace to the frame or panel.

2.19 POST SHORE or POLE SHORES* – Individual vertical member used to support loads. (*These terms may be used synonymously.)

a. Adjustable Timber Single Post Shore – Individual wooden timers used with a fabricated clamp to obtain adjustment and not normally manufactured as a complete unit.

b. Fabricated Single Post Shores – Type I; Single all-metal post, with a fine adjustment screw or device in combination with pin and hole adjustment or clamp.

c. Fabricated Single Post Shores – Type II; Single or double wooden post members adjustable for a metal clamp or screw and usually manufactured as a complete unit.

d. Timber Single Post Shores – Wood timber used as a structural member for shoring support.

2.20 RE-SHORING – A system used during the construction operation in which the original shores are removed and replaced in a sequence planned to avoid any damage to partially cured concrete.

2.21 SAFE LEG LOAD – The load which can safety be directly imposed on the frame leg. (See Allowable Load).

2.22 SCAFFOLD LAYOUT – A design diagram for scaffolding.

2.23 SHOCK LOAD – Impact of material such as the concrete as it is released or dumped during placement.

2.24 SHORE HEADS – Flat or formed metal pieces which are placed and centered on vertical members.

2.25 SHORING LAYOUT – A design drawing prepared prior to erection showing arrangement of equipment for shoring.

2.26 SILL OR MUD SILL* – A footing (usually wood) which distributes the vertical loads to the ground or slab below. (*These terms may be used synonymously.)

2.27 SPAN – The horizontal distance between posts, columns, or upright support members.

2.28 STRINGERS OR LEDGERS* – See Formwork (*These terms may be used synonymously.)

2.29 TESTING APPARATUS OR FIXTURE – A special purpose device fabricated for the express purpose of testing scaffolding and shoring.

2.30 TESTING MACHINE – A compression testing machine of a type usually found in Universities, Colleges, and reputable Testing Laboratories.

2.31 TIMBER STRESSES – Stress-grade lumber conforming to recommended tables in "Wood Structural Design Data Book", by National Lumber Manufacturers Association, Washington, D.C.

2.32 TOWERS – A composite structure of frames, braces and accessories.

2.33 TUBE AND COUPLER EQUIPMENT – An assembly used as a load-carrying structure consisting of tubing or pipe which serves as posts, braces, and ties, a base supporting the posts, and special couplers which serve to connect the uprights and join the various members.

2.34 ULTIMATE LOAD – The maximum load which may be placed on a structure before its failure due to buckling of column members or failure of some component.

3 Calibration of Testing Devices

3.1 The device used to determine loads applied shall be calibrated and certified either immediately before or after the testing by a reputable testing laboratory.

3.2 Testing machines used for compression testing shall be calibrated in accordance with ASTM Specification E4 or current revision during the preceding 12 month period.

4 Test Specimens

4.1 Scaffold and shoring components shall be selected at random from inventory and shall exhibit approximately the same variations in measurements as would be expected from random sampling including mill tolerances on thickness of various members.

4.2 Measurements of specimens. Thickness measurements, when required, shall be made with a suitable micrometer. All other dimensions shall be made with a commercially obtainable measuring tape and all dimensions reported to the nearest 1/16 inch.

5 Procedure of Test

5.1 The scaffold or shoring tower or shore to be tested shall be erected in such a manner as to simulate field conditions and aligned vertically so that it is not out of plumb more than 1/8" in three feet. No greater attempts should be made to adjust the components concentrically then would be expected in actual use.

5.2 The load shall be applied directly on the load bearing member or members by use of load transfer beams or cross head of testing machine; or directly by hydraulic jacks in an approved testing apparatus or fixture.

6 Duration of Test

6.1 The scaffolding tower or shore shall be subject to increasing loads until the ultimate load is reached.

7 Speed of Testing

7.1 The allowable limits for rate of loading on scaffolding towers shall not be less than 5,000 lbs. per minute nor more than 10,000 lbs. per minute.

7.2 The allowable limits for rate of loading on Post Shore shall not be less than 1,000 lbs. per minute nor more than 2,000 lbs. per minute.

7.3 The rate of loading in each test shall remain constant.

8 Types of Tests

8.1 Shoring leg loading shall consist of frames erected into towers composed of four (4) vertical legs with a normal bracing, base plates, and/or adjustment screws. When adjustment screws are used they shall be extended equally on the top and equally on the bottom; but top and bottom extensions need not be the same. All four load bearing legs shall be subjected to simultaneous loading until the ultimate load is reached by the weakest leg. Tests according to this method are "A" series tests.

8.2 Post Shores shall be tested individually at their minimum and maximum heights and at every foot throughout their operating range. The shores may be tested in both a braced condition and an unbraced condition with the individual test data displayed as a graph showing allowable load versus overall height with the manner of bracing, if any, clearly indicated. Loading shall be continued until ultimate load is reached. Tests performed according to this method are "B" series tests.

8.3 Extension devices shall be positioned in or upon the legs of the shoring and tested in the manner outlined in paragraph 8 (a). The extension devices may be tested extended from the top, bottom or both ends of the legs, and also in a braced or unbraced condition. The devices shall be tested at their maximum and minimum height and at every foot through their operating range with the individual test data displayed as a graph showing allowable load versus extended length and the manner of bracing, if any, clearly displayed. Tests performed according to this method are "C" series tests.

8.4 Welded Frame Wall Scaffolds shall be tested one (1) frame wide and two (2) cross brace lengths long. The frame scaffold components shall be erected into the above configuration composed of six (6) vertical legs with normal manufacturer's recommended bracing, and with base plates or adjusting screws on the bottom. The six (6) legs of the test configuration are to be loaded to the scaffold's anticipated allowable leg load and held at that value. Loading shall be continued on the two legs of the center frame until the ultimate load of these two legs is reached. The ultimate load of the center frame legs shall be the ultimate leg load for the welded frame scaffold system. Tests performed according to this method are "D" series tests.

8.5 Scaffold Component Testing – All load-bearing individual scaffold components shall be tested independently to obtain their ultimate load. When components are supported by other scaffold members, a test fixture of design similar to these supporting members shall be used to support the components being tested to assure that the test load can be transferred to the adjacent members. All components shall be tested to ultimate load and horizontal being components shall have their ultimate load as well as their total center deflection recorded. Tests performed according to this method are "E" series tests.

9 Witness of Test

9.1 All tests shall be witnessed by a reputable independent testing laboratory, University, or a registered Professional Engineer, who shall attest that the test was performed in accordance with applicable provisions of this standard.

10. Report of Test Results

10.1 Test results shall be reported on the SSFI Compression Testing Test Summary Form (see Appendix A) including drawing of test setup with the following:

(a) Ultimate Total Test Load
(b) Ultimate Leg Load or Component Test Load as Applicable
(c) Type of Test
(d) Laboratory
(e) Witness

Appendix A

COMPRESSION TESTING TEST SUMMARY FORM

1. Tested by________________________ Date of Test ______________

   Test Number________________________ Component Tested __________

2. Type of Test (Machine ____________ or Apparatus______________)

   a. Series (A), (B), (C), (D), (E) b. Number of Tiers ____________

   c. Adjustment Beyond Leg Top _____, _____, _____, _____
      Bottom _____, _____, _____, _____

   d. Extension Beyond Leg Top _____, _____, _____, _____
      Bottom _____, _____, _____, _____

   e. Extension, (Top _____) (Bottom _____)
      (Braced _____ or Unbraced _____)

   f. Ultimate Total Test Load ____________

   g. Ultimate Load – Each Leg _____, _____, _____, _____ (Average _____)

   h. Ultimate Shore Load _____

   i. Ultimate Ledger Load _____, _____

3. Witness to Test

   a. Company Representative________________________________

   b. Independent Representative ____________________________

   c. Other Witness ______________________________________

   d. Other Witness ______________________________________

4. Attach Sketch of Test

5. Certification

   I certify that the above described test was performed in accordance with the applicable provisions of the Procedure for Compression Testing of Welded Frame Scaffolds and Shoring Equipment as published by the Scaffolding, Shoring & Forming Institute, Inc.

   ____

SHORING SECTION

SSFI TECHNICAL BULLETIN

The Difference Between Shoring and Scaffolding

Shoring is not scaffolding and scaffolding is not shoring! The difference is in the *use*. The American National Standards Institute describes scaffolding as "a temporary elevated or suspended work unit and its supporting structure used for supporting worker(s) or materials, or both," while the same agency describes shoring as " The vertical supporting members in a formwork system."

To further clarify the difference, ask the question: "What is this deck being used for?" Is the deck (platform) going to be used so workers can stand on it while doing work to something else, such as a ceiling, or is the deck there primarily as part of the formwork for a cast in place concrete slab? It is incorrectly assumed that since the worker is standing on the deck, for example installing reinforcing steel for a concrete floor, it must be a scaffold and therefore the scaffold standards apply. *The primary function of the system determines its classification.* In other words, if the deck is part of the formwork, it is not scaffolding in spite of the fact that people are standing on that deck.

It is important to determine the primary use of the system so the applicable standards can be implemented. The significant differences include:

1. The safety factor: For scaffolding the factor is 4, and for shoring it ranges from 2 to 4;
2. Fall protection requirements for scaffolding are addressed in Subpart L of the Federal OSHA standards while fall protection requirements for shoring are addressed in Subpart M of the Federal OSHA standards;
3. Manufacturers' guidelines for the use of each type of equipment must be followed.

Some points to consider when determining whether a particular system is scaffolding or shoring include:

1. It is not the common name of the equipment that determines the use: a scaffold frame can be used as a shoring component and a shoring frame can be used as a scaffold component;
2. A shoring tower that is being used to temporarily support plank for erectors who are constructing a shoring system is a scaffold platform and must comply with the scaffold standards;
3. The safety factor does not determine if the deck is scaffolding or shoring;

Although scaffolding is not shoring and shoring is not scaffolding, remember that each system requires proper fall protection, access, and strength. For further information, consult the latest editions of the following ANSI and OSHA standards:

- ANSI A10.8
- 29CFR1926, Subpart L (OSHA scaffold standards)
- 29CFR1926, Subpart Q (OSHA concrete standards)
- 29CFR1926, Subpart M (OSHA fall protection standards)
- ANSI A10.9

This Technical Bulletin was prepared by members of the SSFI Shoring Section.

SSFI is a trade association comprising manufacturers of shoring, scaffolding, forming, and suspended scaffolding. The institute focuses on engineering and safety aspects of scope products.

May 03 This Bulletin is reviewed periodically. Check www.ssfi.org for the latest version.

SHORING SECTION

SSFI RECOMMENDED PROCEDURE

Recommended Procedures for Inspecting Used Welded Tubular Frame Shoring Equipment

Visual inspection of Used Vertical Shoring Equipment is intended for shoring of all current manufactured products of steel and aluminum and not intended for other materials such as wood products.

Visual Inspection

Inspection teams must be thoroughly trained to recognize the following possible defects or unsafe conditions present in used shoring regardless of age and source:

1. Cracked or broken welds
2. Missing members, legs or crossmembers
3. Split structural tube
4. Voids in legs or crossmembers due to cutting or cutting torch activity
5. Evidence of extreme heat applied to the product
6. Extra drill holes
7. Missing or inoperable cross brace lock devices
8. Tubular members out of round or deviations from normal cross section
9. Bent crossmembers or legs, including dents and dimples
10. Squareness or warp of frames, ledger frames and major components
11. Extreme corrosion
    (Corrosion can affect the overall strength of the product due to loss of cross sectional area)

Various jigs and fixtures can be assembled to inspect and check the out of round tube.

Accurate measurement of wall thickness can be accomplished in place on suspect product with extreme corrosion present, usually at the top or bottom of the frame using ultrasonic test equipment. This equipment does not measure air voids, thus measuring only the solid material, either aluminum or steel.

This Recommended Procedure was prepared by members of the SSFI Shoring Section.

SSFI is a trade association comprising manufacturers of shoring, scaffolding, forming, and suspended scaffolding. The institute focuses on engineering and safety aspects of scope products.

This procedure does not purport to be all-inclusive nor to supplant or replace other additional safety and precautionary measures to cover usual or unusual conditions. If this procedure conflicts in any way with a state, local, federal or other government statute or regulation, said statute or regulation shall supersede this procedure and it shall be the responsibility of each user to comply therewith. This procedure has been developed as an aid to users of shoring equipment. SSFI is not responsible for the use of this procedure.

SHORING SECTION

SSFI TECHNICAL BULLETIN

Standards That Apply to Shoring

The successful use of shoring means compliance with applicable United States Federal Occupational Safety and Health (OSHA) standards, various state standards, (such as California (Cal/Osha) standards), American National Standards Institute (ANSI) standards, and American Concrete Institute (ACI) standards. Following are some of the main applicable standards:

- OSHA 29CFR1926, Subpart Q;
- ANSI A10.9, *American National Standard for Construction and Demolition Operations-Concrete and Masonry Work-Safety Requirements*;
- ACI 347, Guide to Formwork for Concrete

A review of each set of standards illustrates their application. For example, the federal OSHA standards emphasize the proper erection of shoring equipment in accordance with the "formwork drawings." The ANSI standards, by comparison, are more comprehensive and address many topics including the proper maintenance of equipment, the proper installation of the equipment, and the proper design of shoring use. The ACI standards emphasize proper application of the shoring so the finished product is satisfactory.

A properly designed and erected shoring system will follow the manufacturer's instructions and will incorporate *all* applicable standards to ensure that the shoring system is safe and provides adequate results. For additional information, visit the following websites:

- Scaffolding, Shoring, and Forming Institute: www.ssfi.org
- American Concrete Institute: www.aci-int.org
- OSHA: www.osha.gov/doc/outreachtraining/htmlfiles/concrete.html
- OSHA: www.osha.gov/pls/oshaweb/owadisp.show_document?p_table=STANDARDS&p_id=10777

This Technical Bulletin was prepared by members of the SSFI Shoring Section.

SSFI is a trade association comprising manufacturers of shoring, scaffolding, forming, and suspended scaffolding. The institute focuses on engineering and safety aspects of scope products.

This bulletin does not purport to be all-inclusive nor to supplant or replace other additional safety and precautionary measures to cover usual or unusual conditions. If this bulletin conflicts in any way with a state, local, federal or other government statute or regulation, said statute or regulation shall supersede this bulletin and it shall be the responsibility of each user to comply therewith. This bulletin has been developed as an aid to users of shoring equipment. SSFI is not responsible for the use of this bulletin.

Appendix H

Most Frequent Citations

Standards Cited for SIC Division C; All sizes; FederalU.S. Department of

Labor

Occupational Safety & Health Administration

www.osha.gov[skip navigational links]Search Advanced Search | A-Z Index

Standards Cited for SIC Division C; All sizes; Federal

Division C Construction

Listed below are the standards which were cited by Federal OSHA for the specified SIC during the period October 2001 through September 2002. Penalties shown reflect current rather than initial amounts. For more information, see definitions.

Standard	#Cited	#Insp	$Penalty	Description
=Total=	44409	14863	38645453.37	
19260451	8377	3285	7631205.69	General Requirements for all types of Scaffolding
19260501	5403	4680	6973313.28	Fall Protection Scope/Applications/Definitions
19260651	2024	1163	2331742.50	Excavations, General Requirements
19261053	1736	1314	809881.02	Ladders
19260100	1594	1585	863652.78	Head Protection
19101200	1428	769	172034.82	Hazard Communication
19260652	1410	1279	4100918.50	Excavations, Requirements For Protective Systems
19260404	1397	1173	676992.72	Electrical, Wiring Design & Protection
19260405	1379	1004	413348.61	Elec. Wiring Methods, Components & Equip,Gen'l Use
19260020	1215	1066	890321.70	Construction, General Safety & Health Provisions
19260503	1191	1106	598583.50	Fall Protection Training Requirements
19260453	1103	975	962356.88	Manually Propelled Mobile Ladder Stands & Scaf'lds
19260502	1074	713	726028.50	Fall Protection Systems Criteria & Practices
19260021	1029	1001	963478.25	Construction, Safety Training & Education
19260454	919	788	376312.31	Training Requirements for all types of Scaffolding
19261052	860	644	354794.60	Stairways
19260403	720	617	423947.35	Electrical, General Requirements
19100134	717	299	187402.04	Respiratory Protection
19260062	702	124	726876.00	Lead
19260452	669	575	415924.79	Additional Requirements for Specific Scaffolding
19261101	573	105	447371.25	Asbestos

19260102	542	533	263274.25	Eye & Face Protection
19260416	433	392	287885.67	Electrical, Safety-Related Work Practices,Gen Rqts
5A0001	421	401	904459.10	General Duty Clause (Section of OSHA Act)
19260602	418	343	318286.25	Material Handling Equipment
19260350	409	292	182623.16	Gas Welding & Cutting
19260550	390	218	589856.00	Cranes & Derricks
19260025	347	341	139857.50	Construction, Housekeeping
19260701	344	342	417470.80	Concrete/Masonry, General Requirements
19260095	309	306	205535.25	Criteria For Personal Protective Equipment
19100178	269	232	121544.90	Powered Industrial Trucks
19260760	254	220	411013.00	
19261060	248	239	43712.50	Stairways & Ladders, Training Requirements
19260150	237	226	41965.27	Fire Protection
19260251	237	151	164162.50	Rigging Equipment For Material Handling
19260300	216	200	144928.36	Hand & Power Tools, General Requirements
19261051	206	203	129673.25	Stairways & Ladders,General Requirements
19260152	179	128	43462.75	Flammable & Combustible Liquids
19260028	140	139	78352.90	Construction, Personal Protective Equipment
19260304	140	100	58124.00	Woodworking Tools
19260050	119	104	47250.00	Medical Services & First Aid
19260153	119	103	40983.67	Liquefied Petroleum Gas
19260201	104	92	98063.25	Signaling
19260200	101	96	90624.00	Accident Prevention Signs & Tags
19260059	99	62	13716.00	Hazard Communication
19260850	95	71	110913.00	Demolition, Preparatory Operations
19260351	93	70	36312.00	Arc Welding & Cutting
19260106	92	50	73875.50	Working Over or Near Water
19260051	86	80	11180.00	Sanitation
19260105	85	84	246278.00	Safety Nets
19260761	81	74	51629.50	
19260302	80	70	35031.61	Power Opeated Hand Tools
19040002	79	79	12177.00	Log & Summary of Occupational Injuries & Illnesses
19030019	72	72	30425.00	
19260055	70	39	56690.00	Gases, Vapors, Fumes, Dusts & Mists
19260601	69	59	117670.00	Motor Vehicles
19260352	67	55	29168.75	Fire Prevention
19260307	61	38	20095.00	Mechanical Power-Transmission Apparatus
19260600	54	53	161763.50	Equipment

19260103	51	36	13214.00	Respiratory Protection
19260303	49	39	39997.50	Abrasive Wheels & Tools
19040008	45	45	45974.00	Fatality/Multiple Hospitalization Accident Reportg
19260752	45	43	15060.00	Bolting, Riveting, Fitting-Up & Plumbing-Up Steel
19260800	45	15	130175.00	Underground Construction
19260035	38	36	1800.00	Construction, Employee Emergency Action Plans
19101018	34	7	13125.00	Inorganic Arsenic
19260034	34	32	12000.00	Construction, Means of Egress
19100147	33	13	19527.00	The Control of Hazardous Energy, Lockout/Tagout
19260706	32	26	22963.00	Masonry Construction
19260703	31	21	77400.00	Concrete/Masonry, Cast-In-Place Concrete
19260252	30	28	10892.50	Disposal of Waste Materials
19100305	29	19	28356.00	Electrical, Wiring Methods, Components & Equipment
19260052	29	24	28047.50	Occupational Noise Exposure
19260955	28	8	131375.00	Power Transmission, Overhead Lines
19100269	27	12	147225.00	Electric Power Generation/Transmission/Distribu.
19260054	27	19	4125.00	Nonionizing Radiation
19100132	26	19	13503.50	Personal Protective Equipment,General Requirements
19260250	26	24	15108.00	Materials Handling, General Req'ts For Storage
19260417	25	23	20205.00	Electrical, Lockout & Tagging of Circuits
19260552	25	15	20175.00	Material Hoists, Personnel Hoists & Elevators
19260757	23	9	7810.00	
19260950	23	15	68525.00	Power Transmission, General Requirements
19100067	21	16	19532.50	Vehicle-Mounted Elevating/Rotating Work Platforms
19260605	21	12	8680.50	Marine Operations & Equipment
19260852	20	18	9805.00	Demolition, Chutes
19100023	19	15	15053.00	Guarding Floor & Wall Openings & Holes
19260750	17	17	22965.00	Steel Erection, Flooring Requirements
19260754	17	15	66450.00	
19260202	16	15	12412.50	Barricades
19100268	15	8	11350.00	Telecommunications
19260151	15	14	6472.50	Fire Prevention
19260353	15	14	59993.25	Ventila & Protection in Welding, Cutting & Heating
19260500	14	10	15125.00	Floor/Wall Openings, Guardrails,Handrails & Covers
19260755	13	12	11387.50	
19260856	13	12	14337.50	Removal of Walls, Floors & Material With Equipment
19040039	12	12	24600.00	
19100142	12	7	5086.00	Temporary Labor Camps

19100146	12	4	4250.00	Permit-Required Confined Spaces
19260057	12	9	11200.00	Ventilation
19260096	12	12	2970.50	Occupational Foot Protection
19040005	11	9	800.00	Annual Summary, Occupational Injuries & Illnesses
19100106	11	8	7622.00	Flammable & Combustible Liquids
19100107	11	3	1275.00	Spray Finishing w/ Flammable/Combustible Materials
19100303	11	9	10889.50	Electrical Systems Design, General Requirements
19260301	11	11	9641.25	Hand Tools
19260432	11	11	6012.00	Electrical, Environmental Deterioration of Equip.
19040007	10	10	6000.00	Access to Records
19100037	10	9	2675.00	Means of Egress, General
19100215	10	8	2670.00	Abrasive Wheel Machinery
19100333	10	5	18030.00	Electrical, Selection & Use of Work Practices
19260101	10	10	2025.00	Hearing Protection
19260104	10	9	15600.00	Safety Belts, Lifelines & Lanyards
19260951	10	6	12575.00	Power Transmission, Tools & Protective Equipment
19100151	9	9	8094.00	Medical Services & First Aid
19100120	8	5	7325.00	Hazardous Waste Operations & Emergency Response
19100141	8	6	5006.09	Sanitation
19100213	8	5	6105.00	Woodworking Machinery Requirements
19100335	8	4	9200.00	Electrical, Safeguards For Personnel Protection
19101020	8	8	1500.00	Access to employees exposure and medical records
19101052	8	2	3275.00	Methylene Chloride
19260554	8	2	31500.00	Overhead Hoists
19260905	8	4	4250.00	Loading of Explosives or Blasting Agents
19260954	8	5	124000.00	Power Transmission, Grounding For Employee Protec.
19100095	7	3	1500.00	Occupational Noise Exposure
19100157	7	6	1652.00	Portable Fire Extinguishers
19100180	7	4	14175.00	Crawler Locomotive & Truck Cranes
19101000	7	4	5450.00	Air Contaminants
19260056	7	7	2750.00	Illumination
19100133	6	6	2695.00	Eye & Face Protection
19100176	6	6	7877.00	Materials Handling, General
19100212	6	4	10138.00	Machines, General Requirements
19100253	6	5	2586.00	Oxygen-Fuel Gas Welding & Cutting
19260024	6	6	4267.50	Construction, Fire Protection & Prevention
19260441	6	2	1500.00	Electrical, Battery Loca's & Battery Charging
19260604	6	4	6725.00	Site Clearing

19100242	2	2	650.00	Hand & Portable Powered Tools & Equipment, General
19100244	2	1	1500.00	Other Portable Tools & Equipment
19100334	2	2	875.00	Electrical, Use of Equipment
19100420	2	2	1750.00	Diving, Safe Practice Manual
19100430	2	2	787.00	Diving, Equipment Procedures & Requirements
19101048	2	2	1100.00	Formaldehyde
19260305	2	2	3000.00	Jacks-Level & Ratchet, Screw & Hydraulic
19260702	2	2	1300.00	Concrete/Masonry, Equipment & Tools
19260758	2	2	3500.00	
19260759	2	2	1500.00	
19260855	2	2	1800.00	Manual Removal of Floors
19260857	2	2	1500.00	Demolition, Storage
19260858	2	2	7525.00	Demolition, Removal of Steel Construction
19260957	2	1	5000.00	Power Transmiss, Construc in Energized Substations
19030002	1	1	0.00	Posting of Notice, Avail. Of Act & Applic. Stands
19040003	1	1	0.00	Period Covered
19040006	1	1	0.00	Retention of Records
19040017	1	1	0.00	
19100025	1	1	656.00	Portable Wood Ladders
19100027	1	1	1500.00	Fixed Ladders
19100029	1	1	450.00	Manually Propelled Mobile Ladder Stnds & Scaffolds
19100119	1	1	4500.00	Process Safety Management, Highly Hazardous Chem's
19100169	1	1	562.50	Compressed Air Receivers
19100179	1	1	750.00	Overhead & Gantry Cranes
19100243	1	1	250.00	Guarding of Portable Powered Tools
19100252	1	1	750.00	Welding, Cutting & Brazing, General Requirements
19100254	1	1	500.00	Arc Welding & Cutting
19100255	1	1	0.00	Resistance Welding
19100272	1	1	1225.00	Grain Handling Facilities
19100332	1	1	1200.00	Electrical, Training
19100410	1	1	3500.00	Diving, Qualifications of Dive Team
19100421	1	1	1219.00	Diving, Pre-Dive Procedures
19100422	1	1	3500.00	Diving, Procedures During Dive
19100423	1	1	788.00	Diving, Post-Dive Procedures
19100425	1	1	3500.00	Diving, Surface-Supplied Air Diving
19101030	1	1	0.00	Bloodborne Pathogens
19260029	1	1	825.00	Construction, Acceptable Certifications
19260354	1	1	0.00	Welding, Cutting & Heating, Preservative Coatings

19100242	2	2	650.00	Hand & Portable Powered Tools & Equipment, General
19100244	2	1	1500.00	Other Portable Tools & Equipment
19100334	2	2	875.00	Electrical, Use of Equipment
19100420	2	2	1750.00	Diving, Safe Practice Manual
19100430	2	2	787.00	Diving, Equipment Procedures & Requirements
19101048	2	2	1100.00	Formaldehyde
19260305	2	2	3000.00	Jacks-Level & Ratchet, Screw & Hydraulic
19260702	2	2	1300.00	Concrete/Masonry, Equipment & Tools
19260758	2	2	3500.00	
19260759	2	2	1500.00	
19260855	2	2	1800.00	Manual Removal of Floors
19260857	2	2	1500.00	Demolition, Storage
19260858	2	2	7525.00	Demolition, Removal of Steel Construction
19260957	2	1	5000.00	Power Transmiss, Construc in Energized Substations
19030002	1	1	0.00	Posting of Notice, Avail. Of Act & Applic. Stands
19040003	1	1	0.00	Period Covered
19040006	1	1	0.00	Retention of Records
19040017	1	1	0.00	
19100025	1	1	656.00	Portable Wood Ladders
19100027	1	1	1500.00	Fixed Ladders
19100029	1	1	450.00	Manually Propelled Mobile Ladder Stnds & Scaffolds
19100119	1	1	4500.00	Process Safety Management, Highly Hazardous Chem's
19100169	1	1	562.50	Compressed Air Receivers
19100179	1	1	750.00	Overhead & Gantry Cranes
19100243	1	1	250.00	Guarding of Portable Powered Tools
19100252	1	1	750.00	Welding, Cutting & Brazing, General Requirements
19100254	1	1	500.00	Arc Welding & Cutting
19100255	1	1	0.00	Resistance Welding
19100272	1	1	1225.00	Grain Handling Facilities
19100332	1	1	1200.00	Electrical, Training
19100410	1	1	3500.00	Diving, Qualifications of Dive Team
19100421	1	1	1219.00	Diving, Pre-Dive Procedures
19100422	1	1	3500.00	Diving, Procedures During Dive
19100423	1	1	788.00	Diving, Post-Dive Procedures
19100425	1	1	3500.00	Diving, Surface-Supplied Air Diving
19101030	1	1	0.00	Bloodborne Pathogens
19260029	1	1	825.00	Construction, Acceptable Certifications
19260354	1	1	0.00	Welding, Cutting & Heating, Preservative Coatings

19260551	1	1	0.00	Helicopters
19260904	1	1	875.00	Storage of Explosives & Blasting Agents
19260909	1	1	0.00	Blasting, Firing the Blast
19260953	1	1	1800.00	Power Transmission, Material Handing
19600008	1	1	0.00	Agency Responsibilities

Last updated on 021105

[Query Standards Cited | SIC Search | SIC Division Structure]

Back to Top www.osha.gov www.dol.gov

Contact Us | Freedom of Information Act | Information Quality | Customer Survey .

Privacy and Security Statement | Disclaimers

Occupational Safety & Health Administration

200 Constitution Avenue, NW

Washington, DC 20210

Appendix I

Calculating Scaffold Planks

Structural Safety, 2 (1984) 47–57
Elsevier Science Publishers B.V., Amsterdam - Printed in The Netherlands

CALCULATING APPARENT RELIABILITY OF WOOD SCAFFOLD PLANKS

David S. Gromala

*U.S. Department of Agriculture, Forest Service Forest Products Laboratory *, Madison, WI 53705 (U.S.A.)*

(Received February 28, 1983; accepted in revised form January 15,1984)

Key Words: Building codes, reliability analysis, scaffold planks, specifications, structural engineering, wood.

ABSTRACT

Current safety requirements of the American National Standards Institute (ANSI A10.8) for wood scaffold planks are rooted in a history of generally good performance. The requirements are prescriptive in nature, making the standard easy to use. However, in its current form the standard prescribes only maximum allowable spans and limits its scope to three plank sizes and three load conditions. It does not specify what constitutes acceptable performance in solid-sawn scaffold planks. Thus, the requirements cannot be applied with uniform reliability to different plank systems. At the request of the ANSI A10.8 Subcommittee, an ad hoc advisory group from the wood industry was formed to address this problem.

This paper relates published requirements for wood scaffold planks in codes and standards to the allowable bending stresses published by lumber grading agencies based on a set of assumptions recommended by the ad hoc group. A methodology is proposed, based on requiring equivalent structural reliability, for designing alternative wood-based scaffold planks.

It is anticipated that this paper will be useful to producers and users of wood scaffold plank in its illustration of the sensitivity of both performance and reliability estimates to assumed material variability and distribution type. It is hoped that this report will generate discussion, both on the ANSI requirements themselves and on the best methodology to translate those requirements into measures of safety.

* Maintained in cooperation with the University of Wisconsin.

0167-4730/84/$03.00

48

INTRODUCTION

Scaffolds are temporary structures. They are intended to provide access to a building during its construction and maintenance for workers and materials. Because the configurations and loadings on scaffolds vary from site to site and because individual scaffolds are rarely in place long enough to develop a performance history, it is difficult to define performance requirements. By contrast, codes and standards for permanent structures have continually refined load and strength requirements.

This paper proposes a method by which strength requirements for one component of the scaffold system, the planks, can be defined in terms of standards used for permanent structures. The specific objectives of the paper are: (1) To define the intent of the current deterministic American National Standards Institute Standard (ANSI) strength requirement, (2) to relate published allowable stresses for wood planks to the ANSI requirement, (3) to calculate the apparent reliability index for commonly used solid-sawn wood planks, and (4) to propose a procedure for design of alternative wood-based planks based on the concept of equivalent reliability.

The National Bureau of Standards (NBS) has defined research needs for the development of improved scaffold standards. Hunt [1] has compiled many of the findings of the NBS study. His report is useful in this examination of wood scaffold planks because it reviews the requirements of many codes and standards, and it questions the lack of any published connection between scaffold standards and conventional wood design procedures. After attempting to compare stresses under loads specified by ANSI A10.8 [2] with an estimate of plank strengths, Hunt concluded that " the general provision of ANSI (and the Occupational Safety and Health Administration (OSHA) [3]), which specifies a 4 to 1 strength factor, reflects incognizance of the existing technical bases by which wood design values are determined." Hunt based this statement on calculations for one span and loading in which changes in assumed load duration and plank moisture content can vary the safety factor on the fifth percentile from 2.7 to 6.9.

An *ad hoc* advisory group to the ANSI subcommittee revising the A10.8 standard was formed to address some of the questions raised by Hunt's report, and to generalize the standard to cover wood scaffold plank other than traditional visually graded solid-sawn plank. This report examines current usage, based on input from the ad hoc advisory group and as specified in codes and standards, as a basis for determining assumed relationships between loads specified by ANSI A10.8 and strength requirements for wood scaffold planks. These relationships can be used to express design requirements for scaffold planks in terms of conventional allowable design stresses.

CURRENT PLANK REQUIREMENTS

Accepted plank usage

Examination of codes and standards shows that nominal 2- by 10- and 2- by 12-inch (actual 1½- by 9¼-in. (38- by 235-mm) and 1½- by 11¼-in. (38- by 285-mm)) scaffold planks are the most widely used sizes. Limited industry production data [4] support this observation. It appears that nominal 2- by 10-inch planks on an 8-foot (2.4-m) span are commonly used and widely accepted. This size is also used on 10-foot (3.0-m) spans, but such usage is only allowed by about half of the codes and standard cited by Hunt [1]. This break between universal acceptability on an 8-foot span and disagreement on a 10-foot span would appear to indicate a potential "boundary region" of safety in the opinions of specification writers.

ANSI specifies the design load due to the weight of a worker carrying hand tools to be 250 pounds (115 kg). This requirement would be excessively conservative if only static loads are considered, as only about 2% of adult males in the United States weigh at least 250 pounds [5]. However, if one multiplies the estimated average static weight (170 pounds, 75 kg) of a worker by a dynamic amplification factor to account for the effects of walking (averaging about 1.5 according to Sentler [6]), the resulting effective load becomes 255 pounds force (1135 N). Thus, this paper will assume that the ANSI 250-pound concentrated load requirement is a good estimate of the effective weight of one worker on a plank when the dynamic effects of walking are considered.

The following example of a nominal 2- by 10-inch plank under the ANSI "one worker" loading serves to illustrate the relationship between span and maximum induced stress. Under this load the induced maximum stress for a 1½- by 9¼-inch (38- by 235-mm) plank on a simple span ranges from 1,730 pounds per square inch (psi) (11,930 kN/m²) on an 8-foot (2.4-m) span to 2,160 psi (14,890 kN/m²) on a 10-foot (3.0-m) span.

So, in progressing from 1,730 to 2,160 psi (11,930 to 14,890 kN/m²), virtually all specifications implicitly accept the former stress, but half of them do not permit the increase to the latter.

ANSI Strength requirements

There is no single requirement that causes as much confusion and controversy in ANSI A10.8 as section 3.4, which states:

> "Scaffolds and their components except as noted herein shall be capable of supporting without failure at least four times the maximum intended load. . . "

The first thing about this section that strikes the technical reader is that its wording violates the laws of statistical probability in explicitly requiring a zero failure rate. Undoubtedly. this was not intended. Consequently, if one is permitted to interpret the intent of this wording, the only problem is defining, in contemporary terms, the relationship between the load and resistance distributions implied by ANSI. If ANSI-specified loads can be assumed to be the average of the load distribution, a rough estimate of where this average (multiplied by 4) intersects the strength distribution can be found by a process of elimination.

Using the previously defined 2 by 10 on an 8-foot span with a maximum induced stress of 1,730 psi (11,930 kN/m²), the "target level" of comparison is 1,730 psi × 4 or over 6,900 psi (47,600 kN/m²). This stress can be compared to three different strength measures: the published allowable stress, the population near-minimum, and the population mean.

Starting with a conservative assumption, one can compare four times the ANSI-specified load to published allowable design stresses such as those found in the National Design Specification (NDS) for Wood Construction [7]. This would be a reasonable interpretation of the term "safety factor.'' This assumption would require the NDS value to be over 6.900 psi (47,600 kN/m²) to satisfy the factor of 4 requirement. Although published allowable stresses for some grades of visually graded lumber exceed 2,000 psi (13,790 kN/m²), they rarely, if ever, exceed 3,000 psi (20,680 kN/m²). Thus, this assumption is not even close to being realistic.

The assumed relationship between load and resistance would be less stringent if one assumes that the aforementioned 6,900 psi (47,600 kN/m²) relates to a near-minimum (fifth percentile) short-term test strength. Even this assumption would require planks to have an average strength of over 13,000 psi for a population with an assumed coefficient of variation of 30%. Planks on a 10-foot span would need an average strength of over 17,000 psi to satisfy the same criterion. As in the

50

previous assumption, such values are not realistic.

A third, somewhat liberal, assumption would be that the stress induced by the ANSI load and the mean short-term strength are separated by a factor of 4. With estimated average strengths of visually graded solid-sawn wood scaffold planks in the range of 6,000 to 9,000 psi (41,370 to 62,050 kN/m²), this assumption is very good. Thus, current usage appears to be fairly close to producing an "average safety factor" of 4. In equation form, this can be expressed as:

$$\sigma_{wL} \leqslant R_m/4 \quad (1)$$

in which σ_{wL} = maximum induced stress at ANSI working load and R_m = mean plank strength. This is the definition of the ANSI safety factor that will be used to relate ANSI requirements to traditional design stresses for lumber.

Allowable stresses for wood

As described more completely in the NDS, allowable stress values for use in design are based primarily on American Society for Testing and Materials (ASTM) standards augmented by additional experimental information. The relationship between allowable stress and the population is roughly this: The allowable stress is an estimate of a near-minimum point, commonly the fifth percentile, of the strength distribution in a short-term test divided by a factor to adjust for the anticipated duration of the design load, and a safety, or accidental overload , factor. ASTM D 2555 [8] provides information on strength of clear wood specimens and ASTM D 245 [9] explains how to adjust clear wood values to estimate design stresses for visually graded structural lumber.

In the ASTM procedure a population of wood specimens is tested, yielding a mean (R_m) and coefficient of variation (V_R) of strength. The fifth percentile of the population ($R_{0.05}$) is then calculated; commonly used normal statistics yield:

$$R_{0.05} = R_m(1 - 1.645V_R) \quad (2)$$

and lognormal statistics yield:

$$R_{0.05} = R_m\, e^{-(0.5Y^2 + 1.645Y)} \quad (3)$$

in which

$$Y^2 = \ln(1 + V_R^2)$$

TABLE 1

Allowable stress modification factors

Parameter	Factor	Recommended by	Comments
Moisture content	1.00	CSA *	Normal scaffold use **
	See CSA 086 [11]	CSA	If wet use **
	1.00	NDS	If dry use
	0.80	NDS	If wet use
Duration of load	1.25	CSA	Assumes 7-day duration
	-	NDS	
Load sharing	1.10	CSA	3-plank minimum **
	1.15	NDS	3-member minimum
Fire retardant treated	0.90	NDS	
	0.82	APA	
	0.75	AITC	

* CSA S269.2-M1980 [10].

** Factor explicitly given for scaffold plank use.

The fifth percentile is divided by a factor to account for load duration effects (approximately 1.62 to correct to 10-year or "normal" duration) and by an additional explicit safety factor (1.3), so:

$$f_b = R_{0.05}/2.1 \quad (4)$$

in which f_b = published allowable bending stress.

In addition to the factors already considered, published grading rules and corresponding allowable stresses must account for load-sharing, strength ratio and size effects, 'and adjustment for moisture content. Final choice of appropriate stress modification factors to account for these effects is the responsibility of the engineer or designer. However, published standards and other literature (Table 1) offer useful guidelines in this area.

Table 1 shows that the Canadian Standards Association (CSA) [10] scaffold standard specifies factors for duration of load (1.25 for 7-day duration) and for load-sharing systems (1.10, 3-plank minimum). It also provides guidance for definition of inservice moisture condition. Although some NDS factors are also shown in the table, NDS does not specifically address scaffold applications. The CSA scaffold standard is of particular interest specifically because it addresses wood scaffold plank in terms familiar to both scaffold users and wood designers. Its basic reference to design of wood scaffold members [11] is the primary design reference for wood structures in Canada. In contrast to the North American approach, the Standards Association of Australia [12,13] specifications for wood scaffold planks are formatted to look like grading rules rather than design specifications. The British Standards Institution [14] specification for timber scaffold boards is formatted much like the Australian standards.

One factor in Table 1, concerning the effects on strength and on toughness of fire-retardant treatments (FRT), is an area currently under active research investigation. The NDS specifies a 10% reduction in allowable bending stress for fire-retardant-treated wood. However, research shows [15] that impact strength is more severely affected by FRT chemicals than is bending strength. Some industry associations in the United States [16,17] are already using larger reductions of up to 25% as normal design practice. As scaffold planks are potentially subjected to many impact or dynamic loads, a similar reduction may be appropriate.

Equations (1) through (4) can be used to find "scaffold use factors" for various assumed load durations and distribution parameters. Table 2 contains factors that must be applied to published allowable stresses if the ANSI-specified safety factor is to be preserved. For example, assuming a 30% coeffi-

TABLE 2

Factors to relate allowable stresses to ANSI scaffold use *

V_R	Assumed distribution type					
	Normal			Lognormal		
	5-Minute basis	7-Day basis	10-Year basis	5-Minute basis	7-Day basis	IO-Year basis
0.20	0.78	0.60	0.48	0.74	0.57	0.46
0.25	0.89	0.69	0.55	0.81	0.63	0.50
0.30	1.04	0.80	0.64	0.89	0.69	0.55
0.35	1.24	0.96	0.77	0.98	0.75	0.60

* $\sigma_{scaffold} \leqslant$ factor $\times f_b$.

52

cient of variation in strength and using the traditional assumption of a normal distribution coupled with the CSA recommended 7-day duration factor yields:

$$(stress)_{ANSI} \leqslant 0.80 \cdot f_b$$

in which 0.80 is the appropriate entry from Table 2.

For simplicity, the plank dead load stress (s_{DL}) was intentionally left out of the derivation above. It can be included if (stress),,,, is defined as:

$$(stress)_{ANSI} = \sigma_{wL} + \sigma_{dL}/4 \quad (5)$$

The dead load stress in Eq. (5) is divided by 4 because it is excluded in the ANSI standard from the required safety factor of 4.

Based on this derivation, and on the judgment of the ad hoc advisory committee to the ANSI A10.8 subcommittee, the factor relating allowable stresses to stresses for scaffold plank will be 0.80 in the revised A10.8 standard.

RELIABILITY OF SCAFFOLD PLANKS.

Computed reliability of solid-sawn

Methods for computing reliability of structural systems abound in the literature [18,19,20]. Most recent applications of reliability-based analyses to codes and standards have settled on "first-order second-moment" methods to calculate a reliability index, β [21].

As the only objective in calculating reliability indices in this paper is to establish a basis for comparison, the calculations are kept simple and brief. Assumptions such as distribution type and relative positions of mean load and resistance must be reasonable but, in the absence of adequate data, they certainly cannot be rigorous.

In the basic calculations, load and resistance are assumed to have the same type of distribution. The calculations were performed both on assumed normal and lognormal distributions, but lognormals were believed to better represent loads and resistances. Both of these distributions can be completely characterized solely by a mean and standard deviation (or coefficient of variation).

Although most distributions in the literature for strength of 'wood members qualitatively resemble the lognormal (right-skewed), recent test data indicate that some high grades of lumber may actually be left-skewed. Further study in this area is recommended.

For purposes of this example, ANSI section 3.4 will be interpreted to intend that planks designed according to ANSI A10.8 have a mean strength equal to four times the mean induced stress. This is the definition of the "average safety factor" of 4 discussed earlier.

In one notable departure from the exact wording of section 3.4 in ANSI A10.8, the author has chosen to substitute "ANSI working load" for "maximum intended load." The substitution is for two reasons; first, the term "maximum intended load" is not defined either in ANSI or anywhere else in the literature for scaffolds, and second, the author believes that the ANSI working loads represent a load more likely at or above the mean of the load distribution than below. Thus, the substitution is probably conservative.

If normal distributions are assumed the reliability index can be expressed as:

$$\beta = \frac{R_m - Q_m}{\sqrt{\sigma_R^2 + \sigma_Q^2}} \quad (6)$$

in which R_m = mean strength, Q_m = mean load effect, s_R, V_R = standard deviation, coefficient of variation in strength, and s_Q, V_Q = standard deviation, coefficient of variation in load effect.

If lognormal distributions are assumed:

$$\beta = \frac{\ln(R_m/Q_m) + (Z^2 - Y^2)/2}{\sqrt{Y^2 + Z^2}} \quad (7)$$

TABLE 3

Calculated reliability indices (β) for strength variability (V_R) range expected for solid-sawn

V_R	Distribution type					
	Normal			Lognormal		
	$V_Q = 0.2$	$V_Q = 0.3$	$V_Q = 0.4$	$V_Q = 0.2$	$V_Q = 0.3$	$V_Q = 0.4$
0.20	3.64	3.51	3.35	4.95	3.98	3.33
0.25	2.94	2.87	2.79	4.35	3.65	3.13
0.30	2.47	2.42	2.37	3.85	3.34	2.93
0.35	2.12	2.10	2.06	3.43	3.05	2.73

in which $Y^2 = \ln(1 + V_R^2)$ and $Z^2 = \ln(1 + V_Q^2$.

Based on the "average safety factor" of 4 as interpreted by this author from ANSI A10.8, by definition:

$$R_m/Q_m = 4 \tag{8}$$

Table 3 shows the calculated reliability indices for normal and lognormal distributions over a range of load and strength variability.

The calculated reliability index, β, does not reflect actual failure probabilities for scaffold plank systems. The index represents the reliability of a population of single-plank scaffold platforms under a hypothetical population of loads. In reality, planks are often part of multiple systems, which tends to increase reliability. In addition, a major intangible that dramatically decreases the probability of overload is the "human factor". The loads on scaffold planks are not truly random or induced by uncontrollable phenomena. Workers place the loads and walk onto the platforms. Few workers would walk onto a plank already deflecting excessively. Thus, the β's calculated here are only used for calibration of the relative reliability of various plank systems.

As shown in Table 3, the calculated reliability index for visually graded solid-sawn plank based on these assumptions ranges from 2.1 to 4.9. The table also shows that the calculated reliability is significantly different for assumed normal versus lognormal distributions, but is not overly sensitive to small changes in assumed coefficient of variation for solid-sawn plank.

Here, the ad hoc advisory group to the ANSI A10.8 subcommittee has decided that a single β should be recommended for calibration. This group recommends that lognormal distributions be assumed for these calculations and that both distributions have an assumed variability of 30%. Thus, the assumed reliability index for visually graded solid-sawn scaffold plank is 3.34.

Equivalent reliability

As wood-based scaffold planks other than visually graded solid-sawn planks enter the marketplace, a rational basis for calculating their design requirements would be to require

TABLE 4

Average safety factor required for equivalent reliability of alternative plank with V_R as shown *

V_R	Average safety factor	V_R	Average safety factor
0.05	2.59	0.20	3.19
0.075	2.64	0.225	3.36
0.10	2.71	0.25	3.55
0.125	2.80	0.275	3.76
0.15	2.91	0.30	4.00
0.175	3.04		

* Assumes lognormal distribution, β = 3.34, and V_Q = 0.3.

equivalent reliability. In this way materials that exhibit lower strength variability could be designed with a lower average safety factor without decreasing their overall reliability.

Equation (7) can be solved to find the relative safety factor, x (defined as R_m/Q_m), needed to obtain equivalent reliability (equal β) for alternative wood-based planks:

$$x = \exp\left[\beta\sqrt{Y^2 + Z^2} - (Z^2 - Y^2)/2\right] \quad (9)$$

Substituting $\beta = 3.34$, this equation yields the average safety factor required for plank populations with coefficients of variation, V_R, to have the same reliability index as solid-sawn (Table 4).

APPLICATION TO SCAFFOLD PLANK DESIGN

Because the application of this procedure may be somewhat confusing, this section of the paper provides examples of calculations for both visually graded solid-sawn planks and for alternative wood-based planks.

Visually graded solid-sawn planks. The published allowable stress must be corrected for moisture condition. depth effects, and scaffold use. Normal scaffold application is assumed to be dry use [10]. Depth corrections are minor, but are mentioned here because grading agencies commonly use a 3-inch (76-mm) depth basis in establishment of allowable stresses. The scaffold use factor (0.80) already includes a 1.25 load duration increase.

Thus, if a hypothetical 1½- by 9¼-inch (38- by 235-mm) scaffold plank had a published wet-use stress of 1.800 psi (12.410 kN/m²), this number could be multiplied by 1.25 if it qualified for dry-use stresses, by $(3/1.5)^{1/9}$ [9] to account for depth effects, and finally by 0.80 to convert to scaffold use:

$$1.800 \times 1.25 \times (3/1.5)^{1/9} \times 0.80$$

$$= 1.940 \text{ psi } (13.380 \text{ kN/m}^2)$$

The resulting "factored" allowable stress is then compared to stresses induced by ANSI loads. For a one worker loading on an 8-foot (2.4-m) span:

$$(stress)_{ANSI} = \sigma_{wL} + \sigma_{dL}/4$$

$$= 1{,}730 + 90/4$$

$$= 1{,}752 < 1{,}940, \text{ acceptable.}$$

Alternative wood-based planks. Derivation of allowable design stresses for planks other than visually graded solid-sawn are based on the premise that the mean and coefficient of variation of the plank population are known. It is beyond the scope of this paper to recommend either a minimum number of tests or a specific test configuration. However, because the choice of relative safety factor from Table 4 is highly dependent on the coefficient of variation, this parameter must be determined in a statistically valid manner.

The allowable stress for scaffold use is determined by dividing the population mean by the appropriate factor from Table 4 and correcting to 7-day duration:

Scaffold allowable stress

= [(mean of short-term test)/factor]

× (1.25/1.62)

in which "factor" = relative safety factor from Table 4.

For a hypothetical population of planks with a short-term mean MOR of 7,000 psi (48,260 kN/m²) and a coefficient of variation of 15%, the factor (from Table 4) is 2.91 and the scaffold allowable stress is:

$$(7{,}000/2.91) \times (1.25/1.62)$$

$$= 1{,}860 \text{ psi } (12{,}820 \text{ kN/m}^2)$$

When derived in this manner, the scaffold allowable stress is similarly compared to stresses induced by ANSI loads. For the same one-worker loading on an 8-foot (2.4-m) span:

$$(stress)_{ANSI} = \sigma_{wL} + \sigma_{dL}/(\text{factor})$$

$$= 1{,}730 + 90/2.91$$

$$= 1{,}760 < 1{,}860, \text{ acceptable.}$$

DISCUSSION OF ACTUAL RELIABILITY

As in all analytical exercises, the calculations in this paper consider only a limited number of physical parameters and contain certain assumptions (Appendix A). Thus some discrepancy between calculated or "apparent" reliability and actual "in-use" reliability can be expected.

Two factors come to mind that would increase actual reliability in the field—worker experience and wood's "special" material behavior. Worker experience is the biggest intangible influence on actual field reliability. Strength distributions of lumber are based on testing all pieces in a given sample, but if the same sample appeared on a job site, workers would select planks for various uses, and discard some, based on how they "feel." This "feel," based on experience with many planks, may include consideration of such factors as density, ductility, and possibly even some dynamic properties of the plank. Some material properties inherent to wood are its ability to withstand impact loads substantially greater than its long-term strength and its characteristic "crackling noise" as failure approaches. Both properties could provide additional measures of safety in typical field use.

Conversely some scaffold planks are proof-loaded in the field prior to use, often to multiples of design load, in an attempt to eliminate weaker boards. In fact, this practice was actually recommended by the National Safety Council in 1937 [22]. As discussed by Freas [23] with respect to similar practices on ladders, such testing, if improperly done, can damage the plank to the point of eliminating its ductile failure mode. Thus, such onsite proof testing can actually decrease reliability.

For the reasons mentioned above, it is anticipated that the actual reliability of wood scaffold planks would be higher than the calculated reliability in this paper if worker experience and ductile material behavior are accounted for. However, this additional safety could be eliminated if a plank exhibits a brittle failure mode. Thus, neither the definition of safety factor (average = 4) nor the calculated reliability ($\beta = 3.34$) given in this paper should be extrapolated to nonwood planks, or even to wood-based planks that do not exhibit the aforementioned characteristics of wood behavior under load.

SUMMARY AND RECOMMENDATIONS

This paper has reviewed the safety requirements for scaffold planks in ANSI A10.8 and has related those requirements to published allowable stresses based on ASTM D 245.

A methodology is proposed by which wood-based scaffold planks that are different from traditional visually graded solid-sawn planks can be evaluated on an equivalent reliability basis.

All calculations and discussions herein are' based on many assumptions, most of which are explicitly or implicitly part of current ANSI or ASTM standards. The calculations show that solid-sawn scaffold planks can be designed using published allowable stresses with an added safety factor to satisfy ANSI requirements.

The exercises performed in this paper are valid indicators of actual safety of wood scaffold planks only if loadings on planks are truly represented by ANSI load requirements and if the strength of full-size scaffold planks can be deduced through ASTM procedures. To verify both load and resistance the following research is recommended:

(1) Onsite load surveys to determine the true distribution of plank loads.

(2) Surveys of plank usage—recording plank sizes, grades, species, moisture content, and spans.

(3) Detailed study of plank failures.

(4) Experimental determination of the strength distribution of full-size planks.

(5) Study of jobsites to determine whether

56

planks are re-sorted or otherwise selectively used.

Completion of research in all five areas is necessary before actual onsite reliability of scaffold planks can be determined.

ACKNOWLEDGEMENTS

Without the recommendations of the *ad hoc* advisory group to the ANSI A10.8 subcommittee, this report would be little more than an academic exercise. The contributions of John Sebelius, Ray Todd, Brad Dempsey, Paul Myhaver, and especially Sherman Nelson. the chairman, are gratefully acknowledged.

Also acknowledged are Fred Petersen, chairman of the A10.8 subcommittee, under whose guidance the standard is being revised and Bill Galligan, for his extensive review of the initial draft of this paper.

REFERENCES

1 B.J.P. Hunt, Development of a Research Program for Scaffolding Standards. NBS GCR 80-255, National Bureau of Standards, Washington, DC, 1980.
2 Safety Requirements for Scaffolding (ANSI A10.8-1977). American National Standards Institute, New York, 1977.
3 Occupational Safety and Health Administration, Construction Safety and Health Standards. Federal Register Part 1926, U.S. Department of Labor, Washington, DC, 1974.
4 J.A. Sebelius, Personal communication, 1982.
5 Weight and Height of Adults 18–74 Years of Age: United States 1971–74. U.S.D.H.E.W. Vital and Health Statistics, Series 11-Number 211, U.S. Department of Health, Education, and Welfare, Washington, D.C., n.d.
6 L. Sentler, Live Load Surveys: a Review with Discussions. Report No. 78, Division of Building Technology, Lund, Sweden, 1976.
7 National Design Specification for Wood Construction. National Forest Products Association, Washington, DC, 1982.
8 Standard Methods for Establishing Clear Wood Strength Values (ASTM D 2555-78). American Society for Testing and Materials. Philadelphia, PA, 1978.
9 Standard Methods for Establishing Structural Grades and Related Allowable Properties for Visually Graded Lumber (ASTM D 245-74). American Society for Testing and Materials. Philadelphia, PA, 1974.
10 Access scaffolding for Construction Purposes (CSA S269.2-M1980). Canadian Standards Association, Rexdale, Ontario, 1980.
11 Code for the Engineering Design of Wood (CSA 086-1976). Canadian Standards Association, Rexdale, Ontario, 1976.
12 Solid Timber Scaffold Planks (AS 1577-1974). Standards Association of Australia, North Sydney, Australia, 1974.
13 Laminated Timber Scaffold Planks (AS 1578-1974). Standards Association of Australia, North Sydney, Australia, 1974.
14 Specification for Timber Scaffold Boards (B.S.2482: 1981). British Standards Institution, London, 1981.
15 C.C. Gerhards, Effect of Fire-Retardant Treatment on Bending Strength of Wood. USDA For. Serv. Res. Pap. FPL 145, Forest Products Laboratory, Madison, WI, 1970.
16 Design, Laminating Specifications (AITC 117-82). American Instititute of Timber Construction, Englewood, CO, 1982.
17 Plywood Design Specification. American Plywood Association, Tacoma, WA, 1980.
18 B. Ellingwood, Reliability of wood structural elements. J. Struct. Div. ASCE, 107 (ST1) (1981) 73–87.
19 J.R. Goodman, Z. Kovacs and J. Bodig, Code comparisons of factor design for wood. J. Struct. Div. ASCE, 107 (ST8) (1981) 1511–1527.
20 W.W.C. Siu, S.R. Parimi and N.C. Lind, Practical approach to code calibration. J. Struct. Div. ASCE, 101 (ST7) (1975) 1469–1480.
21 Probability-Based Design of Wood Transmission Structures (EPRI EL-2040). Project RP 1352-1, Electric Power Research Institute, Pala Alto, CA, 1981.
22 Wood Scaffolds. Safe Practices Pamphlet No. 12, National Safety Council, Chicago, IL, 1937.
23 A.D. Freas, Don't test that ladder!, International Fire Fighter, 32 (10) (1949) 15–16.

APPENDIX A: ASSUMPTIONS

Assumptions implied in ANSI A10.8

(1) Working loads are realistic and reflect the approximate "average" load for a given

use. (Based on Hunt's limited load survey the area loadings appear to be underestimated; based on the weight distribution of workers, the concentrated loadings are extremely conservative if only static loads are considered, but more realistic if dynamic effects are included.)

(2) The spans specified for wood plank are currently used, and are generally satisfactory.

(3) The specified safety factor of 4 can be interpreted as an "average safety factor"—that is, the average strength equals four times the average induced stress.

Additional assumptions

(1) The published allowable stress for lumber represents the fifth percentile of the strength distribution divided by a factor that includes a "safety factor" and a "load duration factor". (For bending, $SF \times LDF = 2.1$.)

(2) The load duration effect, as defined in the NDS [5], is reasonably accurate for high grades of lumber such as scaffold plank,

(3) Based on the CSA Standard [10], the appropriate duration factor for scaffold use is "normal" duration × 1.25, representing an assumed design load duration of 7 days.

(4) Reliability equations can be used to establish the apparent reliability of current systems.

(5) Normal or lognormal distributions can adequately characterize load and strength distributions.

(6) The coefficient of variation of the load distribution can be adequately characterized at about 30% for each load condition. This value is cited in reference [20] and is estimated to be slightly conservative, especially for the concentrated load cases.

Appendix J

OSHA-Approved Safety and Health Plans

Alaska
Alaska Department of Labor and Workforce Development
Commissioner (907) 465-2700 Fax: (907) 465-2784
Program Director (907) 269-4904
Fax: (907) 269-4915

Arizona
Industrial Commission of Arizona
Director, ICA (602) 542-4411 Fax: (602) 542-1614
Program Director (602) 542-5795
Fax: (602) 542-1614

California
California Department of Industrial Relations
Director (415) 703-5050 Fax:(415) 703-5114
Chief (415) 703-5100 Fax: (415) 703-5114
Manager, Cal/OSHA Program Office
(415) 703-5177 Fax: (415) 703-5114

Connecticut
Connecticut Department of Labor
Commissioner (860) 566-5123 Fax: (860) 566-1520
Conn-OSHA Director (860) 566-4550
Fax: (860) 566-6916

Hawaii
Hawaii Department of Labor and Industrial Relations
Director (808) 586-8844 Fax: (808) 586-9099
Administrator (808) 586-9116 Fax: (808) 586-9104

Indiana
Indiana Department of Labor
Commissioner (317) 232-2378 Fax: (317) 233-3790
Deputy Commissioner (317) 232-3325
Fax: (317) 233-3790

Iowa
Iowa Division of Labor
Commissioner (515) 281-6432 Fax: (515) 281-4698
Administrator (515) 281-3469 Fax: (515) 281-7995

Kentucky
Kentucky Labor Cabinet
Secretary (502) 564-3070 Fax: (502) 564-5387
Federal-State Coordinator (502) 564-3070 ext.240
Fax: (502) 564-1682

Maryland
Maryland Division of Labor and Industry
Commissioner (410) 767-2999 Fax: (410) 767-2300
Deputy Commissioner (410) 767-2992
Fax: 767-2003
Assistant Commissioner, MOSH (410) 767-2215
Fax: 767-2003

Michigan
Michigan Department of Consumer and Industry Services
Director (517) 322-1814 Fax: (517)322-1775

Minnesota
Minnesota Department of Labor and Industry
Commissioner (651) 296-2342 Fax: (651) 282-5405
Assistant Commissioner (651) 296-6529
Fax: (651) 282-5293
Administrative Director, OSHA Management Team
(651) 282-5772 Fax: (651) 297-2527

Nevada
Nevada Division of Industrial Relations
Administrator (775) 687-3032 Fax: (775) 687-6305
Chief Administrative Officer (702) 486-9044
Fax:(702) 990-0358
[Las Vegas (702) 687-5240]

New Jersey
New Jersey Department of Labor
Commissioner (609) 292-2975 Fax: (609) 633-9271
Assistant Commissioner (609) 292-2313
Fax: (609) 1314
Program Director, PEOSH (609) 292-3923
Fax: (609) 292-4409

New Mexico
New Mexico Environment Department
Secretary (505) 827-2850 Fax: (505) 827-2836
Chief (505) 827-4230 Fax: (505) 827-4422

New York
New York Department of Labor
Acting Commissioner (518) 457-2741
Fax: (518) 457-6908
Division Director (518) 457-3518
Fax: (518) 457-6908
OSHA Approved Safety and Health Plans

North Carolina
North Carolina Department of Labor
Commissioner (919) 807-2900 Fax: (919) 807-2855
Deputy Commissioner, OSH Director
(919) 807-2861 Fax: (919) 807-2855
OSH Assistant Director (919) 807-2863
Fax:(919) 807-2856

Oregon
Oregon Occupational Safety and Health Division
Administrator (503) 378-3272 Fax: (503) 947-7461
Deputy Administrator for Policy (503) 378-3272
Fax: (503) 947-7461
Deputy Administrator for Operations (503)
378-3272 Fax: (503) 947-7461

Puerto Rico
Puerto Rico Department of Labor and Human Resources
Secretary (787) 754-2119 Fax: (787) 753-9550
Assistant Secretary for Occupational Safety and Health
(787) 756-1100, 1106 / 754-2171
Fax: (787) 767-6051
Deputy Director for Occupational Safety and Health
(787) 756-1100, 1106 / 754-2188
Fax: (787) 767-6051

South Carolina
South Carolina Department of Labor, Licensing, and Regulation
Director (803) 896-4300 Fax: (803) 896-4393
Program Director (803) 734-9644
Fax: (803) 734-9772

Tennessee
Tennessee Department of Labor
Commissioner (615) 741-2582 Fax: (615) 741-5078
Acting Program Director (615) 741-2793
Fax: (615) 741-3325

Utah
Utah Labor Commission
Commissioner (801) 530-6901 Fax: (801) 530-7906
Administrator (801) 530-6898 Fax: (801) 530-6390

Vermont
Vermont Department of Labor and Industry
Commissioner (802) 828-2288 Fax: (802) 828-2748
Project Manager (802) 828-2765
Fax: (802) 828-2195

Virgin Islands
Virgin Islands Department of Labor
Acting Commissioner (340) 773-1990
Fax: (340) 773-1858
Program Director (340) 772-1315
Fax: (340) 772-4323

Virginia
Virginia Department of Labor and Industry
Commissioner (804) 786-2377 Fax: (804) 371-6524
Director, Office of Legal Support (804) 786-9873
Fax: (804) 786-8418

Washington
Washington Department of Labor and Industries
Director (360) 902-4200 Fax: (360) 902-4202
Assistant Director [PO Box 44600] (360) 902-5495
Fax: (360) 902-5529
Program Manager, Federal-State Operations
[PO Box 44600]
(360) 902-5430 Fax: (360) 902-5529

Wyoming
Wyoming Department of Employment
Safety Administrator (307) 777-7786
Fax: (307) 777-3646
OSHA Approved Safety and Health Plans

Nebraska
(402) 471-4717
(402) 471-5039 FAX

Nevada
(702) 486-9140
(702) 990-0362 FAX

New Hampshire
(603) 271-2024
(603) 271-2667 FAX

New Jersey
(609) 292-3923
(609) 292-4409 FAX

New Mexico
(505) 827-4230
(505) 827-4422 FAX

New York
(518) 457-2238
(518) 457-3454 FAX

North Carolina
(919) 807-2905
(919) 807-2902 FAX

North Dakota
(701) 328-5188
(701) 328-5200 FAX

Ohio
1-800-282-1425 or 614-644-2631
614-644-3133 FAX

Oklahoma
(405) 528-1500
(405) 528-5751 FAX

Oregon
(503) 378-3272
(503) 378-5729 FAX

Pennsylvania
(724) 357-2396
(724) 357-2385 FAX

Puerto Rico
(787) 754-2171
(787) 767-6051 FAX

Rhode Island
(401) 222-2438
(401) 222-2456 FAX

South Carolina
(803) 734-9614
(803) 734-9741 FAX

South Dakota
(605) 688-4101
(605) 688-6290 FAX

Tennessee
(615) 741-7036
(615) 532-2997 FAX

Texas
(512) 804-4640
(512) 804-4641 FAX
OSHCON Request Line: 800-687-7080

Utah
(801) 530-6901
(801) 530-6992 FAX

Vermont
(802) 828-2765
(802) 828-2195 FAX

Virginia
(804) 786-6359
(804) 786-8418 FAX

Virgin Islands
(340) 772-1315
(340) 772-4323 FAX

Washington
(360) 902-5638
(360) 902-5459 FAX

West Virginia
(304) 558-7890
(304) 558-9711 FAX

Wisconsin (Health)
(608) 266-8579
(608) 266-9383 FAX

Wisconsin (Safety)
(262) 523-3040 1-800-947-0553
(262) 523-3046 FAX

Wyoming
(307) 777-7786
(307) 777-3646 FAX

Appendix K

MultiEmployer Directive

U.S. Department of Labor
Occupational Safety & Health Administration

CPL 2-0.124 - Multi-Employer Citation Policy.

Directives - Table of Contents

- Record Type: Instruction
- Directive Number: CPL 2-0.124
- Title: Multi-Employer Citation Policy.
- Information Date: 12/10/1999

DIRECTIVE NUMBER:CPL 2-0.124

EFFECTIVE DATE: December 10, 1999

SUBJECT: Multi-Employer Citation Policy

ABSTRACT

Purpose: To Clarify the Agency's multi-employer citation policy

Scope: OSHA-wide

References: OSHA Instruction CPL 2.103 (the FIRM)

Suspensions: Chapter III, Paragraph C. 6. of the FIRM is suspended and replaced by this directive

State Impact: This Instruction describes a Federal Program Change. Notification of State intent is required, but adoption is not.

Action Offices: National, Regional, and Area Offices

Originating Office: Directorate of Compliance Programs

Contact: Carl Sall (202) 693-2345
Directorate of Construction
N3468 FPB
200 Constitution Ave., NW
Washington, DC 20210

By and Under the Authority of
R. Davis Layne
Deputy Assistant Secretary, OSHA

TABLE OF CONTENTS

Purpose

Scope

Suspension

References

Action Information

Federal Program Change

Force and Effect of Revised Policy

Changes in Web Version of FIRM

Background

Continuation of Basic Policy

No Changes in Employer Duties

Multi-employer Worksite Policy

Multi-employer Worksites

The Creating Employer

The Exposing Employer

The Correcting Employer

The Controlling Employer

Multiple Roles

Purpose. This Directive clarifies the Agency's multi-employer citation policy and suspends Chapter III. C. 6. of OSHA's Field Inspection Reference Manual (FIRM).

Scope. OSHA-Wide

Suspension. Chapter III. Paragraph C. 6. of the FIRM (CPL 2.103) is suspended and replaced by this Directive.

References. OSHA Instructions:

CPL 02-00.103; OSHA Field Inspection Reference Manual (FIRM), September 26, 1994.

ADM 08-0.1C, OSHA Electronic Directive System, December 19,1997.

Action Information

Responsible Office. Directorate of Construction.

Action Offices. National, Regional and Area Offices

Information Offices. State Plan Offices, Consultation Project Offices

Federal Program Change. This Directive describes a Federal Program Change for which State adoption is not required. However, the States shall respond via the two-way memorandum to the Regional Office as soon as the State's intent regarding the multi-employer citation policy is known, but no later than 60 calendar days after the date of transmittal from the Directorate of Federal-State Operations.

Force and Effect of Revised Policy. The revised policy provided in this Directive is in full force and effect from the date of its issuance. It is an official Agency policy to be implemented OSHA-wide.

Changes in Web Version of FIRM. A note will be included at appropriate places in the FIRM as it appears on the Web indicating the suspension of Chapter III paragraph 6. C. and its replacement by this Directive, and a hypertext link will be provided connecting viewers with this Directive.

Background. OSHA's Field Inspection Reference Manual (FIRM) of September 26, 1994 (CPL 2.103), states at Chapter III, paragraph 6. C., the Agency's citation policy for multi-employer worksites. The Agency has determined that this policy needs clarification. This directive describes the revised policy.

Continuation of Basic Policy. This revision continues OSHA's existing policy for issuing citations on multi-employer worksites. However, it gives clearer and more detailed guidance than did the earlier description of the policy in the FIRM, including new examples explaining when citations should and should not be issued to exposing, creating, correcting, and controlling employers. These examples, which address common situations and provide general policy guidance, are not intended to be exclusive. In all cases, the decision on whether to issue citations should be based on all of the relevant facts revealed by the inspection or investigation.

No Changes in Employer Duties. This revision neither imposes new duties on employers nor detracts from their existing duties under the OSH Act. Those duties continue to arise from the employers' statutory duty to comply with OSHA standards and their duty to exercise reasonable diligence to determine whether violations of those standards exist.

Multi-employer Worksite Policy. The following is the multi-employer citation policy:

Multi-employer Worksites. On multi-employer worksites (in all industry sectors), more than one employer may be citable for a hazardous condition that violates an OSHA standard. A two-step process must be followed in determining whether more than one employer is to be cited.

Step One. The first step is to determine whether the employer is a creating, exposing, correcting, or controlling employer. The definitions in paragraphs (B) - (E) below explain and give examples of each. Remember that an employer may have multiple roles (see paragraph H). Once you determine the role of the employer, go to Step Two to determine if a citation is appropriate (NOTE: only exposing employers can be cited for General Duty Clause violations).

Step Two. If the employer falls into one of these categories, it has obligations with respect to OSHA requirements. Step Two is to determine if the employer's actions were sufficient to meet those obligations. The extent of the actions required of employers varies based on which category applies. Note that the extent of the measures that a controlling employer must take to satisfy its duty to exercise reasonable care to prevent and detect violations is less than what is required of an employer with respect to protecting its own employees.

The Creating Employer

Step 1: Definition: The employer that caused a hazardous condition that violates an OSHA standard.

Step 2: Actions Taken: Employers must not create violative conditions. An employer that does so is citable even if the only employees exposed are those of other employers at the site.

Example 1: Employer Host operates a factory. It contracts with Company S to service machinery. Host fails to cover drums of a chemical despite S's repeated requests that it do so. This results in airborne levels of the chemical that exceed the Permissible Exposure Limit.

Analysis: Step 1: Host is a creating employer because it caused employees of S to be exposed to the air contaminant above the PEL. Step 2: Host failed to implement measures to prevent the accumulation of the air contaminant. It could have met its OSHA obligation by implementing the simple engineering control of covering the drums. Having failed to implement a feasible engineering control to meet the PEL, Host is citable for the hazard.

Example 2: Employer M hoists materials onto Floor 8, damaging perimeter guardrails. Neither its own employees nor employees of other employers are exposed to the hazard. It takes effective steps to keep all employees, including those of other employers, away from the unprotected edge and informs the controlling employer of the problem. Employer M lacks authority to fix the guardrails itself.

Analysis: Step 1: Employer M is a creating employer because it caused a hazardous condition by damaging the guardrails. Step 2: While it lacked the authority to fix the guardrails, it took immediate and effective steps to keep all employees away from the hazard and notified the controlling employer of the hazard. Employer M is not citable since it took effective measures to prevent employee exposure to the fall hazard.

The Exposing Employer

Step 1: Definition: An employer whose own employees are exposed to the hazard. See Chapter III, section (C)(1)(b) for a discussion of what constitutes exposure.

Step 2: Actions taken: If the exposing employer created the violation, it is citable for the violation as a creating employer. If the violation was created by another employer, the exposing employer is citable if it (1) knew of the hazardous condition or failed to exercise reasonable diligence to discover the condition, and (2) failed to take steps consistent with its authority to protect is employees. If the exposing employer has authority to correct the hazard, it must do so. If the exposing employer lacks the authority to correct the hazard, it is citable if it fails to do each of the following: (1) ask the creating and/or controlling employer to correct the hazard; (2) inform its employees of the hazard; and (3) take reasonable alternative protective measures. In extreme circumstances (e.g., imminent danger situations), the exposing employer is citable for failing to remove its employees from the job to avoid the hazard.

Example 3: Employer Sub S is responsible for inspecting and cleaning a work area in Plant P around a large, permanent hole at the end of each day. An OSHA standard requires guardrails. There are no guardrails around the hole and Sub S employees do not use personal fall protection, although it would be feasible to do so. Sub S has no authority to install guardrails. However, it did ask Employer P, which operates the plant, to install them. P refused to install guardrails.

Analysis: Step 1: Sub S is an exposing employer because its employees are exposed to the fall hazard. Step 2: While Sub S has no authority to install guardrails, it is required to comply with OSHA requirements to the extent feasible. It must take steps to protect its employees and ask the employer that controls the hazard - Employer P - to correct it. Although Sub S asked for guardrails, since the hazard was not corrected, Sub S was responsible for taking reasonable alternative protective steps, such as providing personal fall protection. Because that was not done, Sub S is citable for the violation.

Example 4: Unprotected rebar on either side of an access ramp presents an impalement hazard. Sub E, an electrical subcontractor, does not have the authority to cover the rebar. However, several times Sub E asked the general contractor, Employer GC, to cover the rebar. In the meantime, Sub E instructed its employees to use a different access route that avoided most of the uncovered rebar and required them to keep as far from the rebar as possible.

Analysis: Step 1: Since Sub E employees were still exposed to some unprotected rebar, Sub E is an exposing employer. Step 2: Sub E made a good faith effort to get the general contractor to correct the hazard and took feasible measures within its control to protect its employees. Sub E is not citable for the rebar hazard.

The Correcting Employer

Step 1: Definition: An employer who is engaged in a common undertaking, on the same worksite, as the exposing employer and is responsible for correcting a hazard. This usually occurs where an employer is given the responsibility of installing and/or maintaining particular safety/health equipment or devices.

Step 2: Actions taken: The correcting employer must exercise reasonable care in preventing and discovering violations and meet its obligations of correcting the hazard.

Example 5: Employer C, a carpentry contractor, is hired to erect and maintain guardrails throughout a large, 15-story project. Work is proceeding on all floors. C inspects all floors in the morning and again in the afternoon each day. It also inspects areas where material is delivered to the perimeter once the material vendor is finished delivering material to that area. Other subcontractors are required to report damaged/missing guardrails to the general contractor, who forwards those reports to C. C repairs damaged guardrails immediately after finding them and immediately after they are reported. On this project few instances of damaged guardrails have occurred other than where material has been delivered. Shortly after the afternoon inspection of Floor 6, workers moving equipment accidentally damage a guardrail in one area. No one tells C of the damage and C has not seen it. An OSHA inspection occurs at the beginning of the next day, prior to the morning inspection of Floor 6. None of C's own employees are exposed to the hazard, but other employees are exposed.

Analysis: Step 1: C is a correcting employer since it is responsible for erecting and maintaining fall protection equipment. Step 2: The steps C implemented to discover and correct damaged guardrails were reasonable in light of the amount of activity and size of the project. It exercised reasonable care in preventing and discovering violations; it is not citable for the damaged guardrail since it could not reasonably have known of the violation.

The Controlling Employer

Step 1: Definition: An employer who has general supervisory authority over the worksite, including the power to correct safety and health violations itself or require others to correct them. Control can be established by contract or, in the absence of explicit contractual provisions, by the exercise of control in practice. Descriptions and examples of different kinds of controlling employers are given below.

Step 2: Actions Taken: A controlling employer must exercise reasonable care to prevent and detect violations on the site. The extent of the measures that a controlling employer must implement to satisfy this duty of reasonable care is less than what is required of an employer with respect to protecting its own employees. This means that the controlling employer is not normally required to inspect for hazards as frequently or to have the same level of knowledge of the applicable standards or of trade expertise as the employer it has hired.

Factors Relating to Reasonable Care Standard. Factors that affect how frequently and closely a controlling employer must inspect to meet its standard of reasonable care include:

The scale of the project;

The nature and pace of the work, including the frequency with which the number or types of hazards change as the work progresses;

How much the controlling employer knows both about the safety history and safety practices of the employer it controls and about that employer's level of expertise.

More frequent inspections are normally needed if the controlling employer knows that the other employer has a history of non-compliance. Greater inspection frequency may also be needed, especially at the beginning of the project, if the controlling employer had never before worked with this other employer and does not know its compliance history.

Less frequent inspections may be appropriate where the controlling employer sees strong indications that the other employer has implemented effective safety and health efforts. The most important indicator of an effective safety and health effort by the other employer is a consistently high level of compliance. Other indicators include the use of an effective, graduated system of enforcement for non-compliance with safety and health requirements coupled with regular jobsite safety meetings and safety training.

Evaluating Reasonable Care. In evaluating whether a controlling employer has exercised reasonable care in preventing and discovering violations, consider questions such as whether the controlling employer:

Conducted periodic inspections of appropriate frequency (frequency should be based on the factors listed in G.3.);

Implemented an effective system for promptly correcting hazards;

Enforces the other employer's compliance with safety and health requirements with an effective, graduated system of enforcement and follow-up inspections.

Types of Controlling Employers

Control Established by Contract. In this case, the Employer Has a Specific Contract Right to Control Safety: To be a controlling employer, the employer must itself be able to prevent or correct a violation or to require another employer to prevent or correct the violation. One source of this ability is explicit contract authority. This can take the form of a specific contract right to require another employer to adhere to safety and health requirements and to correct violations the controlling employer discovers.

(1) Example 6: Employer GH contracts with Employer S to do sandblasting at GH's plant. Some of the work is regularly scheduled maintenance and so is general industry work; other parts of the project involve new work and are considered construction. Respiratory protection is required. Further, the contract explicitly requires S to comply with safety and health requirements. Under the contract GH has the right to take various actions against S for failing to meet contract requirements, including the right to have non-compliance corrected by using other workers and back-charging for that work. S is one of two employers under contract with GH at the work site, where a total of five employees work. All work is done within an existing building. The number and types of hazards involved in S's work do not significantly change as the work progresses. Further, GH has worked with S over the course of several years. S provides periodic and other safety and health training and uses a graduated system of enforcement of safety and health rules. S has consistently had a high level of compliance at its previous jobs and at this site. GH monitors S by a combination of weekly inspections, telephone discussions and a weekly review of S's own inspection reports. GH has a system of graduated enforcement that it has applied to S for the few safety and health violations that had been committed by S in the past few years. Further, due to respirator equipment problems S violates respiratory protection requirements two days before GH's next scheduled inspection of S. The next day there is an OSHA inspection. There is no notation of the equipment problems in S's inspection reports to GH and S made no mention of it in its telephone discussions.

Analysis: Step 1: GH is a controlling employer because it has general supervisory authority over the worksite, including contractual authority to correct safety and health violations. Step 2: GH has taken reasonable steps to try to make sure that S meets safety and health requirements. Its inspection frequency is appropriate in light of the low number of workers at the site, lack of significant changes in the nature of the work and types of hazards involved, GH's knowledge of S's history of compliance and its effective safety and health efforts on this job. GH has exercised reasonable care and is not citable for this condition.

(2) Example 7: Employer GC contracts with Employer P to do painting work. GC has the same contract authority over P as Employer GH had in Example 6. GC has never before worked with P. GC conducts inspections that are sufficiently frequent in light of the factors listed above in (G)(3). Further, during a number of its inspections, GC finds that P has violated fall protection requirements. It points the violations out to P during each inspection but takes no further actions.

Analysis: Step 1: GC is a controlling employer since it has general supervisory authority over the site, including a contractual right of control over P. Step 2: GC took adequate steps to meet its obligation to discover violations. However, it failed to take reasonable steps to require P to correct hazards since it lacked a graduated system of enforcement. A citation to GC for the fall protection violations is appropriate.

(3) Example 8: Employer GC contracts with Sub E, an electrical subcontractor. GC has full contract authority over Sub E, as in Example 6. Sub E installs an electric panel box exposed to the weather and implements an assured equipment grounding conductor program, as required under the contract. It fails to connect a grounding wire inside the box to one of the outlets. This incomplete ground is not apparent from a visual inspection. Further, GC inspects the site with a frequency appropriate for the site in light of the factors discussed above in (G)(3). It saw the panel box but did not test the outlets to determine if they were all grounded because Sub E represents that it is doing all of the required tests on all receptacles. GC knows that Sub E has implemented an effective safety and health program. From previous experience it also knows Sub E is familiar with the applicable safety requirements and is technically competent. GC had asked Sub E if the electrical equipment is OK for use and was assured that it is.

Analysis: Step 1: GC is a controlling employer since it has general supervisory authority over the site, including a contractual right of control over Sub E. Step 2: GC exercised reasonable care. It had determined that Sub E had technical expertise, safety knowledge and had implemented safe work practices. It conducted inspections with appropriate frequency. It also made some basic inquiries into the safety of the electrical equipment. Under these circumstances GC was not obligated to test the outlets itself to determine if they were all grounded. It is not citable for the grounding violation.

Control Established by a Combination of Other Contract Rights: Where there is no explicit contract provision granting the right to control safety, or where the contract says the employer does not have such a right, an employer may still be a controlling employer. The ability of an employer to control safety in this circumstance can result from a combination of contractual rights that, together, give it broad responsibility at the site involving almost all aspects of the job. Its responsibility is broad enough so that its contractual authority necessarily involves safety. The authority to resolve disputes between subcontractors, set schedules and determine construction sequencing are particularly significant because they are likely to affect safety. (NOTE: citations should only be issued in this type of case after consulting with the Regional Solicitor's office).

(1) Example 9: Construction manager M is contractually obligated to: set schedules and construction sequencing, require subcontractors to meet contract specifications, negotiate with trades, resolve disputes between subcontractors, direct work and make purchasing decisions, which affect safety. However, the contract states that M does not have a right to require compliance with safety and health requirements. Further, Subcontractor S asks M to alter the schedule so that S would not have to start work until Subcontractor G has completed installing guardrails. M is contractually responsible for deciding whether to approve S's request.

Analysis: Step 1: Even though its contract states that M does not have authority over safety, the combination of rights actually given in the contract provides broad responsibility over the site and results in the ability of M to direct actions that necessarily affect safety. For example, M's contractual obligation to determine whether to approve S's request to alter the schedule has direct safety implications. M's decision relates directly to whether S's employees will be protected from a fall hazard. M is a controlling employer. Step 2: In this example, if M refused to alter the schedule, it would be citable for the fall hazard violation.

(2) Example 10: Employer ML's contractual authority is limited to reporting on subcontractors' contract compliance to owner/developer O and making contract payments. Although it reports on the extent to which the subcontractors are complying with safety and health infractions to O, ML does not exercise any control over safety at the site.

Analysis: Step 1: ML is not a controlling employer because these contractual rights are insufficient to confer control over the subcontractors and ML did not exercise control over safety. Reporting safety and health infractions to another entity does not, by itself (or in combination with these very limited contract rights), constitute an exercise of control over safety. Step 2: Since it is not a controlling employer it had no duty under the OSH Act to exercise reasonable care with respect to enforcing the subcontractors' compliance with safety; there is therefore no need to go to Step 2.

Architects and Engineers: Architects, engineers, and other entities are controlling employers only if the breadth of their involvement in a construction project is sufficient to bring them within the parameters discussed above.

(1) Example 11: Architect A contracts with owner O to prepare contract drawings and specifications, inspect the work, report to O on contract compliance, and to certify completion of work. A has no authority or means to enforce compliance, no authority

to approve/reject work and does not exercise any other authority at the site, although it does call the general contractor's attention to observed hazards noted during its inspections.

Analysis: Step 1: A's responsibilities are very limited in light of the numerous other administrative responsibilities necessary to complete the project. It is little more than a supplier of architectural services and conduit of information to O. Its responsibilities are insufficient to confer control over the subcontractors and it did not exercise control over safety. The responsibilities it does have are insufficient to make it a controlling employer. Merely pointing out safety violations did not make it a controlling employer. NOTE: In a circumstance such as this it is likely that broad control over the project rests with another entity. Step 2: Since A is not a controlling employer it had no duty under the OSH Act to exercise reasonable care with respect to enforcing the subcontractors' compliance with safety; there is therefore no need to go to Step 2.

(2) Example 12: Engineering firm E has the same contract authority and functions as in Example 9.

Analysis: Step 1: Under the facts in Example 9, E would be considered a controlling employer. Step 2: The same type of analysis described in Example 9 for Step 2 would apply here to determine if E should be cited.

Control Without Explicit Contractual Authority . Even where an employer has no explicit contract rights with respect to safety, an employer can still be a controlling employer if, in actual practice, it exercises broad control over subcontractors at the site (see Example 9). NOTE: Citations should only be issued in this type of case after consulting with the Regional Solicitor's office.

(1) Example 13: Construction manager MM does not have explicit contractual authority to require subcontractors to comply with safety requirements, nor does it explicitly have broad

contractual authority at the site. However, it exercises control over most aspects of the subcontractors' work anyway, including aspects that relate to safety.

Analysis: Step 1: MM would be considered a controlling employer since it exercises control over most aspects of the subcontractor's work, including safety aspects. Step 2: The same type of analysis on reasonable care described in the examples in (G)(5)(a) would apply to determine if a citation should be issued to this type of controlling employer.

Multiple Roles

A creating, correcting or controlling employer will often also be an exposing employer. Consider whether the employer is an exposing employer before evaluating its status with respect to these other roles.

Exposing, creating and controlling employers can also be correcting employers if they are authorized to correct the hazard.

Occupational Safety & Health Administration
200 Constitution Avenue, NW
Washington, DC 20210

Appendix L

OSHA Regional Offices

REGION I: (CT,* ME, MA, NH, RI, VT*)
JFK Federal Building, Room E340
Boston, MA 02203
(617) 565-9860

REGION II: (NJ,* NY,* PR,* VI*)
201 Varick Street, Room 670
New York, NY 10014
(212) 337-2378

REGION III: (DE, DC, MD,* PA,* VA,* WV)
The Curtis Center
170 S. Independence Mall West
Suite 740 West
Philadelphia, PA 19106-3309
(215) 861-4900

REGION IV: (AL, FL, GA, KY,* MS, NC,* SC,* TN*)
Atlanta Federal Center
61 Forsyth Street, SW, Room 6T50
Atlanta, GA 30303
(404) 562-2300

REGION V: (IL, IN,* MI,* MN,* OH, WI)
230 South Dearborn Street, Room 3244
Chicago, IL 60604
(312) 353-2220

REGION VI: (AR, LA, NM,* OK, TX)
525 Griffin Street, Room 602
Dallas, TX 75202
(214) 767-4731 or 4736 x224

REGION VII: (IA,* KS, MO, NE)
City Center Square
1100 Main Street, Suite 800
Kansas City, MO 64105
(816) 426-5861

REGION VIII: (CO, MT, ND, SD, UT,* WY*)
1999 Broadway, Suite 1690
Denver, CO 80202-5716
(303) 844-1600

REGION IX: (American Samoa, AZ,* CA,* Guram, HI,* NV*)
71 Stevenson Street, Room 420
San Francisco, CA 94105
(415) 975-4310

REGION X : (AK,* ID, OR,* WA*)
1111 Third Avenue, Suite 715
Seattle, WA 98101-3212
(206) 553-5930

In case of emergency, please call 1-800-321-OSHA.

*States with State Plans.

Please note, that New Jersey, New York and Connecticut state plans cover state and local government only. Workers in those states who are not state and local government workers are covered by Federal OSHA.

Appendix M

Glossary

Adjustable suspension scaffold - a suspension scaffold equipped with a hoist(s) that can be operated by an employee(s) on the scaffold.

Bearer (putlog) - a horizontal transverse scaffold member (which may be supported by ledgers or runners) upon which the scaffold platform rests and which joins scaffold uprights, posts, poles, and similar members.

Boatswains' chair - a single-point adjustable suspension scaffold consisting of a seat or sling designed to support one employee in a sitting position.

Body belt (safety belt) - a strap with means both for securing it about the waist and for attaching it to a lanyard, lifeline, or deceleration device.

Body harness - a design of straps which may be secured about the employee in a manner to distribute the fall arrest forces over at least the thighs, pelvis, waist, chest, and shoulders, with means for attaching it to other components of a personal fall arrest system.

Brace - a rigid connection that holds one scaffold member in a fixed position with respect to another member, or to a building or structure.

Bricklayers' square scaffold - a supported scaffold composed of framed squares that support a platform.

Carpenters' bracket scaffold - a supported scaffold consisting of a platform supported by brackets attached to building or structural walls.

Catenary scaffold - a suspension scaffold consisting of a platform supported by two essentially horizontal and parallel ropes attached to structural members of a building or other structure. Additional support may be provided by vertical pickups.

Chimney hoist - a multi-point adjustable suspension scaffold used to provide access to work inside chimneys. See Multi-point adjustable suspension scaffold.

Cleat - a structural block used at the end of a platform to prevent the platform from slipping off its supports. Cleats are also used to provide footing on sloped surfaces such as crawling boards.

Competent person - one who is capable of identifying existing and predictable hazards in the surroundings or working conditions which are unsanitary, hazardous, or dangerous to employees, and who has authorization to take prompt corrective measures to eliminate them.

Continuous run scaffold (Run scaffold) - a two-point or multi-point adjustable suspension scaffold constructed using a series of interconnected braced scaffold members or supporting structures erected to form a continuous scaffold.

Coupler - a device for locking together the tubes of a tube and coupler scaffold.

Crawling board (chicken ladder) - a supported scaffold consisting of a plank with cleats spaced and secured to provide footing, for use on sloped surfaces such as roofs.

Deceleration device - any mechanism, such as a rope grab, rip-stitch lanyard, specially-woven lanyard, tearing or deforming lanyard, or automatic self-retracting lifeline lanyard, which dissipates a substantial amount of energy during a fall arrest or limits the energy imposed on an employee during fall arrest.

Double pole (independent pole) scaffold - a supported scaffold consisting of a platform(s) resting on cross beams (bearers) supported by ledgers and a double row of uprights independent of support (except ties, guys, braces) from any structure.

Equivalent - alternative designs, materials, or methods to protect against a hazard which the employer can demonstrate will provide an equal or greater degree of safety for employees than the methods, materials, or designs specified in the standard.

Exposed power lines - means electrical power lines which are accessible to employees and which are not shielded from contact. Such lines do not include extension cords or power tool cords.

Eye or Eye splice - a loop with or without a thimble at the end of a wire rope.

Fabricated decking and planking - manufactured platforms made of wood (including laminated wood and solid sawn wood planks), metal, or other materials.

Fabricated frame scaffold (tubular welded frame scaffold) - a scaffold consisting of a platform(s) supported on fabricated end frames with integral posts, horizontal bearers, and intermediate members.

Failure - load refusal, breakage, or separation of component parts. Load refusal is the point where the ultimate strength is exceeded.

Float (ship) scaffold - a suspension scaffold consisting of a braced platform resting on two parallel bearers and hung from overhead supports by ropes of fixed length.

Form scaffold - a supported scaffold consisting of a platform supported by brackets attached to formwork.

Guardrail system - a vertical barrier, consisting of, but not limited to, toprails, midrails, and posts, erected to prevent employees from falling off a scaffold platform or walkway to lower levels.

Hoist - a manual- or power-operated mechanical device to raise or lower a suspended scaffold.

Horse scaffold - a supported scaffold consisting of a platform supported by construction horses (saw horses). Horse scaffolds constructed of metal are sometimes known as trestle scaffolds.

Independent pole scaffold - see Double pole scaffold.

Interior hung scaffold - a suspension scaffold consisting of a platform suspended from the ceiling or roof structure by fixed length supports.

Ladder jack scaffold - a supported scaffold consisting of a platform resting on brackets attached to ladders.

Ladder stand - a mobile, fixed-size, self-supporting ladder consisting of a wide flat tread ladder in the form of stairs.

Landing - a platform at the end of a flight of stairs.

Large area scaffold - a pole scaffold, tube and coupler scaffold, systems scaffold, or fabricated frame scaffold erected over substantially the entire work area. For example: a scaffold erected over the entire floor area of a room.

Lean-to scaffold - a supported scaffold that is kept erect by tilting it toward and resting it against a building or structure.

Lifeline - a component consisting of a flexible line that connects to an anchorage at one end to hang vertically (vertical lifeline), or that connects to anchorages at both ends to stretch horizontally (horizontal lifeline), and which serves as a means for connecting other components of a personal fall arrest system to the anchorage.

Lower levels - areas below the level where the employee is located and to which an employee can fall. Such areas include, but are not limited to, ground levels, floors, roofs, ramps, runways, excavations, pits, tanks, materials, water, and equipment.

Masons' adjustable supported scaffold - see Self-contained adjustable scaffold.

Masons' multi-point adjustable suspension scaffold - a continuous run suspension scaffold designed and used for masonry operations.

Maximum intended load - the total load of all persons, equipment, tools, materials, transmitted loads, and other loads reasonably anticipated to be applied to a scaffold or scaffold component at any one time.

Mobile scaffold - a powered or unpowered, portable, caster or wheel-mounted supported scaffold.

Multi-level suspended scaffold - a two-point or multi-point adjustable suspension scaffold with a series of platforms at various levels resting on common stirrups.

Multi-point adjustable suspension scaffold - a suspension scaffold consisting of a platform(s) that is suspended by more than two ropes from overhead supports and equipped with means to raise and lower the platform to desired work levels. Such scaffolds include chimney hoists.

Needle beam scaffold - a platform suspended from needle beams.

Open sides and ends - the edges of a platform that are more than 14 inches (36 cm) away horizontally from a sturdy, continuous, vertical surface (such as a building wall) or a sturdy, continuous horizontal surface (such as a floor), or a point of access. Exception: For plastering and lathing operations the horizontal threshold distance is 18 inches (46 cm).

Outrigger - the structural member of a supported scaffold used to increase the base width of a scaffold in order to provide support for and increased stability of the scaffold.

Outrigger beam (Thrustout) - the structural member of a suspension scaffold or outrigger scaffold that provides support for the scaffold by extending the scaffold point of attachment to a point out and away from the structure or building.

Outrigger scaffold - a supported scaffold consisting of a platform resting on outrigger beams (thrustouts) projecting beyond the wall or face of the building or structure, the inboard ends of which are secured inside the building or structure.

Overhand bricklaying - the process of laying bricks and masonry units such that the surface of the wall to be jointed is on the opposite side of the wall from the mason, requiring the mason to lean over the wall to complete the work. It includes mason tending and electrical installation incorporated into the brick wall during the overhand bricklaying process.

Personal fall arrest system - a system used to arrest an employee's fall. It consists of an anchorage, connectors, a body belt or body harness and may include a lanyard, deceleration device, lifeline, or combinations of these.

Platform - a work surface elevated above lower levels. Platforms can be constructed using individual wood planks, fabricated planks, fabricated decks, and fabricated platforms.

Pole scaffold - see Single-pole scaffold and Double (independent) pole scaffold.

Power operated hoist - a hoist that is powered by other than human energy.

Pump jack scaffold - a supported scaffold consisting of a platform supported by vertical poles and movable support brackets.

Qualified - one who, by possession of a recognized degree, certificate, or professional standing, or who by extensive knowledge, training, and experience, has successfully demonstrated his/her ability to solve or resolve problems related to the subject matter, the work, or the project.

Rated load - the manufacturer's specified maximum load to be lifted by a hoist or to be applied to a scaffold or scaffold component.

Repair bracket scaffold - a supported scaffold consisting of a platform supported by brackets that are secured in place around the circumference or perimeter of a chimney, stack, tank, or other supporting structure by one or more wire ropes placed around the supporting structure.

Roof bracket scaffold - a rooftop-supported scaffold consisting of a platform resting on angular-shaped supports.

Runner (ledger or ribbon) - the lengthwise horizontal spacing or bracing member that may support the bearers.

Scaffold - any temporary elevated platform (supported or suspended) and its supporting structure (including points of anchorage), used for supporting employees or materials or both.

Self-contained adjustable scaffold - a combination supported and suspension scaffold consisting of an adjustable platform(s) mounted on an independent supporting frame(s) not a part of the object being worked on, and which is equipped with a means to permit the raising and lowering of the platform(s). Such systems include rolling roof rigs, rolling outrigger systems, and some masons' adjustable supported scaffolds.

Shore scaffold - a supported scaffold that is placed against a building or structure and held in place with props.

Single-point adjustable suspension scaffold - a suspension scaffold consisting of a platform suspended by one rope from an overhead support and equipped with means to permit the movement of the platform to desired work levels.

Single-pole scaffold - a supported scaffold consisting of a platform(s) resting on bearers, the outside ends of which are supported on runners secured to a single row of posts or uprights, and the inner ends of which are supported on or in a structure or building wall.

Stair tower (Scaffold stairway/tower) - a tower comprised of scaffold components and which contains internal stairway units and rest platforms. These towers are used to provide access to scaffold platforms and other elevated points such as floors and roofs.

Stall load - the load at which the prime-mover of a power-operated hoist stalls or the power to the prime-mover is automatically disconnected.

Step, platform, and trestle ladder scaffold - a platform resting directly on the rungs of stepladders or trestle ladders.

Stilts - a pair of poles or similar supports with raised footrests, used to permit walking above the ground or working surface.

Stonesetters' multi-point adjustable suspension scaffold - a continuous run suspension scaffold designed and used for stonesetters' operations.

Supported scaffold - one or more platforms supported by outrigger beams, brackets, poles, legs, uprights, posts, frames, or similar rigid support.

Suspension scaffold - one or more platforms suspended by ropes or other non-rigid means from an overhead structure(s).

System scaffold - a scaffold consisting of posts with fixed connection points that accept runners, bearers, and diagonals that can be interconnected at predetermined levels.

Tank builders' scaffold - a supported scaffold consisting of a platform resting on brackets that are either directly attached to a cylindrical tank or attached to devices that are attached to such a tank.

Top plate bracket scaffold - a scaffold supported by brackets that hook over or are attached to the top of a wall. This type of scaffold is similar to carpenters' bracket scaffolds and form scaffolds and is used in residential construction for setting trusses.

Tube and coupler scaffold - a supported or suspended scaffold consisting of a platform(s) supported by tubing, erected with coupling devices connecting uprights, braces, bearers, and runners.

Tubular welded frame scaffold - see Fabricated frame scaffold.

Two-point suspension scaffold (swing stage) - a suspension scaffold consisting of a platform supported by hangers (stirrups) suspended by two ropes from overhead supports and equipped with means to permit the raising and lowering of the platform to desired work levels.

Unstable objects - items whose strength, configuration, or lack of stability may allow them to become dislocated and shift and therefore may not properly support the loads imposed on them. Unstable objects do not constitute a safe base support for scaffolds, platforms, or employees. Examples include, but are not limited to, barrels, boxes, loose brick, and concrete blocks.

Vertical pickup - a rope used to support the horizontal rope in catenary scaffolds.

Walkway - a portion of a scaffold platform used only for access and not as a work level.

Window jack scaffold - a platform resting on a bracket or jack which projects through a window opening.

Index

ANSI *see* The American National Standards Institute
AEGCP *see* Assured Equipment Grounding Program
BLS *see* Bureau of Labor Statistics
FECA *see* Federal Employees Compensation Act
GFCI *see* Ground Fault Current Interrupter
NIOSH *see* National Institute of Occupational Safety and Health
NTOF *see* National Traumatic Occupational Fatalities
OSH *see* Occupational Safety and Health Act
OSHA *see* Occupational Safety and Health Administration

A

Access, 155 – 157, 237
Access requirements, 61
Accidents, causes of scaffold, 104
Active fall protection, 123 – 127
  active solutions and equipment, 119 – 123
Adjustable multi-point suspension scaffolds, 223
Aerial lift requirements, 63 – 64
Aerial work platforms, 239 – 242
  common hazards and precautions, 254 – 261
    with bucket trucks and boom lifts, 258 – 259
    with scissor lifts, 260
  specifications, 246 – 267
Alarms, 146 – 147
American National Standards Institute (ANSI), 3
Annual inspections, 253
Assured Equipment Grounding Program (AEGCP), 133 – 134,135 – 136

B

Boatswains' chair *see* Single-point suspension scaffolds
Boom lifts, 262
Bucket trucks, 262
Bureau of Labor Statistics (BLS), 4

C

Capacity, 54
Casters, 234 – 235
Catenary scaffold, 220 – 222
Citations, listing of, 33 – 34
Collapses/tipovers, 257
Competent Person, 72, 205 – 206
Construction and safety, 233 – 234
Construction legislation, history of, 1 – 3
Contract, 142 – 143

D

Descending ladders, 334
Designations, competent, qualified, and authorized, 72 – 73
  on the jobsite, 75 – 83

Dismantlers, erectors and, 205 – 206

E

Electrical power tools, use of on scaffolds, 133 – 136
Electrocution on scaffolds, 129 – 130
  case studies, 137 – 138
  hazards, 254 – 255
  precautions, 255
  preventing contact with electrical lines, 132 – 133, 154 – 155
Elevated structures, scaffolds on, 94 – 99
Emergency response and rescue, 141 – 142
  Possible scenarios, 143 – 145
End frames, 169 – 170
Engineering design analysis, 15
Equipment, 37 – 40
Erectors and dismantlers, 205 – 206
Evacuation, 147 – 148
Evaluating condition of scaffold planking, 98 – 99
Evaluating the span of a scaffold plank, 99

F

Failure to understand and/or comply with regulatory requirements, 20
  case studies, 20 – 25
Fall hazards, 255 – 256
Fall precautions for bucket trucks and boom lifts, 256
Fall precautions for scissor lifts, 256 – 257
Fall protection, 115, 202 – 206, 261 – 262, 309, 335 – 336
  criteria for, 158 – 160
  solutions, 118 – 122
Fall, reasons scaffolds fall, 14 – 16
  case studies, 16 – 20
Falling object protection, 63, 102 – 104
Federal Employees Compensation Act (FECA), 1
Fiber rope, 108 – 113
Fixed ladders, 317
  fall protection, 335 – 336
Float scaffold, 219 – 220
Form scaffolds, 193 – 196
Forming systems, 193 – 196
Foundation/support, 86 – 94
Frequent inspections, 253

G

Ground Fault Current Interrupter (GFCI), 133 – 135
Guardrails, 190, 160 – 161
  systems, 204 – 205
Guy wires, 152

H

Half-ladder, 165 – 167
Hazards of elevated work platforms, 7 – 8
Hazards, identifying, 115 – 118
  common hazards and precautions, 254 – 261, 305 – 308
  of boom lifts and bucket trucks, 257 – 258
  with scissor lifts, 259 – 260
Hoisting devices, 201 – 202
Hoists, 60 – 61

I
Industry standards, 246, 303 – 304, 320
Inspection checklists, 330
Inspection requirements, 253 – 254
Interior hung scaffold, 217 – 219

J
Jacks, 170
Job Hazard Analysis, 36 – 37
  form, 115 – 118
Job-made ladders, 314 – 315
Job-manufactured scaffolds, 193 – 196

L
Ladder jack scaffolds, 187, 190 – 191
Ladder selection, 321
Ladders, 313
  care, 328 – 329
  common hazards, 330 – 331
  common precautions, 332
  training requirements, 329 – 330
  types of, 313 – 314
  typical malfunctions and injuries, 337 – 340
Leveling and stabilizing the ladder, 323 – 325
Load capacity, 248 – 249
Loads, 200 – 201
Lumber grading, 96 – 98

M
Malfunctions and injuries, typical, 262 – 267
Manually propelled rolling towers, 233
Mason's frames, 163 – 165
Masons multi-point scaffolds, 223 – 225
Maximum platform height, 247
Mechanics, training of, 252
Mending plates, 189
Moving a scaffold, 235 – 236
Moving ladders, 322
Multiple point suspension scaffolds, 217

N
National Institute of Occupational Safety and Health (NIOSH), 9
National Traumatic Occupational Fatalities (NTOF), 9
Needle beam scaffolds, 227 – 228

O
Occupational Safety and Health Act (OSH), 2
Occupational Safety and Health Administration (OSHA), 1, 3, 6
  regulations, 172, 242 – 245, 301 – 303, 317 – 318
  standards, 40 – 64
Operator training, 251 – 252
Outrigger beam scaffolds, 229 – 230
Outrigger brackets, 173 – 174
Outrigger type scaffolding, 227
Outriggers, 236

P
Pad, 170
Passive solutions, 118 – 119

Planks, 94 – 99
Platforms, 99 – 102, 211 – 213
Poles, 189 – 190
Power lines, determining location of, 131 – 133
  working in the vicinity of, 138 – 140
Pre-bid process, 36
Pre-erection inspection, 171 – 172
Planning, 170 – 171
Portable ladders, 314 – 317
  installation and removal, 320 – 329
  fall protection, 336
Power source, 247
Powered platform, 216
Pre-operation inspection, 253 – 254
Preplanning, 35
Pump jack scaffolds, 187
  structure, 189 – 190
  structure and poles, importance of, 187 – 189

R
Removing ladders, 328
Rental/lease and purchase, 250
Rescue plans, 126 – 27
Resolution, 148 – 149
Rigid braces, 153 – 154
Roof brackets, 196 – 197

S
Safety factors, 130 – 131
Safety features, 249
Scaffold platform construction, 54 – 55
Scaffold sway, preventing, 210 – 211
Scaffold tags, 73 – 75
Scaffolds
  proper assembly of, 26
    case studies, 26 – 33
Scissor lifts, 262
Self-supporting portable ladders, 315 – 317
  setting up, 325 – 328
Setting up all portable ladders, 322
Setting up straight and extension ladders, 322 – 325
Single-point suspension scaffolds, 207, 213 – 216
Soil bearing capacities, 85 – 86
Square and level, 158
Stability, maintaining, 152 – 154
  and weight, 200 – 201
Stairways, 301
  common hazards and precautions, 305 – 308
  fall protection, 309
  installation, 304
  typical malfunctions and injuries, 309 – 310
Stall force, 202
Statistics, 9 – 14
Stepping on/off a ladder at height, 334
Stone setters *see* Masons multi-point scaffolds
Structural Work Act, 5 – 6
Structure attachment points, 200
Subpart L, 3 – 5
Subpart M, 7
Subpart R, 7
Support, 199 – 200
Supported scaffolds, 151 – 152
  criteria, 55 – 56
  materials and methods, 85

Suspended scaffolds
and welding, 136 – 137
criteria, 57 – 60
fall protection, 202 – 206
general information, 199

T

Tie wires, 153
Tipovers, collapses/, 257
Tire type, 247
Training, 65 – 72
materials, 65 – 72
specific requirements, 65, 250 – 252, 305
Tube and coupler scaffolding, 175 – 178
standard tables for minimum construction, 179 – 181
Tubular welded-frame scaffolds, 163 – 165 *see* also Mason's Frames
climbing of, 172 – 174
Two-point suspension scaffolds, 207 – 210

U

Use requirements, 62
Uses, 249 – 250

W

Walk-through frame, 167 – 168
Weight and stability, 200 – 201
Welding, 136 – 137, 138
Wheel base width, 248
Window-jack scaffolds, 230 – 231
Wire suspension ropes, 104 – 108
Wood pole scaffolds, 183 – 185
Working from a ladder, 332 – 334
Workplace emergencies, 145 – 146